VISIONS OF CELL BIOLOGY

T0138204

CONVENING SCIENCE:
DISCOVERY AT THE MARINE BIOLOGICAL LABORATORY

A Series Edited by Jane Maienschein

I am delighted to serve as editor for the new University of Chicago Press series Convening Science: Discovery at the Marine Biological Laboratory. These books will highlight the ongoing role the Marine Biological Laboratory plays in the creation and dissemination of science, in its broader historical context, as well as current practice and future potential. Each volume is anchored at the MBL and includes work about the MBL and its science and scientists; work by those scientists; work that begins with workshops, research, or courses at the MBL; collaborations made possible by the MBL; and so on. Books by, about, with, for, inspired by, and otherwise related to the MBL will capture the spirit of discovery by the community of MBL scientists and students. Some will be monographic, while others will be collaborative coherent collections.

We look forward to discovering new ideas and approaches that find their way into volumes of this series. I first did summer research, with a small NSF grant, as a graduate student in 1976; it led to my first edited volume inspired by this special place. Many other people have been similarly inspired, and this new series invites us to bring together our works into a collection of reflections on the MBL's role in promoting discovery through its exceptional role in convening science at the seaside.

JANE MAIENSCHEIN
Series Editor
University Professor and Director of the Center for Biology and Society, Arizona State University
Fellow, Marine Biological Laboratory

VISIONS OF CELL BIOLOGY

REFLECTIONS INSPIRED BY COWDRY'S *GENERAL CYTOLOGY*

Edited by

KARL S. MATLIN, JANE MAIENSCHEIN,

and MANFRED D. LAUBICHLER

THE UNIVERSITY OF CHICAGO PRESS

Chicago and London

The University of Chicago Press, Chicago 60637
The University of Chicago Press, Ltd., London
© 2018 by The University of Chicago
All rights reserved. No part of this book may be used or reproduced in any manner whatsoever without written permission, except in the case of brief quotations in critical articles and reviews. For more information, contact the University of Chicago Press, 1427 E. 60th St., Chicago, IL 60637.
Published 2018
Printed in the United States of America

27 26 25 24 23 22 21 20 19 18   1 2 3 4 5

ISBN-13: 978-0-226-52048-3 (cloth)
ISBN-13: 978-0-226-52051-3 (paper)
ISBN-13: 978-0-226-52065-0 (e-book)
DOI: 10.7208/chicago/9780226520650.001.0001

Library of Congress Cataloging-in-Publication Data

Names: Matlin, Karl S., editor. | Maienschein, Jane, editor. | Laubichler, Manfred Dietrich, editor. | Cowdry, E. V. (Edmund Vincent), 1888–1975. General cytology.
Title: Visions of cell biology : reflections inspired by Cowdry's General cytology / edited by Karl S. Matlin, Jane Maienschein, and Manfred D. Laubichler.
Other titles: Convening science.
Description: Chicago ; London : The University of Chicago Press, 2018. | Series: Convening science
Identifiers: LCCN 2017026678 | ISBN 9780226520483 (cloth : alk. paper) | ISBN 9780226520513 (pbk. : alk. paper) | ISBN 9780226520650 (e-book)
Subjects: LCSH: Cytology. | Cells. | Cowdry, E. V. (Edmund Vincent), 1888–1975. General cytology.
Classification: LCC QH581.2.V58 2018 | DDC 571.6—dc23
LC record available at https://lccn.loc.gov/2017026678

♾ This paper meets the requirements of ANSI/NISO Z39.48-1992 (Permanence of Paper).

CONTENTS

CHAPTER 1

INTRODUCTION

Karl S. Matlin, Jane Maienschein, and Manfred D. Laubichler

Cell biology as many scientists know it today is generally considered to have arisen after World War II and is often associated historically with particular technical developments, most prominently the electron microscope and cell fractionation. Despite its extraordinary success in describing both the structure and function of cells, modern cell biology tends to be overshadowed by molecular biology, a field of inquiry that developed in the same period. Nevertheless, cell biology, which considers both the molecular aspects of cells and cell form, is often more effective than approaches focused only on molecules in explaining biological phenomena at the cellular level.

The investigation of cells began, of course, much earlier than the postwar period. As a number of recent studies have explored, the cellular conception of life emerged gradually within a rich context of cultural trends, philosophical claims, changing epistemologies and aesthetic preferences, political debates, and institutional settings (Duchesneau 1987; Harris 1999; Parnes and Vedder 2008; Rheinberger and Müller-Wille 2007; Weigel 2005). Cells were identified in the seventeenth century, and during the eighteenth century and into the beginning of the nineteenth century, observations of mainly plant but also animal cells accumulated. By the beginning of the nineteenth century, improved light microscopes with significantly reduced chromatic aberration became widely available, and the pace of new discoveries about cells accelerated. Building on the work of Henri Dutrochet and François-Vincent Raspail in France, among others, Matthias Schleiden and Theodor Schwann produced widely read works in the 1830s suggesting that all organisms are composed of cells. Shortly thereafter, Robert Remak and Rudolf Virchow emphasized that all new cells are formed by division of old cells (Harris 1999).

As the nineteenth century progressed, cytologists increased our understanding of cell structures through morphological observations of embryos and other preparations, while also developing the protoplasm concept that began to address questions of cell chemistry (Geison 1969; see also Reynolds, chap. 3 in this volume). These observations also contributed to ideas about the theoretical foundation of biology as the fundamental science of life (Driesch 1893; Hartmann 1925 and 1933; Hertwig 1892; 1898; and 1906; Laubichler 2006; Schaxel 1919; Verworn 1895).

Further studies led to the formulation of hypotheses about the relative functions of the nucleus and cytoplasm, and the mechanism of cell division (Laubichler and Davidson 2008). Then, at the beginning of the twentieth century, cytologists correlated the distribution of Mendelian characters into dividing cells with the separation of chromosomes, and cytology became, for a time, closely linked with the study of heredity through the new field of genetics (see Allen, chap. 8 in this volume; and Sutton 1903). This link was grounded in a conceptual understanding of inheritance that predates the subsequent split into thinking in terms of genetics (transmission) and development (expression), a split triggered both by methodology and by new conceptual orientations (Laubichler 2014; Maienschein and Laubichler 2014). A casualty of this split into two experimental disciplines was the apparent loss of importance of cells to each.

Cytology, in the meantime, had become an increasingly experimental field based upon manipulations of early embryonic cells by Hans Driesch, Theodor Boveri, Wilhelm Roux, and others (Maienschein 1991). Although cytology still relied on the light microscope, biologists generally agreed that the processes within cells also depended on chemical reactions that researchers could not directly observe. There was, however, a dilemma. The living substance of the cell was the protoplasm, a gel-like mass that contained within it the nucleus and other "formed elements." Life was viewed as the consequence of protoplasmic organization, but the disruption of protoplasm believed to be necessary to study its chemistry was thought to render suspect the biological relevance of any reactions uncovered by these manipulations (Wilson 1896, 238; Geison 1969).

Cowdry's *General Cytology*

It is within this context that a group of American biologists decided to initiate a new project, the creation of a comprehensive cytology textbook with individual chapters from many of the leaders in the field, most of whom already interacted on a regular basis during summers at the Marine Biological Laboratory (MBL) in Woods Hole, Massachusetts. When *General Cytology* was published in 1924, the volume sought to treat cytology comprehensively, but also to go beyond what the authors saw as the usual morphological considerations. Chapters focused on the chemical and physical activities of cells, and new techniques, such as cellular microsurgery and tissue culture, joined more traditional observational and experimental methods (Cowdry

1924). Edmund Beecher Wilson, one of the leading synthesizers of all knowledge of the cell up to then, noted in his introduction to the volume that *General Cytology* represented a new era of multi-perspectival cell biology.

The idea for *General Cytology* originated at a meeting of scientists working at the MBL in Woods Hole in September 1922. At the suggestion of Edward Conklin, the cytologist and anatomist Edmund V. Cowdry was asked to edit the volume. After accepting, Cowdry began assembling contributors, starting with the core group from the MBL (see Maienschein, chap. 2 in this volume). Among the individuals on this original list was Jacques Loeb, the great promoter of mechanistic views of the cell and formerly an MBL regular, with the suggested topic "Physical chemistry of the cell with special reference to proteins" (see fig. 1.1). Loeb later dropped off this list, presumably due to ill health (he died in 1924) (Pauly 1987). Robert Bensley, an anatomist from the University of Chicago and Cowdry's thesis advisor, was also proposed tentatively as author of two chapters on secretion and methods of fixation and staining. In the end, neither chapter by Bensley appeared in the final book, nor did a chapter by the botanist and physiologist Winthrop J. V. Osterhout on cellular permeability or a proposed "historical resume" by Fielding H. Garrison (see fig. 1.1). Other authors, however, eventually stepped in to fill these gaps.

A subsequent meeting determined the final title, list of authors, and suggestions for a publisher. The original working title was *Cellular Physiology*, indicating perhaps the desire to avoid what the group saw as the morphological connotations of the term cytology (fig. 1.1). By the time of the meeting, however, *General Cytology* was set as the final title. Wilson, who had been asked to write the introduction, planned to include historical material largely drawn from the upcoming edition of his book *The Cell in Development and Inheritance*, making a separate chapter on history superfluous (Wilson 1925). Merkel Jacobs, at the time responsible for directing the noted physiology course at the MBL, replaced Osterhout on cell permeability, and Thomas Hunt Morgan was added to write about the experimental analysis of the chromosome theory of heredity to pair with Clarence E. McClung's presumably more comprehensive chapter on the same topic (see fig. 1.2; see also Maienschein, chap. 2 in this volume, for more details).

While it seemed that Cowdry had preempted the others by speaking to Appleton Publishers about the book, in the end Frank Lillie, the MBL director and University of Chicago professor, approached the University of Chicago Press. On December 23, 1922, the contract for the book was completed (fig. 1.2). After that, adjustments to the content were minor but telling:

Cellular Physiology

	Pages text	Pages Figs.
1. Introduction - E. B. Wilson?	5	
2. Historical Resume - Fielding H. Garrison?	25	3
3. Physical chemistry of the cell with special reference to proteins. J. Loeb?	30	1
4. Qualitative chemistry of the cell. A. P. Mathews?	40	2
5. Physical structure of protoplasm as determined by micro-dissection. Robert Chambers.		
6. Methods of fixation and staining. C. E. McClung and R. R. Bensley?	25	2
7. Cytoplasmic inclusions (Metochondria, Golgi apparatus and chromidial substance) E. V. Cowdry	30	3
8. Secretion - R. R. Bensley?		
9. Cell behavior in vitro - W. H. Lewis and M. R. Lewis		
10. Permeability and conductivity - W. J. V. Osterhout?	40	3
11. Irritability and stimulation - R. S. Lillie		
12. Fertilization - F. R. Lillie		
13. Growth and differentiation - E. G. Conklin		
14. Material basis of heredity - C. E. McClung		
	195	14
	(405)	(36)
	500	50

Figure 1.1. An early proposal of contributors and topics for the volume that became *General Cytology*. Note the provisional title, *Cellular Physiology*. The handwritten notes may be those of Frank Lillie, since this document was found in his archives at the University of Chicago. Lillie submitted the proposal for the book on Edmund Cowdry's behalf to the University of Chicago Press. From the Special Collections Research Center, University of Chicago Library, Frank Lillie Archives, box 2, folder 23.

Morgan changed his title to "Mendelian Heredity in Relation to Cytology," and Lillie added Ernest Everett Just as a coauthor. Just, an African American scientist and one of Lillie's former graduate students, was struggling to get recognition for his work, and Lillie may well have wanted to support him by giving him this opportunity while, at the same time, easing the burden of writing on himself (see fig. 2.1 in Maienschein, chap. 2 in this volume).

MEMORANDUM OF AGREEMENT, made this twenty-third day of December, nineteen hundred twenty-two (1922), between THE UNIVERSITY OF CHICAGO, a corporation organized under the laws of the State of Illinois, of Chicago, Illinois, hereinafter referred to as the University, its successors and assigns, party of the first part, and E. V. COWDRY, of New York City, hereinafter referred to as the Editor, his heirs, executors, administrators and assigns, party of the second part, WITNESSETH:

In consideration of the covenants and agreements of the Editor as herein contained, the University covenants and agrees with the Editor as follows:

1. To publish at its own expense, through its University Press in such style or styles and at such prices as it deems best suited to its sale, a work entitled "A Textbook of General Cytology," to be composed of eleven (11) chapters with authors and pages of each chapter substantially as follows:

1.	Introduction	E.B.Wilson	25	pages
2.	Chemistry and metobolism of the cell	A.P.Mathews	70	"
3.	Permeability of the cell to diffusing substances	M.H.Jacobs	55	"
4.	Reactivity of protoplasm	R.S. Lillie	60	"
5.	Physical structure of protoplasm as determined by microdissection and microinjection	R. Chambers	55	"
6.	Mitochondria, Golgi apparatus, and chromidial substance	E.V. Cowdry	~~70~~ 65	" (and M. R. Lewis)
7.	Cell behavior in ~~vitro~~ tissue culture	W.H. Lewis	45	"
8.	Fertilization	F.R. Lillie	60	"
9.	Cellular differentiation	E.G. Conklin	70 ~~65~~	"
10.	Chromosome theory of heredity	C.E. McClung	70	"
11.	Experimental analysis of the chromosome theory of heredity	T.H. Morgan	35	"

it being understood that the University isto make separate contracts with each of the said chapter authors; in the event of any change in the names of chapters of or authors of chapters or of the number of pages, it is understood that only such change as shall be recommended by the Editor and accepted by the University shall become effective under this contract.

2. To pay to the Editor upon the submission of the manuscript in accordance with the terms of this contract, in full of all editorial royalties, two and one-half per cent (2 1/2%) of the retail published price of twenty-five hundred (2500) copies of the said book.

Figure 1.2. The first part of the formal agreement between the University of Chicago Press and Edmund Cowdry to publish *General Cytology*, dated December 23, 1922. Note the final list of contributors, including Thomas Hunt Morgan. As this document is from the press archives, handwritten notations are likely those of an editor. From the Special Collections Research Center, University of Chicago Library, University of Chicago Press Archives, box 126, folder 6.

When published in 1924, *General Cytology* was very well received, and the book became somewhat of a best seller, especially considering that it ran to more than seven hundred pages and was intended for a specialized audience. As an edited volume put together by a collection of experts, the book was a concrete illustration that the scope of cytology had expanded beyond the capacity of an individual biologist, as Wilson noted in his introduction.

But was *General Cytology*, despite its popularity and stellar set of contributors, successful in transforming cytology into the modern, interdisciplinary science that the authors envisioned? That is less clear. It *was* forward-looking and *did* emphasize the importance of chemistry and even physics to understanding cellular phenomena. At the time, however, this was certainly not unique, since even earlier textbooks such as Allan MacFadyen's *The Cell as a Unit of Life*, published in Britain in 1908, as well as other works, also addressed the chemical aspects of cell function. In addition, by 1924 biochemists like Frederick Gowland Hopkins had already mounted an attack on the protoplasm concept, proposing to substitute enzyme specificity as the foundation on which the chemical organization of the cell was built (Needham 1949). Furthermore, Cowdry's book failed to predict the dramatic advances achieved just a short time later. In the 1930s Robert Bensley and, particularly, Albert Claude at the Rockefeller Institute began to disrupt current ideas about the cell and protoplasm, eventually leading to new epistemic strategies to explain cellular phenomena and to the merger of morphology, physics, and chemistry in cell studies (see Matlin, chap. 11 in this volume, and Bechtel 2006). Nevertheless, what *General Cytology* did accomplish was to mark the beginning of a transition in cell studies to an era in which the barriers constraining progress in cytology were overcome by new technologies and by collaborative and multi-perspectival approaches.

Reflecting on the Past, Present, and Future of Cell Biology

In October 2014 another group of leading scientists, historians, and philosophers of biology came together at the MBL in Woods Hole to reflect on the Cowdry volume from the perspective of the twenty-first century (table 1.1). Among the scientists were not only individuals who clearly identified as cell biologists, but also those more focused on gene expression and its regulation, topics many would consider more properly as molecular biology. The historians and philosophers were also an eclectic group. In presentations, several focused on events in cytology that preceded and produced the scientific context of the Cowdry volume, while others looked at what

Table 1.1. Participants in the first workshop, "Updating Cowdry at the MBL"

Jane Maienschein / Arizona State University and the MBL
    Revisiting Cowdry's *General Cytology*: Looking Back to See Ahead
Laura Otis / Emory University
    The Evolution of Imagined Cells
Andrew Reynolds / Cape Breton University
    Updating the Metaphors of Cytology and Cellular Biology since Cowdry 1924
Michael Dietrich / Dartmouth College
    Finding the Pulse of Protoplasm: The Search for Mechanisms of Cytoplasmic Streaming
Kai Simons / MPI-CBG Dresden
    Cell Membranes: How to Cope with Complexity
Benjamin Glick / University of Chicago
    Self-Organization and Maturation of Endomembrane Compartments
Jutta Schickore / Indiana University
    Pitfalls of Carmine and Canada Balsam: Accounts of Sources of Error in Cowdry's *General Cytology*
Rudolf Oldenbourg / MBL
    Shinya Inoué and the Reemergence of Light Microscopy
Karl Matlin / University of Chicago
    On the Relationship between Morphology and Molecular Explanation
Daniel Nicholson / University of Exeter
    Mechanism versus Organicism: Two Views of the Cell
Garland Allen / Washington University
    The Chromosome Theory of Heredity and the Cell: A Fruitful Convergence, 1910–1930
Hannah Landecker / UCLA
    From Information to Conformation: Chromatin and Cell Biology Then and Now
Jason Lieb / University of Chicago
    Genomics and the Cell Biology of the Nucleus
Manfred Laubichler / Arizona State University and the MBL
    The Regulatory Genome from Boveri to Davidson
Eric Davidson / Caltech
    A Formal View of the Regulatory System of the Animal Cell
James Nelson / Stanford University
    Challenges and Solutions to Probing Adhesive Mechanisms
Clare Waterman / NIH
    The Dynamic, Nano-Scale 3D Molecular Clutch
Michael Bennett / Albert Einstein University
    From *General Cytology* to Cellular Biology of Neurons: Modes of Synaptic Transmission
Fridolin Gross / Humboldt University
    Updating Cowdry with Computational Models
Ed Munro / University of Chicago
    Getting at Cellular Dynamics

(*continues*)

Table 1.1. (continued)

Bill Bechtel / UCSD
 The Evolving Understanding of Mechanisms
William Summers / Yale
 Cellular Pathogenesis: Virus Inclusions and Histochemistry
Bill Aird / Harvard Medical School
 Seeing the Endothelium: A Story of Blind Spots, Blind Turns, and Blind Alleys
Lijing Jiang / Princeton
 How Aging Became a Part of Cellular Life
Gary Borisy / Forsyth Institute
 Missing Microbes

---

occurred afterward, and even explored future strategies to address cellular complexity.

On the basis of discussions at the meeting and at a subsequent workshop one year later, some of the original participants plus a few others who were not in attendance contributed chapters to this volume, *Visions of Cell Biology*. Although some contributors focused on Cowdry and his book, most used Cowdry's *General Cytology* and the imagined atmosphere of the MBL in 1924 as points of departure to try to understand not only how studying the cell has changed historically, but also how cell studies have affected and been affected by developments in the twentieth and even twenty-first centuries.

Our intention was neither to review the history of cell biology before and after Cowdry comprehensively, nor to attempt to directly relate the proposals articulated by the authors of *General Cytology* to later developments. Instead, Cowdry's book was used to inspire us to think about the study of cells from the microscopic observations of Schleiden and Schwann in the 1830s to the dynamic imaging of living cells today. In the end, we believe that our analysis establishes that cell biology is neither a discipline that arose in the modern, postwar period, nor one whose time has passed. Instead it is clear that the study of cells today exists within the mainstream of a historical continuum going back to the very origins of biology, a key part of the scientific study of life that was ushered in by the microscope and the ensuing recognition that, to paraphrase Wilson, everything alive began its existence as a single cell (Wilson 1925, 1).

Visions of Cell Biology is about how biologists attempted and continue to attempt to understand cells, not only before and after the appearance of Cowdry's *General Cytology*, but also in the present, when cellular complexity is beginning to yield to computationally based strategies. Technology is part of this story. Jutta Schickore (in chap. 4 of this volume) maps the use of the light microscope in the nineteenth and early twentieth centuries as it changed from a tool for simple observation to an experimental instrument. As microscopes improved, their magnification increased, and different forms of illumination and chemical dyes were added to the scientific repertoire to allow previously invisible structures to be seen. These manipulations created problems of interpretation, but also opened up possibilities to conduct real analysis of cells through the microscope. Karl Matlin (chap. 11 in this volume) picks up this thread by describing the introduction of electron microscopy as well as cell fractionation to cell biology in the 1940s and 1950s, while Rudolf Oldenbourg (chap. 12 in this volume) relates the reemergence of light microscopy in the modern period, exemplified by the work of the great microscopist Shinya Inoué and his application of sophisticated optics and digital imaging to living cells.

Another part of understanding cells is how the very idea of the cell has changed and continues to change. Andrew Reynolds (chap. 3 in this volume) traces emerging concepts of cellular organization, from the original use of the term *cell* that emphasized the cell wall and not the space within, to the focus by the end of the nineteenth century on the protoplasm, the critical living substance filling the cell's interior. As he makes clear, it was understood at the time that protoplasmic function was governed by chemistry. Yet even when *General Cytology* appeared, cytologists did not see a way to apply the emerging science of biochemistry to this problem, despite the pleadings of the great enzymologist Frederick Gowland Hopkins from Cambridge (Needham 1949). Instead, as Reynolds describes, they grasped mechanical and physical metaphors to try to conceptualize protoplasmic organization and find their way forward.

In her contribution, Jane Maienschein (chap. 2) looks most directly at the Cowdry volume itself and reviews the cell concept from that time and after, comparing views of the cell as an independent living unit and as a component responsible for the growth and differentiation of complex living systems. Against this background, she also details the creation of *General Cytology* in the 1920s, exploring the credentials of the biologists chosen as contributors, as well as the reception of the book after its publication and its impact on future conceptions of cells.

William Summers (chap. 5) adds to this by reminding us that the cell can be a source of a disease. He accomplishes this by following Edmund Cowdry's career both before and after *General Cytology*. During this time,

Cowdry became an expert on intracellular pathogenesis and cellular inclusions caused by viruses and bacteria. As Summers points out, while Cowdry's histochemical approach was eventually superseded by molecular analysis and tissue culture, aspects of the classification schemes that he developed remain important. Lijing Jiang (chap. 6) follows another aspect of Cowdry's diverse career, the study of cell aging. Here the original belief that cells were immortal was replaced by the concept of a cellular lifespan called the Hayflick limit, eventually explained by telomere shortening.

A different kind of perspective on cells is provided by Beatrice Steinert and Kate MacCord (chap. 7). They describe pictorial representation in *General Cytology*, suggesting that the forms that these representations take can reflect evolving epistemic strategies in the field of cytology itself. Realistic drawings of cells harken back to simple morphological descriptions designed to convey as accurately as possible what can be seen through the microscope. However, as they point out, additional drawings depict experimental manipulations of cells that point in directions beyond straightforward observation, consistent with the transcendent goals of *General Cytology*. Still others illustrate proposed explanations of biological phenomena or theories, abstractions that now have little to do with real representation of cells as physical entities but focus instead on ideas about cellular function derived from data that is not necessarily microscopic.

Bill Bechtel (chap. 13) explores the use of representations and images in the modern period by examining the roles of diagrams in mechanistic explanations of biological phenomena. He notes that diagrams are traditionally used to illustrate recomposed mechanisms, enabling the "mental rehearsal" of mechanistic steps, whereas the complex mechanisms recognized in current biology require different sorts of diagrams that are intuitively out of reach but can be deciphered through computational simulation and graph theoretic network analysis.

Another way that biologists try to understand cells is through genetics and the expression of genes. The discipline of genetics is rooted in the cytological study of chromosomes. After the rediscovery in 1900 of Mendel's work on peas, Walter Sutton and Theodore Boveri linked chromosomes directly to Mendel's characters (Laubichler and Davidson 2008; Sutton 1903). Boveri then developed a compelling conceptual framework that focused on the role of the hereditary material as a structured system of causal agents that controlled development and, by implication, evolution. This was the beginning of a trajectory that finds its most recent expression in the concept of gene regulatory networks, as Eric Davidson's work and investigative pathway, which began more than six decades ago at the MBL, demonstrates so well (Laubichler and Davidson 2008).[1]

After Boveri and Sutton, but well before any thought of gene regulatory networks, genetics took off through the study of model organisms, most prominently *Drosophila melanogaster*, and gradually separated itself from cytology. Nevertheless, in *General Cytology*, the importance of chromosomes and Mendelian genetics to cytologists led to inclusion of the chapters by McClung and Morgan. In *Visions of Cell Biology*, Garland Allen reviews Morgan's contribution to explain why Morgan, at the late date of 1924, felt it necessary to focus on Mendelian genetics. He reports that, in fact, at that time aspects of what Allen calls the Mendelian-chromosome paradigm of heredity were still controversial, justifying Morgan's vigorous defense, while also showing how the contributions from both Morgan and McClung led to the firm establishment of cytogenetics as a lasting part of cell biology.

Inheritance through genes on chromosomes is not, however, the only form of inheritance important to cell biology. Jan Sapp, in chapter 9, reminds us that nonchromosomal inheritance, or epigenetics, has been a constant theme in the history of cytology and remains of great significance today, despite narrowing definitions and continued confusion about the term *epigenetics* itself. Sapp explores the debates about cytoplasmic versus nuclear inheritance, along with inconsistencies in the logic suggesting that a one-dimensional code can give rise to a three-dimensional, spatially differentiated cell, by relating classical studies of Tracy Sonneborn on cortical inheritance in *Paramecia*. At the same time, he traces the path to our current definition of epigenetics as non-sequence-based changes in chromosomal DNA from proposals of Joshua Lederberg in the 1950s.

In the modern period, of course, the way we try to understand cells is through molecular mechanisms. Daniel Liu, in his contribution to *Visions of Cell Biology*, describes how chemists' acquisition of the capacity to visualize molecules, specifically the heads and tails of lipids, enabled them to conceive of a model for the lipid bilayer (chap. 10). The lipid bilayer is the basis of the biological membrane enclosing cells and cellular compartments, and is perhaps the most iconic element of the cell's three-dimensional form. As related by Karl Matlin in his chapter, one of the most significant next steps in chemically characterizing cells was to disrupt the cell membrane and separate the resulting membrane-bound compartments by centrifugation, the process of cell fractionation. This enabled determination of the biochemical identity of cellular organelles and, through parallel electron microscopy of the whole cell and its parts, also made functional inferences possible. Eventually, as he describes, this approach led to an almost comprehensive molecular understanding of certain biological processes within the cell.

The problem, though, is that the demands of trying to understand cells at the molecular level ultimately require a confrontation with cellular complexity. We now know many of the molecular parts of cells and their cellular locations because of high-throughput technologies, advanced microscopy, and bioinformatics. However, this knowledge has also demonstrated that the relationship between the molecular parts and cellular functions is decidedly nonlinear. As noted earlier, Bechtel describes an approach to this challenge by highlighting diagrams as they began to portray cellular processes only accessible through computational approaches, including the current focus on networks as a way of understanding biological mechanisms. Fridolin Gross (chap. 14) also describes the transition to computational modeling, certainly one of the most important developments in modern cell biology. He explores the different heuristics used by "experimental" versus "computational" cell biologists as seen from their analysis of cell cycle dynamics at finer and finer resolution. A take-home message from both Bechtel and Gross is that computational modeling of biological systems is an element of the epistemic strategy of cell biology that will continue to develop. Whether it will finally resolve debates about the relation of the parts (molecules) to the whole, which reaches back to the origins of biology itself, remains to be seen.

In the call for our workshops we emphasized a focus on the past, present, and future of cell biology as part of our reflections. While many of the scientists at the meetings gave detailed overviews of the present state in their various subdisciplines, and historians and philosophers provided detailed accounts in the form of specific case studies and thematic reflections, the future of cell biology remains somewhat unspecified. In light of Yogi Berra's declaration ("The future ain't what it used to be.") this is not that surprising. Nevertheless, as with the Cowdry volume that initially inspired us to come together, *Visions of Cell Biology* demonstrates that historians, philosophers, and scientists bring different perspectives and approaches to their reflections. Cowdry and his group recognized in 1924 that such a confluence provides something more valuable than one contributor alone can offer. Perhaps, as Wilson noted about cell biology in 1924, it is no longer possible for any one scholar to write a history of cell biology. As a remedy, we offer *Visions of Cell Biology*, in which once again the MBL has brought people together to explore new directions.

Notes

1 Davidson gave a special lecture at the first workshop at the MBL that led to the development of this volume, and he and Manfred Laubichler intended to contribute a chapter. Unfortunately, Davidson's untimely death prevented this from happening.

References

Bechtel, William. 2006. *Discovering Cell Mechanisms*. Cambridge: Cambridge University Press.

Cowdry, Edmund V., ed. 1924. *General Cytology: A Textbook of Cellular Structure and Function for Students of Biology and Medicine*. Chicago: University of Chicago Press.

Driesch, H. 1893. *Die Biologie als selbstsständige Grundwissenschaft*. Leipzig: Wilhelm Engelmann.

Duchesneau, F. 1987. *Genese de La Theorie Cellulaire*. Paris: Librarie Philosophique J. Vrin.

Geison, G. L. 1969. "The Protoplasmic Theory of Life and the Vitalist-Mechanist Debate." *Isis* 60 (3): 272–92.

Harris, H. 1999. *The Birth of the Cell*. New Haven, CT: Yale University Press.

Hartmann, M. 1925. "Aufgaben, Ziele und Wege der Allgemeinen Biologie." *Klinische Wochenschrift* 4:2229–34.

———. 1933. *Allgemeine Biologie. Eine Einführung in die Lehre vom Leben*. Jena: Gustav Fischer.

Hertwig, O. 1892. *Die Zelle und die Gewebe. Grundzüge der Allgemeine Anatomie und Physiologie. Erster Theil*. Jena: Gustav Fischer.

———. 1898. *Die Zelle und die Gewebe. Grundzüge der Allgemeinen Anatomie und Physiologie. Zweiter Theil*. Jena: Gustav Fischer.

———. 1906. *Allgemeine Biologie*. Jena: Gustav Fischer.

Laubichler, M. D. 2006. "Biologie als selbstständige Grundwissenschaft und die allgemeinen Grundlagen des Lebens." In *Der Hochsitz des Wissens; Das Allgemeine als wissenschaftlicher Wert*, edited by M. Hagner and M. D. Laubichler, 185–205. Zürich: Diaphanes Verlag.

———. 2014. "Gene Regulatory Networks." In *Discoveries in Modern Science: Exploration, Invention, Technology*, edited by J. Trefil, 393–401. Farmington Hills, MI: Macmillan.

———, and Eric H Davidson. 2008. "Boveri's Long Experiment: Sea Urchin Merogones and the Establishment of the Role of Nuclear Chromosomes in Development." *Developmental Biology* 314 (1): 1–11. doi: 10.1016/j.ydbio.2007.11.024.

Maienschein, J. 1991. *Transforming Traditions in American Biology*. Baltimore: Johns Hopkins University Press.

———, and M. D. Laubichler. 2014. "Exploring Development and Evolution on the Tangled Bank." In *Evolutionary Biology: Conceptual, Ethical, and Religious Issues*, edited by P. Thompson and D. Walsh, 151–171. Cambridge: Cambridge University Press.

Needham, J., ed. 1949. *Hopkins and Biochemistry 1861–1947*. Cambridge: W. Heffer and Sons.

Parnes, O., and U. Vedder. 2008. *Das Konzept der Generation: Eine Wissenschafts- und Kulturgeschichte*. Berlin: Suhrkamp.

Pauly, P. J. 1987. *Controlling Life*. New York: Oxford University Press.
Rheinberger, H.-J., and S. Müller-Wille. 2007. *Heredity Produced: At the Crossroads of Biology, Politics, and Culture, 1500–1870*. Cambridge, MA: MIT Press.
Schaxel, J. 1919. *Grundzüge der Theorienbildung in der Biologie*. Jena: Gustav Fischer.
Sutton, W. S. 1903. "The Chromosomes in Heredity." *Biological Bulletin* 4 (5): 231–50.
Verworn, M. 1895. *Allgemeine Physiologie: ein Grundriss der Lehre vom Leben*. Jena: Gustav Fischer.
Weigel, S. 2005. *Genea-Logik: Generation, Tradition und Evolution zwischen Kultur- und Naturwissenschaften*. Freiberg: Fink.
Wilson, E. B. 1896. *The Cell in Development and Inheritance*. New York: Macmillan.
———. 1925. *The Cell in Development and Heredity*. 3rd ed. New York: Macmillan. Orig. pub. as *The Cell in Development and Inheritance* in 1896; 2nd ed., 1900.

CHAPTER 2

CHANGING IDEAS ABOUT CELLS AS COMPLEX SYSTEMS

Jane Maienschein

As Edmund Beecher Wilson finished writing the third and final edition of his *The Cell in Development and Heredity*, he noted that in the future probably no single author could write such a cell biology text. The subject had become too complex and required too many different kinds of expertise to do it justice. About the same time, Wilson joined Edmund Cowdry and other leading biologists in a workshop at the Marine Biological Laboratory (MBL), where the distinguished group divided up the topics to write the collective 1924 volume *General Cytology* (Wilson 1925; Cowdry 1924). Cowdry then convened a larger and even more diverse group to produce two volumes on *Special Cytology* (Cowdry 1928; Cowdry 1932).

Far from being of merely antiquarian interest, these volumes reveal underlying assumptions that both reflected and informed the directions of scientific research. The 1924 Cowdry volume focused on the architecture and activities of individual cells, not primarily as building blocks of living organisms but also as the fundamental units that were themselves living, and the authors emphasized the value of studying these living cells in detail. The group clearly saw the beginning of a new field of cell biology emerging, one that might lose the coherence of a single approach but gain from different points of view, using different techniques to ask different questions about complex cells and their activities.

This chapter looks at the context in which Cowdry's volume appeared, a context constructed on the foundation of the first cell theory of 1839 and subsequent developments. The story leads to questions about what the Cowdry volume tells us about the science of understanding cells more generally. It looks at what the 1924 volume offers in seeing cells as having gained autonomy, integrity, and biological importance as complex living systems in their own right. In addition, some of the chapters and reviews focus on understanding that the individual cells work together to make up complex organisms, such that organization arises through their connections. Yet the chapters mostly remain focused on the individual cells themselves rather than on how they communicate with each other and work as a whole. The Cowdry volume presents an American story, focused on Woods

Hole at a time after World War I when scientists in the United States aspired to scientific leadership. To gain perspective on the contributions of Cowdry's volume, it is useful to start briefly at a prominent discussion at the centennial of the cell theory for a broader view looking back, and then to move on to *General Cytology* itself.

Reflecting on the First Century of Cell Theory

In his introduction to the centennial volume entitled *The Cell and Protoplasm* in 1940, editor Forest Ray Moulton noted that the American Association for the Advancement of Science was publishing the book as part of a series. It grew out of a symposium, held in 1939, to celebrate the centennial of Matthias Schleiden and Theodor Schwann's introduction of the scientific cell theory. Because of the rich history of thinking about cells up to that time, Moulton felt that "in a sense the Cell Theory is not new." Yet, he continued, "In another sense the Cell Theory is always new, for every discovery respecting this primary and essential unit of living organisms, both plant and animal, has raised more questions than it has answered and has always widened the fields of inquiry" (Moulton 1940, "Foreword"). The volume set out to show both what was old and well established and what was new a century after the original idea of cells.

By 1940, discussion of cells usually separated plant, microbial, and animal cells. For plants, discussion typically involved looking at such predictable topics as cell walls, while discussion of animal cells looked more closely at delineation of individual cells; contents of cells, including nucleus, cytoplasm, and organelles; and environments, both internal and external to each cell. Along the way came considerations of biochemistry and cell physiology. More surprising in the 1940 volume are the less standard chapters on microbiology, viruses, enzymes, hormones, and vitamins. The choice of topics and of contributors makes clear just how much remained in 1940 to be discovered about cells and especially about the ways they interact with each other as well as with their environment. The contributors realized that they still knew relatively little about how the individual cells add up to an organized, whole, complex organism, though they recognized that the process of making coordinated combinations of parts was key to understanding living systems and organisms.

We see this emphasis on individual cells in textbooks of the time as well. For example, Lester W. Sharp's very widely used *Fundamentals of Cytology* of 1943 (as well as other editions) laid out the way that cytoplasm and nucleus work in the cell, looked at different kinds of cells, and recognized that cells make up organisms, but had relatively little to say about the latter point. Sharp noted that in addition to the cell theory focused on the cells themselves, other researchers had a different view of living organisms that supported an emphasis on each organism as a whole. The two different perspectives have to come together in some way, Sharp recognized, for "in every normal mass of protoplasm, whatever its growth pattern or degree of differentiation, the many diversified activities are so coordinated that it behaves as a consistent whole, or individual, from the beginning of development onward; without such harmony there obviously could be no organism" (Sharp 1943, 20). Yes, but how could this harmony be achieved?

The connections were not yet clear. Some biologists continued to look at organisms as organized living systems that happen to consist of cells, while others looked at aggregations of cells as making up organized living systems. At issue was partly a matter of emphasis, but also partly a matter of causal efficacy. Do cell divisions and actions cause organisms, or does some integrated wholeness cause cells to behave as they do? What drives the integration of the whole organism? These were still questions in the 1940s, and the discussion shows that biology had not embraced a single "cell theory" to explain the basis for all living systems.

Sharp explained that cell structure and function affect or perhaps even effect the organism, but it remained unclear just how that happens. In an earlier picture of cells as structural units or building blocks, it was easier to treat them like bricks or stones that combined into a larger organism through forces outside the cells themselves. But if cells were each living units in their own right, then how do all those separate cells relate to the organism as a whole, and how do they make up that whole? How could new research resolve persistent debates? In particular, by the end of the nineteenth century, in the face of increasing knowledge about protoplasm and internal workings of cells, which theory about life should hold? "The proponents of the cell theory stressed the cell as the primary agent of organization, while adherents of the organismal theory insist upon the primacy of the whole, cells when present being important but subsidiary parts" (Sharp 1943, 21). Furthermore, looking at evolutionary relationships by comparing studies of different organisms suggested that a different kind of protoplasm might serve to connect cells and might thereby help to bring together coordinated whole organisms in somewhat different ways and not necessarily in exactly the same processes and patterns for each kind of organism. Many questions remained in 1940, including questions about what cell theory was and how it had changed over time.

Again, this work of the mid-twentieth century reinforced just how much had been learned about the details of cells, and yet how little that knowledge revealed about the ways cells work together in a coordinated way in more complex animals. It is worth looking at how the science had gotten to this point. The history of cell biology shows a first stage of thinking about cells as the structural units of living organisms, followed by a stage of thinking about cells as themselves more nearly the "agent[s] of organization," as Sharp put it. Reflecting on the development of cell theory took Sharp, Cowdry's group, and takes us as well, to previous studies of cells, starting with Schleiden and Schwann.

In the Beginning

Most of us have heard something about the basic story of Schleiden and Schwann and the cell theory that they invented, which is recounted in textbook after textbook—except that we do not know the historically accurate story, because the textbooks usually get it wrong, or partly wrong. Those accounts tell of these two German innovators, one working on plants and the other working on animals, as coming up with *the* theory that cells are the fundamental units of life. The story goes that these two began to put together the available evidence and reasoning to develop what they called the *Zellenlehre* and what others labeled the *Zellentheorie* that has grounded all of biology since (Schwann 1839).

Everybody likes a good myth, and like most, this one is not completely inaccurate. Matthias Schleiden and Theodor Schwann did, respectively, study plants and animals and did see and describe structural units that they called cells. Their work added to earlier observations by Robert Hooke, Anthony Leeuwenhoek, and others to establish the idea of structural cellular units as bounded by walls and consisting of some internal fluid-like or gel-like substance. The claim that they established cell theory in the sense that the cell is the fundamental unit of life, or the fundamental living unit, is less clear.

In the mid-twentieth century, several biologists looked much more closely at the historical record and at what each contributor to the foundational biological idea had actually done. They began to replace the earlier, oversimplified interpretations. In 1948, the Oxford University cytologist John Randal Baker began a series of essays in the *Quarterly Journal of Microscopical Science*. Under the title "The Cell-Theory: A Restatement, History, and Critique," Baker reported on his close studies of the primary sources that textbooks had so frequently mentioned but seldom studied in any detail. He began his essays with the point that "several zoological text-books published during the last two decades have cast doubts on the validity of the cell-theory." Baker resolved to review the recent attacks, the nature of the evidence, and to establish the current status of the cell theory. He found that different critiques were attacking different aspects of what was lumped into the cell theory, and he found some of the attacks were justified, while others were not—largely because the critics were really talking about different things (Baker 1948, 103).

Baker broke down the larger cell theory into seven propositions, then explored each in turn. They involved claims that (1) most organisms consist of microscopic cells, (2) cells have definable characters, (3) cells usually come from other cells, (4) cells are the living parts of organisms, (5) cells are individuals, (6) cells are like living protists, and (7) many-celled organisms may have resulted from protists coming together. And his discussion focused on the shape, characteristics, origin, development, individuality, and claims about the relationships of multiple cells in multicellular organisms (Baker 1988, 107).

Through his five essays, Baker provided a tremendous service in clarifying what was at issue in discussions of cell biology. He showed that Schleiden and Schwann each, in different ways, made assumptions about how cells originate or about their structure and nature that went beyond their data from what they could observe. In some cases, they worked with inadequate microscopic tools; in other cases, they started with views about what they should see and then somehow became convinced that they had actually seen it—whether it was really there or not. In particular, Schwann was confident that he saw cells forming around a nucleus, much as crystals form from inorganic matter. Baker's essays appeared from 1949 to 1955, and his careful research showed very clearly who had said and thought what, when, and why.

Shortly after, in 1959, Cambridge University anatomist Arthur Hughes published *A History of Cytology*. Like Baker, and probably for some of the same reasons related to questions about the cell theory at the time, Hughes sought to clarify the development of the understanding of cells. Hughes emphasized the cytological methods of investigation alongside the theories, with special emphasis on the nucleus and the cytoplasm (Hughes 1959).

Each of these mid-twentieth century historical studies focused on cell structure and asked how we came to the idea that cells are the fundamental structural units of living organisms. Textbooks of the time reinforced the underlying view that cells do, in fact, play this central role, thereby reinforcing some version of a cell theory. Yet Moulton's question from 1940 remained: Do the cells themselves serve as the drivers for development and organization, or are cells instead the results of the process of organismal development. To put it another way: To what extent, and in what ways, are cells the units of life rather than simply the structural units of living systems? A century after the introduction of the cell theory, this question had not yet been answered.

Cells as Structural Units of Living Organisms

For many who saw Schleiden and Schwann as the beginning for cell theory, those two gave cell theory a name. They declared, first, that cells exist and are constituents of living organisms and, second, that the theory might help to explain the individuality of more complex organisms that consist of multiple cells. The decades following brought a great deal of additional observation as well as interpretation. Those years also brought improvements in both microscopes and microscopic techniques, as Hughes discussed in some detail. In studying cells, what one can see and how well one can see it are of crucial importance; and making sure that others can see the same thing is especially important. Better lenses reduced chromatic aberration, and better fixing, staining, and sectioning methods brought consistency to the preparation of specimens (Bracegirdle 1978). The whole story is much richer than this, and a number of historians of science have taken up aspects of early cell theory. In addition, as Jutta Schickore explains in chapter 4 of this volume, the technical innovations brought additional questions of interpretation. The emphasis here is on the emerging understanding of cells that informed the 1924 Cowdry volume, to help us interpret the contribution and impact of that volume.

In 1834, Karl Ernst von Baer had presented observations of frog cleavage stages, which clearly showed what later biologists saw as cells dividing, each reliably dividing in the same way and following the same basic patterns, or what were later called lineages of cell divisions within the organism. Von Baer's images were also taken as showing that the full material in the initial egg divides into more and more cells, which remain separate. His illustration clearly supported a claim that the collection of cells is what makes up the developing organism—not intercellular connections or noncellular material (von Baer 1834). The images were taken as representing division into separate structural units, and for those who held that those units were cells, they played an important structural role.

Yet not everybody accepted cells as having such a role, especially those who focused on organisms as a whole. Historian Marsha Richmond points to Thomas Henry Huxley as a leading critic, and she argues that he rejected cell theory in part because it seemed to assign the cells a sort of preformationist role, as if the cells themselves cause development and body structure. Huxley held a more epigenetic view. As Richmond notes, Huxley insisted that cells are "not instruments, but indications—that they are no more the producers of the vital phenomena than the shells scattered in orderly lines along the sea-beach are the instruments by which the gravitative force of the moon acts upon the oceans. Like these, the cells mark only where the vital tides have been, and how they have acted" (Richmond 2002, citing Huxley 1853). Richmond further discusses the debates of the time, which make clear that cell theory was not a clearly defined, unified, or universally accepted idea. (See also Whitman 1893 on what he called "the inadequacies of the cell theory.")

In fact, one main alternative idea persisted, affirming that some sort of protoplasm lies outside the cells and connects them. Huxley put forth such ideas. Adam Sedgwick was still invoking this idea through the end of the nineteenth century, as Baker discusses at greater length (Baker 1988, 175; Sedgwick 1894). The idea of a reticulum, or syncytial connections, proved attractive because it seemed to offer an explanation for how the cellular parts might work together as an organismal whole. Physical connections make the parts into a network. The same reasoning held for the nervous system. At the end of the nineteenth century, researchers argued about whether the nervous system is there from the very beginning in a sort of reticulum that then grows larger while maintaining its structure (Billings 1971). This idea of a protoplasmic reticulum could explain how the complex structure arose and was maintained. In contrast, the neuronal theory held that individual neuroblast cells then develop nerve fibers that grow out and make connections. Gradually, they extend, develop connections, and make up the nervous system.

Some researchers, such as Ross Granville Harrison, could easily imagine how such a complex system can arise from the interactions of individual cells. His study of individual cells led him to develop the first successful tissue culture and the first stem cell research, with transplanted neuroblasts (neural cells). Harrison's work was taken by many as having resolved the question in favor of the action of individual cells working together. Certainly Santiago Ramon y Cajal agreed. In contrast, Ramon y Cajal's corecipient of the Nobel Prize, Camillo Golgi, and others never gave up their convictions that the system was inextricably interconnected from the beginning. The discussions were part of persistent debates about whether development is more preformationist, that is laid out from the very beginning in a preformed way, or epigenetic, that is arising only gradually over time. These debates have been discussed in detail elsewhere, and we need not repeat the entire story here (Maienschein 1983; Maienschein 1991).

It is worth remembering that when different people looked at cells, some saw them as newly arisen objects making up an organism, and some saw them more as products of cell division from past cells. Both are partly true, and it really depends on how one does the looking and what one is looking for. It depends, as is so often the case, on perspective.

The former, epigenetic view requires an explanation for how the individual cells arise and how they make up a whole organism. Where does the organization and where does the life come from if the separate and individual cells come together only gradually over time to make a whole? The temptation was strong to invoke some form of teleological or vitalistic principle or force to offer such an explanation and drive the process. Aristotle had given us two and a half millennia of thinking in such gradual, epigenetic terms, and his instincts fit with those of many other observers. While later thinkers discarded Aristotle's entelechy, they had to find explanations for the emergence of form and function in some other way (see Maienschein 2011).

In contrast, if cells arise only from other cells, then the "life" and the beginnings of the form are in some sense already there from the beginning. Thus, when Robert Remak showed that cells divide and give rise to other cells, and Rudolf Virchow famously declared that "*omnis cellula e cellula*," their assertions that cells come from other cells were clear and direct. The claim was neither entirely new, however, nor was it universally accepted at the time. Asserting that cells come from other cells pushes back the explanation of where they come from in the first place as well as the question of how they come to be "alive." If cells are the fundamental unit of life, and more than just the structural building blocks of living things, then how so? And what follows for our understanding of biological processes? (See Maienschein 2014, chap. 1.)

Cells as Living Units

This very quick look at leading ideas brings us to the end of the nineteenth century and to the work of Theodor Boveri, Oscar Hertwig, and Edmund Beecher Wilson. A number of researchers had demonstrated that the cell has structure, with a distinct bounded nucleus, liquid or gel-like cytoplasm, and other structures including the mitochondria and Golgi bodies, with spindle fibers, asters, and centrosomes playing important roles during cell division. Again, historians have covered this period in detail, and the rich historical work of cell biologist Sir Henry Harris at the University of Oxford provides the best modern account in *The Birth of the Cell*, which appeared in 1999.

Harris goes over much of the same ground as Baker and Hughes, but with considerably more interpretive subtlety. He has reread the original sources, and furthermore has the benefit of an additional half-century of biological discovery and reflection on our understanding of cells. Harris frames his work with a selected quotation from the French microscopist François-Vincent Raspail: "Give me an organic vesicle endowed with life and I will give you back the whole of the organized world." Making the claim that the German story came to dominate—and perhaps to distort—the history of cell biology as well as the work of cytology itself, Harris calls for recognizing the alternative point of view held by Raspail and a few others. While most of those studying cells were still focused on establishing all the details of structure, Raspail already by the early nineteenth century saw the cell as a "kind of laboratory" that allowed development of the life of an organism out of the life of the individual cells (Raspail 1833; Harris 1999).

Understanding life also involves sorting out what is going on with heredity and development. By the last quarter of the nineteenth century, Oscar Hertwig provided a widely cited solidification of the accumulating evidence and reasoning about the nature of fertilization, concluding that the nucleus of the egg and sperm come together to make a new nucleus for the zygote. This provided the starting point for a new cell. Efforts to understand whether chromosomes retain their individuality throughout cell divisions, and whether they retain all their material or undergo some sort of reduction division occupied Hertwig's attention and led to greater clarity of what the questions were surrounding fertilization and cell division. As Harris discusses, a number of other researchers also began to ask similar questions directed at understanding the cell as a living functional unit. In particular, they wanted to know how each cell grows, divides, differentiates, and otherwise changes over time in ways that add up to a complex organized organism?

Theodor Boveri provided some answers, looking closely at the contributions of the nucleus. In 1902, for example, Boveri demonstrated that chromosomes are defined structures and, furthermore, that they retain their individuality through cell divisions. They divide in such a way that each of the daughter cells has its own set of chromosomes after divisions, but they retain their individuality nonetheless. Observing carefully both naturally occurring and experimentally derived examples, Boveri added immeasurably to the understanding of cell division with his experimental work.

Manfred Laubichler and Eric Davidson have suggested that Boveri was thinking in very forward-looking terms about the cell and the roles of its parts. They help us see Boveri as a visionary able to imagine something conceptually similar to today's complex organisms guided by gene regulatory networks, even though Boveri thought of them as determinants on chromosomes in the nucleus and did not yet have a concept of genes specifically (Laubichler and Davidson 2008).

Cells as Complex Living Systems: Wilson's *The Cell*

Edmund Beecher Wilson built on the work of Hertwig, Boveri, and many others. For Wilson, the cell plays a foundational role for life and therefore also for the study of life through biology. Wilson's work influenced generations of cell and developmental biologists because of the way he brought together and made sense of so many different pieces of evidence about parts of the cell and its changes over time. In the first 1896 edition of his classic textbook, entitled *The Cell in Development and Inheritance* and dedicated to Boveri, Wilson opened his introduction by pointing to Schleiden and Schwann: "During the half-century that has elapsed since the enunciation of the cell-theory by Schleiden and Schwann, in 1838–39, it has become ever more clearly apparent that the key to all ultimate biological problems must, in the last analysis, be sought in the cell." Furthermore, "No other biological generalization, save only the theory of organic evolution, has brought so many apparently diverse phenomena under a common point of view or has accomplished more for the unification of knowledge. The cell-theory must therefore be placed beside the evolution-theory as one of the foundation stones of modern biology" (Wilson 1896, 1). He saw his task in part as bringing the two together, showing the role of cells in development and heredity, in ways that made evolution possible.

By the third and final edition in 1925, Wilson acknowledged that a great deal had changed—the volume had grown from 371 to 1232 pages and had undergone reconceptualization while seeking to retain its synthetic approach. For that last edition, he opened with a slightly different tone: "Among the milestones of modern scientific progress the cell-theory of Schleiden and Schwann, enunciated in 1838–39, stands forth as one of the commanding landmarks of the nineteenth century." Yet he went on to note that their ideas were just a "rude sketch" and that it nonetheless succeeded in "opening a new point of view for the study of living organisms, and revealing the outlines of a fundamental common plan of organization that underlies their endless external diversity" (Wilson 1925, 1).

In this third edition, Wilson pointed to three rough periods since the inception of the initial idea of cells: the first focused on the basic ideas about cells and their roles; the second looked at development and cell division; and the third brought in the chromosome theory of heredity, which introduced explanations of the causes of cell division. This third period had made heredity more a matter of biochemistry and metabolism, Wilson thought (Wilson 1925, 1114). Wilson pointed to several key phenomena that remained puzzles, concluding that "we are still without adequate understanding of the physiological relations between nucleus and cytoplasm and of the manner in which the nucleus is concerned in the operations of constructive metabolism, of growth and repair, and in the determination of hereditary traits. The same may be said of our present knowledge of development, above all in respect to the problem of localization." Further, he asked, "What determines the appearance of hereditary traits in regular order of space and time? How are the operations of development so coordinated as to give rise to a definitely ordered system?" (Wilson 1925, 1115). And how can a proper understanding in physicochemical terms of the "organization" of the organism push away any temptations toward vitalism that he found in some of his contemporaries?

In the third edition also, even more than in the previous two, he ended by pointing to the successes of recent years in moving forward on all three contributions, while acknowledging that many questions remained. "If we are confronted still," he wrote in the final paragraph, "with a formidable array of problems not yet solved, we may take courage from the certainty that we shall solve a great number of them in the future, as so many have been in the past" (Wilson 1925, 1118).

Senior scholars today recall buying this last edition of Wilson's book and reading it for one or another class, as well as being instructed to keep the volume nearby for reference. With his series of three editions of *The Cell*, Wilson provided a compendium of existing knowledge about cells and the ways that they reflect the processes of life. His message was that each cell is a fundamental living unit, useful for understanding the processes of life as well as the structure of living organisms. Interpreting those processes, though, required bringing together heredity, development, and evolution.

Wilson provided a view of the cell as an individual, complex, living system. The year before his first edition in 1895, he had published *The Atlas of Fertilization and Karyokinesis of the Ovum*, which included a set of beautiful print copies of photographs taken of the early stages of fertilization and cell division. Collaborating with photographer Edward Leaming, Wilson sought to show his readers the complex parts of the cell and how they change during those early stages. By 1896, he provided considerably more detail about later roles of the cell as well. His point in the *Atlas* was precisely to provide an atlas, a sort of collection of maps of structures.

The Cell added function and development. In the process, the cell came alive. Cells still went through stages of development, but Wilson sought to capture more than the standardized stages characteristic of normal tables. He wanted to understand more about what it meant to be alive and especially what it meant to be organized into an individual organism with integrity and autonomy. What Wilson did not quite see yet, despite his clarity of vision and depth of understanding, was the importance of understanding the ways in which cells interact with other cells and the complexity of the regulatory processes that reside within the inherited material but go beyond each individual cell itself.

Edmund Cowdry, *General Cytology*

As Wilson had acknowledged, and despite his attempts to provide a summary update of the field with his own third edition, by 1924 the challenges of understanding the cell had already grown beyond what any single researcher could grasp. Indeed, the very brief summary of ideas taken to be important leading up to 1924 has focused especially on parts of the story about study of cells. Other researchers were looking more intently at physiology, biochemistry, and other areas that fed into the study of cells, especially in later periods. The point here has been to put us at least partly inside the thinking of those who gathered to produce the volume edited by Cowdry.

By 1924, the biologists who gathered to produce the edited volume agreed that it was time for a cooperative approach, which Edmund Cowdry coordinated at the Marine Biological Laboratory in Woods Hole (see also the introduction to this volume). Cowdry noted that, because contributors had worked in the MBL facilities, "the volume, as it stands, is to be considered, to some extent at least, as a contribution from the Marine Biological Laboratory" (Cowdry 1924, v).

The University of Chicago, which published the volume, has a folder of reviews and letters related to the book. One is labeled "17. A Textbook of General Cytology. By Frank R. Lillie, et al." and summarizes the proposal for the book, which Lillie apparently presented to the press. Claiming that the total would not exceed 650 pages (the actual was 754 pages), the press calculated a net investment of $3436.75. The press estimated sales of one thousand copies at $5 each. Those numbers probably look astonishingly low to today's publishers, who would nonetheless be reassured to note that the costs included $1010 of "overhead."

THE CONTRIBUTORS

The contributors to Cowdry's volume all had close ties with the MBL, which was a prime gathering place for biological research by the 1920s (figs. 2.1 and 2.2). They each had a home institution, but they came together at the MBL in the summers to discuss their shared interest in cell biology. It is worth getting a sense of the people involved, and a short biographical sketch of each gives a sense of the group. Yet they each had independent research careers, and so their biographies remain separate and largely not overlapping beyond their collaborations at the MBL.

Edmund Vincent Cowdry

Edmund Vincent Cowdry was born in Alberta, British Columbia, in 1888, the same year that the MBL opened its doors. Cowdry received his Bachelor's degree from the University of Toronto and a PhD from the University of Chicago. He moved to the Johns Hopkins University in anatomy and in 1916 married Alice Hanford Smith, going to the MBL for summer research. A year later, the China Medical Board of the Rockefeller Foundation recruited Cowdry to establish and lead an Anatomy Department of Peking Union Medical College in Beijing, and he and Alice moved there in 1917. With the birth of their first child in 1920, Cowdry returned to the United States to the Rockefeller Institute in New York and studied a range of topics, including anatomy, cytology, parasitic diseases, and aging. He continued to spend many summers at the MBL and to establish his editorial credentials. Cowdry took up an academic position at Washington University in St. Louis in 1930. There, he moved increasingly to studies of aging while continuing to focus on cytology, looking at cell degeneration in particular. As Hyung Wook Park has shown, Cowdry became a leader in gerontology and organized a conference on aging at the MBL in 1937, which was supported by

GENERAL CYTOLOGY

A TEXTBOOK OF CELLULAR STRUCTURE AND FUNCTION FOR STUDENTS OF BIOLOGY AND MEDICINE

By

ROBERT CHAMBERS
EDWIN G. CONKLIN
EDMUND V. COWDRY
MERLE H. JACOBS
ERNEST E. JUST
MARGARET R. LEWIS
WARREN H. LEWIS
FRANK R. LILLIE
RALPH S. LILLIE
CLARENCE E. McCLUNG
ALBERT P. MATHEWS
THOMAS H. MORGAN
EDMUND B. WILSON

Edited by
EDMUND V. COWDRY

THE UNIVERSITY OF CHICAGO PRESS
CHICAGO · ILLINOIS

TABLE OF CONTENTS

vii

Figure 2.1. The title page and table of contents from Cowdry's *General Cytology*, as published in 1924.

the Josiah Macy Jr. Foundation and is considered the first such scientific meeting (Park 2008; 2016).

Edmund Beecher Wilson

Edmund Beecher Wilson was the leading cell biologist of the day and a senior statesman. Born in Geneva, Illinois, in 1856, young Wilson enjoyed learning about natural history. He received a PhB degree from Yale University in 1875 and proceeded to the new Johns Hopkins University for a PhD under William Keith Brooks. After a visit to Germany and the Naples Zoological Station, Wilson spent a year at Williams College, then visited at MIT, where he wrote a biology textbook with his fellow Hopkins graduate William Sedgwick, and in 1885 he took a position as head of the biology department at the new Bryn Mawr College for Women. In 1891 Wilson moved to Columbia University, where he remained for the rest of his career,

Figure 2.2. The contributors to *General Cytology*, as pictured in the 1920s. Edmund Cowdry, the editor, is in the center. Others include (*clockwise from the top left*) Robert Chambers, Edwin Conklin, Merle Jacobs, Ernest Just, Margaret Lewis, Warren Lewis, Frank Lillie, Ralph Lillie, Clarence McClung, Albert Matthews, Thomas Hunt Morgan, and E. B. Wilson. Images reproduced from the website "History of the Marine Biological Laboratory" (http://hpsrepository.asu.edu/).

spending almost all his summers at the MBL. His *The Atlas of Fertilization and Karyokinesis*, and *The Cell* set the standard, as discussed earlier (Wilson 1895; 1896; Morgan 1940).

Albert P. Mathews

Albert P. Mathews was born in 1871 in Chicago. He received a Bachelor's degree from MIT, studying biology under William Sedgwick, and then a PhD from Columbia University. He taught first at Tufts College Medical School, and then in 1901 moved to the University of Chicago, where he remained for fifteen years and was promoted to professor and eventually chairman of the Physiology Department. In 1916, he moved to the University of Cincinnati as Carnegie Professor of Biochemistry, and served as head of the Biochemistry Department until he retired in 1939. Though his work focused on chemistry, he also pursued physics in relation to life and wrote work on gravity, matter, space-time, and such topics. Mathews began his career looking at the biochemistry of secretions, but soon moved to study of living cells. E. Newton Harvey explained that "Woods Hole was just the place for a man of Mathews' broad interests, and the group benefited immensely from his new and stimulating ideas." That group, living near Mathews's house on Buzzards Bay Avenue, included Thomas Hunt Morgan next door, Wilson across the street, Conklin and Jacques Loeb nearby, among many others. Harvey recalled, "I can still see him, walking briskly with great strides along the streets of Woods Hole, with his head held high and a keen penetrating look in his blue eyes, as if he were about to lay bare the secrets of the universe. His convictions were strong and his ideals high" (Harvey 1958a, 744). We see those strong convictions in his highly idiosyncratic essay in the Cowdry volume.

Merkel H. Jacobs

Merkel H. Jacobs was born in 1885 in Harrisburg, Pennsylvania. He received his Bachelor's and PhD degrees from the University of Pennsylvania, and after a year in Berlin, he returned to the university in protozoology. He spent the war years in the Sanitation Corps and then returned to Penn in 1921, remaining there until he retired in 1955. A memorial by Warner E. Love reported that "he was tenacious of purpose, very hard working, high-principles, and kept his own council. He spoke ill of no one. To those around him, he was above all, gentle" (Love 1971, 16). Shortly after Cowdry published his collaborative volume, Jacobs became the third director of the MBL. He had been a member of the MBL Corporation since 1911 and became associate director in 1925–1926 and then director from 1926–1938, while also directing the physiology course. The 1920s was a lively time for the MBL, with significant growth of the physical facilities and increasing numbers attending the courses and carrying out research. In contrast, the Depression of the 1930s brought serious challenges that Jacobs had to navigate to keep the institution afloat. That he nonetheless managed to continue his studies of cell permeability and to inspire many to enter the field is evident from the dedication of a special issue of the *Journal of Cellular and Comparative Physiology* to Jacobs in 1956 (Harvey 1956).

Ralph Stayner Lillie

Ralph Stayner Lillie, was born in Toronto and received his BA from the University of Toronto in 1896 and his PhD from the University of Chicago. After that he worked at the Nela Research Laboratory in Cleveland, the Carnegie Institution of Washington at Johns Hopkins, and then at Clark University as professor of biology, with positions also at the University of Nebraska, Harvard, Johns Hopkins, and the University of Pennsylvania. In 1924, Lillie moved back to the University of Chicago as professor of general physiology until he retired in 1952. Lillie was intrigued by the dynamic activity of organisms, as reflected in his *General Biology and Philosophy of Organism*. Like his brother Frank, Ralph spent many summers at the MBL as a researcher and as a trustee (Ralph S. Lillie Papers).

Robert Chambers

Robert Chambers was born in Erzerum, Turkey, where his parents served as missionaries. He graduated from Robert's College in Istanbul and received an MA degree from Queens University in Kingston, Ontario, Canada. After earning his PhD in Munich, studying cell physiology and embryology, he spent time at the MBL, first as a researcher, then on the teaching staff, and then directing a summer cell biology laboratory at the MBL. His first academic position was in Cincinnati at the Medical College for three years, then at Cornell University Medical College in New York City until he retired in 1949. His student Irene P. Goldring noted that Chambers encouraged her to go to the MBL embryology course, which she did in 1948, and she later wrote, "My introduction to the life of that community enabled me to hear at first-hand, anecdotes and somewhat apocryphal tales of the legend that Robert Chambers had become." Chambers told Goldring that during his own introduction to embryology, he had said, "Dear God, I believe in the chromosomes, I believe in the spindle, I believe in the asters, now help me to find out what they actually are" (Goldring 1979, 1271). Chambers served as an MBL trustee and regarded summers in Woods Hole as the most important time of his year.

Warren Harmon Lewis

Warren Harmon Lewis was born in 1870 in Suffield, Connecticut, and soon moved to Chicago. He received his BS from the University of Michigan and remained as an assistant for a year. In 1896, he entered the still new Johns Hopkins University Medical School, where he became fascinated by anatomy and graduated in 1900. He studied experimental embryology with Jacques Loeb at the MBL. As his biographer George Corner put it, "A summer with Jacques Loeb could not fail to open the eyes of his young associate to the exciting possibilities of experimental cytology" (Corner 1967, 326). At the MBL, he later met Margaret Reed, whom he married in 1910, and they had three children. Margaret and Warren Lewis worked closely in their shared research, carrying out tissue and cell culture studies designed to observe and document cell movements under different conditions. They improved culture media, made videos of developing cells, and continued exploring movements in living cells. In 1917, Warren Lewis moved to the Carnegie Institution of Washington's Department of Embryology, which was on the Johns Hopkins campus. The study of living cells eventually led them to explorations of how cancer cells behave in different culture conditions. As Corner wrote, "Dr. and Mrs. Lewis led a quiet life of devotion to work in their laboratory. They were seldom seen apart. Their vacations were generally spent at Woods Hole, later at the Mt. Desert Island Biological Laboratory, where they varied their work by observations and experiments on marine organisms" (Corner 1967, 342).

Margaret Reed Lewis

Margaret Reed Lewis was born in Kittanning, Pennsylvania, in 1881. She received her BA degree from Goucher College, then studied at Bryn Mawr College, Columbia University, and abroad at Zurich, Paris, and Berlin, though she did not receive a graduate degree. Her studies included regeneration in amphibians and crayfish, and she served as an assistant to Thomas Hunt Morgan at both Bryn Mawr and Columbia. She taught as an assistant in zoology at Bryn Mawr in 1901–2 and at the New York Medical College for Women in physiology, and then served as a lecturer at Barnard College, and later trained nurses at Johns Hopkins. In 1910, she married Warren Lewis and began a long and fruitful career of collaboration in cell biology, embryology, and related studies. That work built on her 1908 visit to Berlin, where she transplanted guinea pig bone marrow into a solution of nutrients for culturing tissue in an experiment later cited as the first successful culture of mammalian cells. The Lewises went on to develop highly successful approaches to culturing tissues in different growth media and produced impressive videos of the process (Landecker 2004). In 1915, Lewis became a researcher at the Carnegie Institution of Washington Department of Embryology.

Frank Rattray Lillie

Frank Rattray Lillie was born in 1870 in Toronto, and was brother to Ralph Stayner Lillie. He received his BA from the University of Toronto, where he became intrigued by embryology. This led him to the MBL where Charles Otis Whitman recruited Lillie for studies of cell lineage in the freshwater mussel *Unio*. Whitman enticed Lillie to graduate study at Clark University in 1891 and then to the University of Chicago, where Whitman moved in 1892 to take up the first directorship of zoology. Lillie received his PhD degree two years later. At the MBL, Lillie became course instructor for the embryology course when it began in 1893. Lillie held positions at the University of Michigan and Vassar College, then returned to the University of Chicago as assistant professor of embryology and remained there throughout his career as professor, chairman of the Department of Zoology, and dean of the Division of Biological Sciences until he retired. He was a talented administrator, who helped shape and sustain both the University of Chicago and the MBL. His research included a textbook on *The Development of the Chick* in 1908, study of marine invertebrates, and study of freemartins in Chicago. His study of fertilization was regarded as his most important contribution, despite disagreements about interpretation. It is this work, developed in *Problems of Fertilization* in 1919 and updated in 1924 with his student Ernest Everett Just, that they present in the Cowdry volume. That work involved what Lillie referred to as "a working hypothesis" that a substance, "fertilizin," contributed to the joining of egg and sperm. Lillie continued to support his students, especially Just (Willier 1957).

Ernest Everett Just

Ernest Everett Just was born in 1883 in Charleston, South Carolina, and was sent to a boarding school, the Kimball Union Academy in Meriden, New Hampshire. He graduated from Dartmouth College in 1907 *magna cum laude* as a Rufus Choate scholar. He began teaching at Howard University in Washington, DC, and soon became chair of the zoology department. Just began going to the MBL in 1909, where he worked with Frank Lillie, studying the process of fertilization as the starting point for individual development. Just received his PhD from the University of Chicago for this work. It is not surprising that Lillie would invite Just to join him as coauthor on the chapter for the volume. Biographer Kenneth R. Manning has written a definitive and provocative interpretation of Just's scientific contributions and his place in the history of biological sciences as well as in the culture and life of academic society more generally (Manning 1983).

Edwin Grant Conklin

Edwin Grant Conklin had a large presence in cell biology and at the MBL, as seen in stories in an interview taped just days before his death (Conklin 1952). Conklin was born in 1863 in Waldo, Ohio. The family lived on a farm, and Conklin worked while attending a country school with one room and one teacher. He studied natural history at Ohio Wesleyan University and received his degree in 1885. While teaching at the missionary college for blacks, Rust University, from 1885 to 1888, he met and married Belle Adkinson. They had three children. Conklin received his PhD from Johns Hopkins in 1891. He studied embryology, cells, and related topics, while also working to reconcile his Methodist convictions with evolutionary biology. While at Johns Hopkins, William Keith Brooks sent Conklin and other students to the US Fish Commission station in Woods Hole. Very quickly, Conklin learned about the MBL just across the street, where he met Whitman and began his own cell lineage studies under Whitman's encouragement. His study of ascidian eggs became his dissertation work, and his completed work made up 226 pages, 9 plates, and 105 colored figures in the *Journal of Morphology*. Conklin enjoyed explaining how his dissertation very nearly bankrupted the journal. His work demonstrated how cells divide, step by step, and acquire differentiation in their different locations within the embryo. Conklin builds on that work in his essay in the Cowdry volume. At first Brooks had been skeptical of Conklin's proposal to study cell lineage, but in the end Brooks said, "Well, we give students degrees for counting words in classics, so I guess we can give you a degree for counting cells" (Harvey 1958).

Clarence E. McClung

Clarence E. McClung was born in 1870 in Clayton, California. He received a Bachelor's and PhD degree from the University of Kansas and then became professor and dean of the medical school there. In 1912, he went to the University of Pennsylvania as director of zoology and remained until he retired in 1940. He also chaired the Division of Biology and Agriculture of the National Research Council through World War I, serving from 1912 to 1921. After retiring, he spent one year at the University of Illinois as acting director of the Department of Zoology and then became the acting head of the Department of Biology at Swarthmore. His obituary in the *New York Times* reported that his one hundred or so students had honored his "profound influence on individuals and organizations concerned with biological research" (*New York Times* 1946). McClung's study of heredity led him to hypothesize that in grasshoppers, the number of X chromosomes determines the sex of an individual organism. Males lack a second X chromosome, which led to the idea of a sex-determining chromosome and provided early evidence that a particular chromosome carries a definable set of hereditary units and thereby shapes inheritance. His discussions of heredity stimulated others, such as Thomas Hunt Morgan.

Thomas Hunt Morgan

Thomas Hunt Morgan was born in 1866 in Lexington, Kentucky, into a family with deep roots in US history. Morgan received his BS degree from the University of Kentucky and his PhD from Johns Hopkins in 1891, following Wilson and along with Conklin. From Hopkins, Morgan followed Wilson to Bryn Mawr College, because Wilson had just left for Columbia. One of his students, Lilian V. Sampson, was especially notable, and he married her in 1904. They worked together on embryological research at Columbia and the MBL. In 1904, Morgan again followed Wilson to Columbia, where he remained until he left for Caltech in 1928. Morgan spent most summers at the MBL as a trustee and an active researcher. Morgan is best known for his work on fruit fly genetics, for which he received a Nobel Prize. Yet his work on regeneration of planarians, earthworms, and hydra, culminating in *Regeneration* in 1901, continues to play a role in stimulating how we interpret regeneration. Morgan studied many different species, and continued to do so even after achieving his reputation with flies (Sturtevant 1959; Allen 1978).

Initially, Cowdry's volume was to include an essay by Jacques Loeb on physical chemistry, with a special focus on proteins. Loeb was one of the luminaries of late nineteenth- and early twentieth-century biology. He asked challenging questions and posed provocative interpretations about the nature of life, and he supported a physicochemical interpretation of the mechanics of organisms. Loeb's work was always stimulating, and the contributors had surely benefited from Loeb's presence at the MBL over the years. Unfortunately, Loeb became ill and died in 1924 (Pauly 1987).

THE VOLUME

In his introduction to the volume, Wilson pointed to three periods of studying cells as he had with his own book, but in somewhat different terms. The first involved the early, largely structural, cell theory, while the second brought in modern cytology and embryology. The third involved Mendelian heredity, and therefore genetic analysis of cell phenomena and more study of the details of the cells. This period required bringing together cell morphology and physiology, biophysics and biochemistry, embryology, and genetics, and all together were leading to a new, "many-sided cellular biology" with increasing cooperation among approaches and among researchers (Cowdry 1924, 10). Wilson noted that "it is hardly possible to arrive at complete unity in a work produced by several collaborators representing widely diverse fields of research. Such a group, however, can at least bring to their task a broader and more critical knowledge of the subject than any single writer can at this day hope to command" (Cowdry 1924, 11). This commentary by the preeminent cell biologist set up the volume in a way that allowed each author considerable individual control over his topic.

Rather than providing a review of each chapter, it is useful to reflect on the approach of some of the chapters and the resulting whole. Many of the chapters seem rather surprising in modern terms. In discussing chemistry, Mathews entitles his first section "Chemistry and Psychism" and discusses the chemistry of "mentality" and vital forces as well as such mystifying ideas as the "psychology of hydrogen." He invokes Sir Oliver Lodge's etherions, and emphasizes the importance of providing an explanation for how the living is created from the nonliving. Despite the rather remarkable quirks in his eighty-page chapter, he also covers a lot of contemporary discussion of molecular chemistry, even if his chemistry colleagues would not have recognized some of it. Mathews did not have the most forward-looking view when he concluded his chapter by noting that little was known about the chemistry of genes and that existing knowledge therefore seemed to weigh against the gene theory. Mathews nonetheless showed that, while the biochemistry of cells might leave much open for question, it was essential for understanding cells.

In contrast, Jacobs's discussion of cell permeability explicitly acknowledged how little was known and yet how important it would be to know about the process of crossing cell membranes. Cells do not contain everything that they will ever need from the beginning, so they must have permeability. But how that permeability works and what controls the diffusion of materials across the membrane remained unknown.

Ralph Lillie asked about how cells react to their environments. Understanding their reactions requires knowing about both the stimuli and the responses. As Lillie's chapter explores possible chemical, mechanical, electrical, and other possible factors influencing reaction, it becomes clear that here, too, remained many open questions. Yet Lillie introduced the idea that cells do react and are not completely autonomous or insular. The ability of cells to react to stimuli from outside is what makes cell-cell interactions, as they were called later, actually work; it allows individual cells to work together as whole coordinated and organized organisms. Ralph Lillie's chapter really just points to the interactions, which became much more important in later decades.

The first four chapters, including Wilson's introduction, all raised many more questions than answers. Clearly, a community of researchers had come to recognize that cells are themselves complex and are also parts of complex systems, requiring diverse kinds of methods and questions to increase understanding.

In his seventy-page chapter, Chambers provided a much more definitive report on the results of microscopic techniques for establishing the physical structure of protoplasm. Chambers acknowledged that there was more to be learned about asters and other details of the cleavage process, but also that researchers had already learned an astonishing amount about both the cytoplasm and nucleus. Whereas Mathews, especially, had veered toward the theoretical and abstract, Chambers grounded his discussion in concrete observations. The same is largely true also of Cowdry's own seventy-page chapter on the cellular parts—mitochondria, Golgi apparatus, and chromidial substance. Thus, we see a diversity of methods and approaches as well as topics.

In their sixty-two-page chapter, Warren and Margaret Lewis introduced experimental approaches, looking at cells in tissue culture. As the acknowledged authorities on this topic, their work focused on laying out the technique and observing how cells behave as a result of being moved to artificial media. Theirs is the only chapter in the volume with photographic plates, which had become a standard way to demonstrate the results of tissue culture experiments. They discussed work with a number of different kinds of cells and concluded the chapter with a short section on cell death in culture. In brief, cells in culture eventually die, they reported, and they did not know why. In fact, it took several more decades before researchers began to sort out factors leading to the death of cells under normal conditions as well as in the artificial conditions of tissue culture.

Fertilization was a more familiar topic by the 1920s, and it is worth noting that Cowdry included Frank Lillie and Just rather than Loeb on the eighty-five-page chapter on the topic. Lillie and Just, on the one hand, and Loeb, on the other, had rather different interpretations of what happens at fertilization and of the extent to which the process is strictly mechanical and chemical-physical (as Loeb said) or involves a special substance called fertilizin (as Lillie maintained) (Pauly 1987; Manning 1983). While Loeb had worked for a number of years at the MBL, Lillie served as the second and long-term director of the MBL. It is therefore not surprising that Cowdry included Lillie's and Just's interpretation and did not discuss the controversies. Perhaps because the debates had continued for a number of years already, this chapter comes across as more specific and established than some of the others.

The same is true of Conklin's forty-eight-page chapter on cellular differentiation. Conklin was a leader in examining the cell lineage in several invertebrate organisms and then pursuing the causes and patterns of differentiation in each case. Differentiation takes cells from a more general to a more specialized state and constitutes development, Conklin explained. The process of differentiation occurs because of changes in both the nucleus and the cytoplasm. For Conklin in 1924, it was not the genes that drive development, however. He held a common view of those focused on cells and embryology that "the genes or Mendelian factors are undoubtedly located in the chromosomes, and they are sometimes regarded as the only differential factors of development, but if this were true these genes would of necessity have to undergo differential division and distribution to the cleavage cells, as Weismann maintained. Since this is not true, it must be that some of the differential factors of development lie outside of the nucleus, and if they are inherited, as most of these early differentiations are, they must lie in the cytoplasm" (Cowdry 1924, 601). That sounds misguided, or at least over-simplified, to us today, but it made sense at the time in the face of existing evidence.

Two chapters followed Conklin's and focused on the nucleus and its contents. McClung's seventy-nine pages on the chromosome theory of heredity and Morgan's forty-two pages on Mendelian heredity focused on the contents of the nucleus in germ cells. Look inside the cell for the driver of living processes, these approaches said. Chromosomes and chromatin carry heredity—somehow. McClung concluded that "the chromosome theory as it stands is logical, consistent, and generally applicable to both plants and animals. Admittedly incomplete, it yet stands as one of the highest achievements in biology and offers the most promising guide to further advances" (Cowdry 1924, 682). Morgan emphasized that even though the nucleus and hereditary genes might contribute to driving what happens in the cell, the cytoplasm remains essential as well. We need, Morgan suggested, more information about the "physiological processes that take place in the chromosomes and in the cytoplasm" (Cowdry 1924, 728). That ended the book.

Throughout, the authors noted the need for more information, more understanding, and more success in putting together a picture of the complex organism and its interacting parts. Cells are organic units, and they are in a real sense alive. But they are not the sole factor in making up living organisms and are also responsive to environmental conditions and to changing internal conditions. Most of the authors went on to further studies that expanded on their summary review approaches here, and some of the ideas here were left behind with time. Yet the overall picture is one of growing understanding of the need for multiple approaches, perspectives, and interpretations of cell structure, function, and interactions. Many questions remained, with many opportunities for further study, some of which are picked up in other chapters in this volume.

REACTIONS TO *GENERAL CYTOLOGY*

Overall, reviewers responded enthusiastically to the edited volume. They recognized the challenges of having thirteen authors with relatively short contributions on each topic (though they seem relatively long as chapters). And, as always with edited volumes, they liked some sections better than others. All acknowledged that the authors were all leaders in their respective fields. And several noted that the volume could not have been written by any one person alone. It took a group to provide what they all acknowledged as an authoritative, comprehensive, and overall very impressive laying out of the contemporary study of cells.

A review in the *Nation* began by noting that "the summer capital of biology in America is at Woods Hole, Massachusetts. A couple of years ago some dozen of the leaders of this scientific convent decided that there should be a new book about cells" (Thone 1925). Another review by Raoul M. May in the history of science journal, *ISIS*, reviewed Cowdry's volume and Wilson's third edition together. He concluded that "too much good cannot be said of these two great contributions to science. While, however, Wilson's book is a milestone, the combined studies of the American investigators which together form *General Cytology* are a stepping-stone. Wilson has mainly elaborated, as in the previous editions of his book, on questions concerning cellular morphology, while *General Cytology* includes a great deal which is physiological in nature. The two books together are a splendid *mise au point* of all that is known concerning that most fundamental of all living structures—the cell." Wilson's was a classic volume, marking the state of a field, while Cowdry's also pointed to new ideas and directions for future study (May 1925, 214). Another review by Arvilla Meek Taylor in the *Chicago Evening Post Literary Review* summarized the book and concluded with a call for more such collaborative projects, "for nothing will do more to advance the cause of science as a whole than such efforts as this" (University of Chicago Archives).

Other reviews offered similar views, though a number of them did find parts of Mathews's chapter on chemistry decidedly odd. Cytologist J. Brontë Gatenby provided a long and detailed review, in which he pointed out what he found missing or misleading in places, though he applauded the volume as a whole. He referred to Mathews's discussion of chemistry and psychism, in particular with regard to understanding how life emerges. It is worth reflecting on this point more closely, because Cowdry, as editor, had allowed this part to remain. The persistence of the views Mathews espoused shows that in 1924 biologists were not yet clear on how cells gain life, nor how they make up living organisms. Mathews wrote, "It is in fact the luminiferous ether which has made things alive, for ether is the storehouse of energy; it is itself nothing else than space and time; energy and time" (Cowdry 1924, 185). Today, Gatenby's response seems reserved in its critique: "It is impossible for a working cytologist adequately to comment on such passages. They may mean something to the metaphysician, but one cannot help feeling that Prof. Mathews' views on the relationship between cell lipins and cell proteins, or on the biochemistry of development, would have been more useful" (Gatenby 1925, 186).

A few critics went further, suggesting that Mathews's chapter ought not to have been included at all. Wilder B. Bancroft, writing a six-page review for the *Journal of Physical Chemistry* noted that the Mathews chapter was decidedly the weakest. After quoting passages related to psychism and the soul of atoms, Bancroft concluded that "this sort of speculative metaphysics may be justifiable in a popular article; but it should not have been allowed in a book like this" (Bancroft 1925, 107). Fortunately, all the reviewers seem to have agreed that the other chapters ranged from very useful to excellent.

Indeed! Yet what Mathews shows, alongside the collection of chapters, is the range of ideas about cells that were available in 1924. While researchers had learned a great deal, much remained to be learned. Cowdry's volume was, in fact, a stepping stone. And the steps forward lead to more study of areas that are only hinted at in the Cowdry volume but became increasingly important, such as cell-cell interaction, cell signaling, gene transcription and regulation, and so on, including the range of topics discussed in other essays in our volume.

Deeper understanding of the cells themselves, if not of their interactions, began to appear in the 1928 and 1932 editions of *Special Cytology* (of two and three volumes, respectively) that Cowdry edited. Here, too, Cowdry brought together a collection of authors. He explained that the purpose was to present in more detail knowledge about the different kinds of cells. "The book," he explained, "is to be regarded as supplementary to an earlier volume called *General Cytology*." There, the authors looked at "the fundamental principles of architecture and activity which cells possess in common" (Cowdry 1928, vii). *Special Cytology* looked instead at the characteristics of specialized cells. The thirty-seven chapters allowed room for a variety of types of cells as well as some of the methods used to study them.

Conclusion

This brings us back to the 1940 celebration of a century of cells in *The Cell and Protoplasm*. That 205-page volume included relatively short chapters by very distinguished researchers on cells, protoplasm, cell walls, chromosomes and genes, enzymes, molecular structure, plant hormones, vitamins, differentiation, physiology, viruses, microorganisms, techniques, and a chapter by Charles Kofoid on "Cells and Organisms."

While these researchers had acquired more knowledge about details by this 1940 symposium and volume, it is striking how many of the papers again include acknowledgements about how much remained to be learned. Conklin noted that "the mystery of mysteries is not the mechanism of evolution, but the evolution of the mechanism by which cells and protoplasm came to have the organization that has resulted in 'the promise and potency of all life.' This is the great problem which is sure to occupy increasingly the attention of biologists in the future" (Moulton 1940, 18). Richard Goldschmidt pointed to the need to get past thinking in terms of individual particulate genes and to focus instead on connections and chromosomes, even though this work may be harder and not fully understood yet. But the volume showed that "nothing is gained by hiding the head in the sand, or by erecting sign-boards 'Verboten' or by calling names" (Moulton 1940, 66). Other contributions pointed to the lack of completeness, or to remaining unaddressed questions. As in 1925, cytology in 1940 was still in its early stages.

In 1959 Jean Brachet and Alfred E. Mirsky edited *The Cell: Biochemistry, Physiology, and Morphology*, which grew to five large volumes and showed how much, and in what ways, the field had expanded. They recognized the tremendous recent advances in molecular biology and genetics, and the ways that understanding the complex interactions of morphological and physiological factors, grounded in biochemistry, had truly revolutionized the understanding of cells. It is clearly true that the knowledge available had expanded, and that the way researchers understood cells and their roles had changed.

By the 1960s, researchers began to discover the details of and reasons for cell cycles, finding that cells go through predictable stages following molecular triggers. Lee Hartwell, Paul Nurse, and Timothy Hunt are credited with having observed that cycles occur and having worked to understand underlying mechanisms. They shared the 2001 Nobel Prize in Physiology or Medicine for their respective contributions, and the study of cyclin and cell cycles became a core way to interpret cell division (Nobel Prize 2001). Along with a growing understanding of cell death, the cell cycle work helped reinforce the idea of a cell that is itself alive, matures and specializes, and then dies.

Many other contributions added to our understanding of how each cell works, how they interact, and how that interaction makes up a whole organism that changes and responds to its environment. The regulation of that process, its timing, and the factors that influence it all come from the environment of other neighboring cells as well as from within the individual cell itself. Cells behave in part in response to their neighbors. They are therefore living units themselves, yes, and also parts of larger, whole, integrated complex systems. Finally, we have moved closer to integrating the different ideas about the roles of cells that Cowdry's group and others were trying to grasp. Yet, as those researchers all noted, there remains much more to be learned.

Acknowledgments
Thanks to the National Science Foundation, Arizona State University, and the Webster Foundation for generous and timely support, and to Karl Matlin, Richard Creath, and Kate MacCord for valuable suggestions.

References

Allen, Garland E. 1978. Thomas Hunt Morgan, the Man and His Science. Princeton, NJ: Princeton University Press.

Baer, Karl Ernst von. 1834. "Die Metamorphose des Eies der Batrachier vor der Erscheinung des Embryo und Folgerungen aus ihr für die Theorie der Erzeugung." *Müller's Archiv für Anatomie, Physiologie, und wissenschaftliche Medizin*, 481–509.

Baker, John Randal. 1948. "The Cell-Theory: A Restatement, History, and Critique–Part I." *Quarterly Review of Microscopical Science* , series 3, 89 (1): 103–25. (All parts are reprinted in Baker 1988.)

———. 1988. *The Cell-Theory: A Restatement, History, and Critique.* New York: Garland Publishing.

Bancroft, Wilder B. 1925. "Review of *General Cytology.*" *Journal of Physical Chemistry* 29 (1): 106–11.

Billings, Susan M. 1971. "Concepts of Nerve Fiber Development, 1839–1930." *Journal of the History of Biology* 4: 275–305.

Bracegirdle, Brian. 1978. *A History of Microscopical Technique.* Ithaca, NY: Cornell University Press.

Conklin, Edwin Grant. 1952. "Reminiscences of Dr. Edwin Grant Conklin (1863–1952), Biologist." Parts 1 and 2. Deposited in the American Philosophical Society Archives. Available online at http://history.archives.mbl.edu/archives/formats/audio.

Corner, George W. 1967. "Warren Harmon Lewis, 1870–1964." In *Biographical Memoirs*, 323–58. National Academy of Sciences. Washington, DC: National Academies Press.

Cowdry, Edmund V., ed. 1924. *General Cytology: A Textbook of Cellular Structure and Function for Students of Biology and Medicine.* Chicago: University of Chicago Press.

———. ed. 1928. *Special Cytology.* New York: Paul B. Hoeber; 2nd ed., Chicago: University of Chicago Press, 1932.

Gatenby, J. Brontë. 1925. "*General Cytology*: A Textbook of Cellular Structure and Function for Students of Biology and Medicine." *Nature* 115 (2884): 185–87.

Goldring, Irene P. 1979. "Robert Chambers, 1881–1957." *American Zoologist* 19: 1271–73.

Harris, Henry. 1999. *The Birth of the Cell.* New Haven, CT: Yale University Press.

Harvey, E. Newton. 1956."Merkel Henry Jacobs and the Study of Cell Permeability." *Journal of Cellular and Comparative Physiology* 47: 5–10.

———. 1958a."Albert Prescott Mathews, Biochemist." *Science* 127: 743–744.

———. 1958b. "Edwin Grant Conklin, 1863–1952." In *Biographical Memoirs*, 54–91. National Academy of Sciences. Washington, DC: National Academies Press.

Hughes, Arthur. 1959. *A History of Cytology.* London: Abelard-Shuman.

Huxley, T. H. 1853. "The Cell-Theory." *British Foreign Medico-Chirurgical Review* 12: 285–314.

Landecker, Hannah. 2004. "The Lewis Films: Tissue Culture and 'Living Anatomy,' 1919–1940." In Jane Maienschein, Marie Glitz, and Garland Allan, eds., *Centennial History of the Carnegie Institute Department of Embryology*, 117–44. Cambridge, MA: Cambridge University Press.

Laubichler, Manfred D., and Eric H. Davidson. 2008."Boveri's Long Experiment: Sea Urchin Merogones and the Establishment of the Role of Nuclear Chromosomes in Development." *Developmental Biology* 314 (1): 1–11.

Lillie, Frank R. *Problems of Fertilization*. 1919. Chicago: University of Chicago Press.

———. 1908. *The Development of the Chick: An Introduction to Embryology*. New York: Henry Holt.

Lillie, Ralph S. 1945. *General Biology and Philosophy of Organism*. Chicago: University of Chicago Press.

———. Various papers. Ralph S. Lillie Papers. University of Chicago Archives. Online at https://www.lib.uchicago.edu/e/scrc/findingaids/view.php?eadid=ICU.SPCL .CRMS58).

Love, Warner E. 1971. "Markel Henry Jacobs" in the report for 1970. *Biological Bulletin*. 141(1): 15–17.

Maienschein, Jane. 1983. "Experimental Biology in Transition: Harrison's Embryology, 1895–1910." *Studies in History of Biology* 6:107–27

———. 1991. *Transforming Traditions in American Biology, 1880–1915*. Baltimore, MD: Johns Hopkins University Press.

———. 2011. "'Organization' as Setting Boundaries of Individual Development." *Biological Theory* 6 (1): 73–79.

———. 2014. *Embryos Under the Microscope: Diverging Meanings of Life*. Cambridge, MA: Harvard University Press.

Manning, Kenneth R. 1983. *Black Apollo of Science: The Life of Ernest Everett Just*. New York: Oxford University Press.

May, Raoul M. 1926. "Book Review: *General Cytology*, E. V. Cowdry; *The Cell in Development and Heredity*, Edmund B. Wilson." *Isis* 8 (1): 213–15.

Morgan, Thomas Hunt. 1940. "Edmund Beecher Wilson, 1856–1939." In *Biographical Memoirs*, 315–42. National Academy of Sciences. Washington, DC: National Academies Press.

Moulton, Forest Ray, ed. 1940. *The Cell and Protoplasm*. Washington, DC: American Association for the Advancement of Science, The Science Press.

New York Times. 1946. Obituary for Clarence E. McClung, January 19.

Nobel Prize. 2001. "The Nobel Prize in Physiology or Medicine." Online at http://www .nobelprize.org/nobel_prizes/medicine/laureates/2001/.

Park, Hyung Wook. 2008. "Edmund Vincent Cowdry and the Making of Gerontology as a Multidisciplinary Scientific Field in the United States." *Journal of the History of Biology* 41: 529–72.

———. 2016. *Old Age, New Science: Gerontologists and Their Biosocial Visions, 1900–1960*. Pittsburgh, PA: University of Pittsburgh Press.

Pauly, Philip J. 1987. *Controlling Life: Jacques Loeb and the Engineering Ideal in Biology*. New York: Oxford University Press.

Raspail, François-Vincent. 1833. *Nouveau système de chimie organique, fondé sue des methodes nouvelles d'observation*. Paris: Ballière. Quoted in Harris 1999, 32.

Richmond, Marsha. 2002. "Thomas Henry Huxley's Developmental View of the Cell." *Nature Reviews Molecular Cell Biology* 3: 61–65.

Schwann, Theodor. 1839. Published in English as Schwann, Theodor. 1847. *Microscopical Researches into the Accordance in the Structure and Growth of Animals and Plants*. Translated by Henry Smith. London: Printed for the Sydenham Society. Available

online at http://www.biodiversitylibrary.org/ia/microscopicalres47schw#page/9/mode/1up.
Sedgwick, Adam. 1894. "On the Inadequacy of the Cellular Theory of Development, and on the Early Development of Nerves, Particularly of the Third Nerve and of the Sympathetic in Elasmobranchii," *Quarterly Journal of Microscopical Science*, series 2, 37 (145): 87–112.
Sharp, Lester W. 1943. *Fundamentals of Cytology*. New York, McGraw-Hill.
Sturtevant, A. H. 1959. "Thomas Hunt Morgan, 1866–1945." In *Biographical Memoirs*, 283–325. National Academy of Sciences. Washington, DC: National Academies Press.
Thone, Frank. 1925. "A Monument in Biology." *Nation*, April 1, 361.
University of Chicago Archives, Special Collections of the University of Chicago Library, box 127, folder 1. Thanks to Karl Matlin for finding and sharing this information.
Whitman, Charles Otis. 1893. "The Inadequacy of the Cell-Theory of Development," *Journal of Morphology* 8: 639–58.
Willier, B. H. 1957. "Frank Rattray Lillie, 1870–1947." *Biographical Memoirs*, 179–236. National Academy of Sciences. Washington, DC: National Academies Press.
Wilson, E. B. 1895. *An Atlas of Fertilization and Karyokinesis of the Ovum*. New York: Macmillan. Available online at http://www.biodiversitylibrary.org/bibliography/6244#/summary.
———. 1896. *The Cell in Development and Inheritance*. New York: Macmillan.
———. 1925. *The Cell in Development and Heredity*. 3rd ed. New York: Macmillan. Orig. pub. as *The Cell in Development and Inheritance* in 1896; 2nd ed., 1900.

CHAPTER 3

IN SEARCH OF CELL ARCHITECTURE

GENERAL CYTOLOGY AND EARLY TWENTIETH-CENTURY CONCEPTIONS OF CELL ORGANIZATION

Andrew Reynolds

In chapter 2, Jane Maienschein discusses how biologists grappled with the conceptual difficulties concerning the cell's status in relation to the living organism as a whole: was the cell merely a structural unit of life, a part subordinate in status to the organism? Or was it the fundamental and primitive unit of life, a living whole primary in some sense to multicellular organisms? This chapter is concerned with another development in the conceptualization of the cell in the late nineteenth and early twentieth centuries: this is the description of the living cell as an *organized system of heterogeneous parts.*

By the first two decades of the twentieth century, many biologists had become convinced that the cell's vital properties were to be ascribed neither to some mysterious property of protoplasm (construed as a homogenous and simple substance) nor to any more elementary living molecule or entity hypothesized to exist within the cell. If the cell truly is the fundamental unit of life, then life must be a property of the cell as a whole that arises from the systemic organization of all its component parts. This meant proposing that the chemical molecules of living cellular protoplasm were organized in some fashion. The chief difficulty was that, aside from a few notable structures (organelles) embedded in the cytoplasm, this internal organization, or cell "architecture," could not be seen with the light microscope, though something of its nature could be inferred on the basis of biochemical and biophysical investigations.

In this chapter I describe how researchers working in the period of Cowdry's *General Cytology* (roughly the first three decades of the twentieth century) went about feeling their way toward an understanding of what this sub-visible organization might be like. Most of these biologists adopted biochemical and biophysical approaches to cellular structure and function, but for reasons explained by Karl Matlin in chapter 11, these techniques were not able to reveal clearly the internal organization of the cell's protoplasm. In order to help them think through how the living matter might be spatially arranged so as to achieve such things as the temporally ordered processes of metabolism or the creation and transmission of electrical charge, these "architects of the cell" (if we might call them that) drew comparisons to familiar examples of organized systems, such as laboratories, batteries, and machines.

Biologists were, then, somewhat "in the dark" when it came to the cell's internal organization, but by means of analogy and metaphor they grasped their way toward a conception of the cell as something more than a biological atom, a primitive speck of living jelly, or even an elementary organism (Kyne and Crowley 2016). They began sketching a picture or blueprint of the cell as a complex, dynamic, and highly ordered system, one whose structural and functional architecture is today understood in the modern technological terms of electronic computer circuitry and networks of signaling pathways. Our modern understanding of the cell as a complex system of molecular components with a logical architecture of genetic circuits and subroutines was traced out in an earlier time and on the model of earlier technology. A closer look at these developments in our understanding of the cell reveals both how much has changed and how much has remained the same since the 1920s.

General Cytology: Uniting Structural and Functional Considerations of the Cell

At the beginning of the twentieth century, several distinct and specialized lines of investigation into the cellular basis of living organisms had developed. These included the long-established approach of descriptive morphology ("microscopical anatomy") and the newer, more experimental and physiologically focused approaches of biochemistry, experimental embryology, and cellular genetics. As these separate ventures made progress individually and their conceptual, theoretical, and methodological principles were more fully articulated, the need for a more synthetic and integrated picture of the cell as a whole became apparent. This was the task set by a group of leading researchers which met at the Marine Biological Laboratory at Woods Hole in the summer of 1922 under the editorial direction of Edmund V. Cowdry of the Rockefeller Institute for Medical Research in New York City.

Cowdry (1924b, v) explained that the objective was to "present briefly for the first time within the scope of a single volume data concerning the cell fundamental, alike, to the sciences of botany, zoology, physiology, and pathology [and] . . . to emphasize the results obtained in different lines of work bearing upon the cell, as the fundamental unit in health and disease."

Achieving this, he noted, would involve a "close rapprochement between physicochemical and morphological points of view." The concrete result of these efforts was the book *General Cytology: A Textbook of Cellular Structure and Function for Students of Biology and Medicine* (University of Chicago Press, 1924). A more abstract result was the eponymous field of general cytology.

The chief instruments of cytologists were the optical microscope and selective dyes with which they studied cell constituents like the nucleus, centrosome, mitochondria, and the protoplasmic substance within whole cells. Cell physiologists, on the other hand, studied the chemical activities of extracts and "juices" squeezed from broken cells. Both, however, were thinking their way toward a conception of intracellular organization. Cowdry, for instance, talks of cellular "architecture" in the preface to the later volume *Special Cytology*: "In 'General Cytology' the fundamental principles of architecture and activity which cells of different kinds possess in common were discussed by a group of workers chiefly recruited from the biological sciences. This involved, primarily, a rapprochement between physicochemical and morphological points of view, which is one of the most recent and profitable departures in cytology" (Cowdry [1928] 1932, ix).

To speak of the cell as having an "architecture" is to say more than that it contains various structures (such as chromosomes, Golgi apparatus, etc.); it is to suggest that these visible structures are arranged or organized within the cell in a specific manner. And as several of the chapters in *General Cytology* reveal, it is also to suggest that protoplasm and its ultramicroscopic constituents (the enzymes and other chemically active molecules) are likely spatially organized as well. The chief difficulty was in discerning what this internal cell organization might look like.

A review of the volume observed that "its chief value lies in the exposition of cellular physiology, and in the fact that it presents the cell as a dynamic whole, a viewpoint which has been largely ignored in books of this type up to now" (May 1926, 214). In his introduction Edmund B. Wilson (1924, 10) spoke of how the earlier tradition of morphological cytology had entered into closer cooperation with more experimentally focused biophysics and biochemistry, creating a new and integrated "cellular biology," which he hoped would become a systemic and organized disciplinary whole. But achieving this goal required a conception of the cell itself as an integrated, organized, and systemic whole. This would be a holistic perspective, in the sense that it was the *organization* of the cell's microscopic and ultramicroscopic structures that made of it a system with the capacity for life. So despite the strategy of many cytologists to investigate particular cell constituents and to tease out what, if any, was their function, or of biochemists to grind up cells so as to explore the chemical activity of their molecular components, the cell remained a fundamental scientific concept because it was believed that only from the cellular organization of physical and chemical elements could life arise as an emergent, system-level property. But to get to this point, biologists worked through several radically distinct conceptions of what they understood a cell to be. The next section traces some of the important developments in the cell concept as a means of describing the conceptual context in which the field of general cytology was created.

History of Distinct Conceptions of the Cell

"The word cytology denotes the study of cells," E. B. Wilson wrote in his introductory essay to *General Cytology* (Wilson 1924, 3). Wilson and Cowdry, like many biologists (but by no means all), considered the cell to be the "fundamental unit of life" and thus the starting point for the investigation and understanding of life, health, and disease.[1]

But why *cell* and why *cytology*? Why these terms? These terms—*cell* from the Latin *cellula*, *cytology* from the Greek *cytos*, and similarly for the German equivalent *Zelle*—all denote an empty space, a vessel, chamber, booth, or box. Robert Hooke first used the term *cell* in 1665 to describe the compartmentalized structure of dead plant tissue as it appeared to him under the microscope (Hooke 1665). It reminded him of the hexagonal units of a bee's honeycomb, which were called *cells* in metaphorical reference to a storeroom or chamber. Hooke also described these and similar structures visible only with aid of the microscope as "pores," "boxes," "bladders," "bubbles" and "caverns." Others would refer to these and similar structures visible in other (typically plant) tissues also as "vesicles" and "chambers." Use of the term *cell* seems to have caught on more widely because the German scientists Matthias Schleiden and Theodor Schwann used its German equivalent (*Zelle*) in their essays, which were considered foundational of the cell theory in the late 1830s. These essays were translated into English in 1847 using the term *cell* (Schleiden 1847; Schwann 1847). *Cytology* then became the proper Greek term for the study of, or science specially devoted to, cells, but not, as Wilson (1924, 3) explains, until the 1870s, when improvements in microscope design and techniques for fixing and staining material (see Schickore, chap. 4 in this volume) made the internal structure of cells more accessible.

Even by the middle of the nineteenth century, researchers were finding fault with the term *cell* as an adequate description of what was supposed to be a living unit of life, let alone the fundamental unit of all life. Even if it were the case, which turned out not to be so, that all those things to be denoted by the term *cell* had the characteristic structural feature which had inspired the term originally—namely, a well-defined outer wall—this morphological feature could not be considered the fundamentally living element. Many animal "cells" and those of the more "primitive" unicellular forms of life were found to lack a cell wall altogether. What all living units did seem to share was the semifluid and jellylike substance known variously as "sarcode" or "protoplasm," and within it a nucleus. So the cell was redescribed by investigators like Max Schultze as "a naked speck of protoplasm containing a nucleus" (Schultze 1861), and soon T. H. Huxley (who was never sold on the cell theory to begin with) was professing protoplasm to be the "physical basis of life" (Huxley [1868] 1968). But if the term *cell* focused on an inessential structural feature, the cell theory was saved by the practice (traceable back to Schleiden and Schwann) of referring to the cell as a "little organism." In an influential paper of 1861, the physiologist Ernst Brücke asked "what should we understand the term "cell" to mean?" and answered "an elementary organism" (Brücke 1861). Afterward, the focus shifted from an inessential morphological feature to a more essential, physiological one. Attention could now move from the "box," "vessel," or "prison" to its occupant, the living stuff itself.

For the purpose of arguing for the plausibility of evolution (just recently made scientifically respectable by Darwin in 1859), it paid to portray the cell as a rather homogenous blob of protoplasm (and for this role, the amoeba cell was frequently given star billing), making it appear reasonable that such a simple system might evolve quite naturally from more primitive physicochemical materials. This was a conception frequently employed by Ernst Haeckel in his popular writings on evolution (e.g., Haeckel 1876). For the purpose of actually understanding cell physiology and inheritance, however, many agreed with Brücke that, if the cell is indeed a little organism, it must possess an internal organization of its own of even smaller parts by which it is able to carry out all the essential functions of life.[2]

Investigations into the chemical composition of protoplasm successfully revealed the elements of which it is typically composed, but early observations with the light microscope could not conclusively determine what, if any, structure it possessed essentially. Fundamentally, there were two chief ideas: either protoplasm was essentially a contractile solid, with fluid interspersed throughout a structure composed of either a connected, netlike reticulum or a filamentous arrangement of separate fibers; or protoplasm was essentially a viscid fluid, with a foamlike or alveolar structure running throughout. The protozoologist Otto Bütschli favored the latter and published the results of his experiments in creating artificial systems from suspensions of oil and soap in water and other materials that exhibited features similar in some respects to living protoplasm (Bütschli 1894). By the 1920s the idea that protoplasm consists of a colloidal suspension of large molecules in a viscid fluid was gaining in popularity, as were descriptions of the cell as a complex, polyphasic (exhibiting solid, fluid, and gaseous properties) colloidal system (more on this below).[3]

To all concerned, but especially to the physiologists, it was clear that inside cells there is lots of chemistry. And chemical activity involves molecules and molecular structure. Pasteur discovered in the nineteenth century that many molecules display chirality, a spatial property he believed peculiar to living organisms. This revealed that biological activity is tied to the structural features of molecules. Pasteur also believed that fermentation was a peculiarly vital process requiring the presence of intact and live yeast cells. This was refuted in 1897 when Eduard Buchner exhibited cell-free fermentation in a test tube, spurring the development of what have been called the "grind and find" techniques of enzymology (Welch and Clegg 2010). These exceptionally disruptive techniques of the newly emerging biochemistry raised concerns (even among some of its practitioners) that the cell was being treated as a mere bag of enzymes.[4] However, among the pioneers of biochemistry, there was recognition that cell structure and organization were required in order to achieve the levels of efficiency and productivity that were unattainable to them in their structureless test-tube solutions. Within a complete and spatially organized cell, enzymes and their substrates would not be left to diffuse about randomly and to interact by chance alone.

Franz Hofmeister (who coined the term *biochemistry*) speculated that the cell must be a highly differentiated and orderly space that provided favorable environments for the meeting of enzymes and their specific substrates and for the orderly arrangement of the stepwise nature of complex chemical reactions involved in metabolism. "Is the cell as a whole one vessel, filled with a homogenous solution, in which the collection of chemical processes occurs, or does it contain a number of definite separate vessels, in which the undisturbed sequence of individual reactions are secured next to one another in an approximately sequential fashion?" (Hofmeister 1901, 24).[5] Hofmeister compared the internal environment of the cell to an organized chemical laboratory (*Laboratorium*) or factory (*Betrieb*), wherein the proper materials and tools were arranged in close spatial proximity to one another on colloidal structures like machine tools (*Handwerkszeug*; *Zahnrad*; *Räderwerk*) on an assembly line for maximum efficiency (Hofmeister 1901, 11–12, 19, 28, 29).

This conception of the cell as an organized and heterogeneous space was also promoted by the influential Cambridge biochemist Frederick Gowland Hopkins in the early part of the twentieth century. Hopkins opposed the idea that the activities of life are carried out by protoplasm *en masse* rather than in stages under the catalytic influence of a multitude of specific enzymes. "It is clear," he wrote, "that the living cell as we now know it is not a mass of matter composed of a congregation of like molecules, but a highly differentiated system" (Hopkins 1913, 715). He believed that the conception of protoplasm as a homogenous living substance was inhibitory to a proper understanding of the cell's physiology. "The use of the term protoplasm may be morphologically justified but chemically it denotes an abstraction. It is sure that it is made up of parts in which the influence of molecular structure as understood by the chemist is all potent" (Hopkins 1949a, 310).

Experimental studies suggested that metabolism occurs through the agency of intracellular enzymes operating in localized "compartments" in the cell environment (Hopkins 1913, 714–15). Hopkins suspected that the temporal procession of metabolism in orderly stages could be traced to another level of organization lying beneath the limits of microscopical detection, one involving the physical structure of the molecules involved and their spatial localization within the polyphasic colloid environment of the cell. "For the dynamic chemical events which happen within the cell, these colloid complexes yield a special *milieu*, providing, as it were, special *apparatus* and an *organized laboratory*" (Hopkins 1913, 715; emphases added). The specific affinity of enzymes to catalyze particular substrates he attributed to their structural chemistry, following the earlier "lock and key" metaphor of Emil Fischer. In this way their ability to function still in cell-free environments would be due to their structural properties, just like a wrench or other workshop tool, removed from the factory. And if, as Hopkins suspected, some enzymes were attached to membranes or other molecular surfaces, this would create localized compartments of activity where distinct chemical processes could occur simultaneously and with greater efficiency than would be possible if the interaction between an enzyme and its specific substrate were left to chance meeting by random diffusion in the cell's aqueous internal environment. Elsewhere, he invoked a different spatial metaphor to complement his mechanical language when he declared that "the cell, too, has a *geography*," he wrote, "and its reactions occur in colloidal *apparatus*, of which the form, and the catalytic activity of its manifold surfaces, must efficiently contribute to the due guidance of chemical reactions" (Hopkins 1949b, 236; emphasis added).[6]

In summary, in addition to the largely morphological emphases of cytology at the microscopic level, it became apparent that, from a biochemical perspective, the cell cytoplasm must contain a good deal of organization and structure at the molecular or chemical level. That some enzyme activity could be shown to occur in broken or cell-free systems spoke to this internal structure, the enzymes operating as Hofmeister's "machines" or "gears," capable of performing their various chemical work independently of the larger structure and organization of the cell as a whole. On the other hand, it might also suggest that the cell itself as a morphological unit is not an essential feature of the chemical activities associated with life. It is worth noting, however, that although Hopkins defended the techniques of the biochemist which involved destroying the intact cell, he insisted that life as it is commonly understood can only be ascribed to the cell as a whole organized system, not to any of the parts individually (Hopkins 1913, 715).

This emphasis on internal cell structure and organization was an important shift from the idea of protoplasm as "the physical basis of life" to a more mechanistic picture of the cell in terms of the coordinated activity of multiple components. Scientists began to stress the importance of organization within the cell—that function requires both structure and organization.[7] The "tools," or "enzyme equipment" as they came to be known at this time, must be properly arranged next to one another in order to carry out the work required in a timely and efficient manner, and processes involving distinct types of chemical activities (catabolic and anabolic) must be isolated one from the other so that the cell can carry out all the different processes it must to remain alive.[8]

Cells, Systems, and Machines

Biologists in the dawning years of the twentieth century became increasingly aware of the cell's complexity. The discovery by Ernest Starling and Sir William Bayliss of hormones in 1902 and Hopkins's discovery of vitamins in 1912 added to the list of enzymes and more conspicuous structures (organelles) now known to be simultaneously active within the tiny space of the cell. Although it took some time to associate various biochemical reactions with distinct organelles and membranes, biologists at this time were keen to discern some kind of organization within the cell, to see it not just as a random jumble of molecular interactions, but as a coordinated arrangement of activities. Talk of the cell as a kind of "system" becomes increasingly common in the early twentieth century, and although description of living things as machines was in no way new, the practice now provided biologists with a useful model of an organized system composed of multiple parts.[9] This section discusses how biologists employed the concepts of "system" and "machine" (and their offshoot, "mechanism") to confront the problem of cell organization.

An illustration of this shift is evident in the comparative descriptions of the cell in the first and second editions of E. B. Wilson's influential text *The Cell in Development and Inheritance*. In the first edition of 1896, the cell is described as both a "mass of protoplasm containing a nucleus" (*sensu* Schultze) (Wilson 1896, 14) and an "elementary organism" (*sensu* Brücke) (Wilson 1896, 211). But in the second revised edition released just four years later, Wilson added that "life can only be properly regarded as a property of the cell-system as a whole" (Wilson 1900, 29).

The Oxford English Dictionary defines the term *system* thus: "—n. a complex whole, set of connected things or parts, organized body of material or immaterial things; the established political or social order" (*Concise Oxford Dictionary*, 6th ed. 1976, s.v. "system"). This emphasizes the notion of organization. The *Chambers Etymological English Dictionary* provides the following suggestive definition: "—n. anything formed of parts placed together to make a regular and connected whole working as if one machine" (Anonymous, London: 1966, s.v. "system"). Despite the later provenance of this definition, we will see that many early twentieth-century biologists also regarded machines as instructive examples of an organized system.

The third and final edition of Wilson's *The Cell* appeared the year after publication of *General Cytology*. One of the revisions in this much-expanded version was the inclusion of a section titled "The Cell a Chemical Machine." Here Wilson stated that, "physiologically, . . . the cell may be regarded as an apparatus for the transformation and application of chemical energy. In the phrase of Loeb, it is a chemical machine. . . . We assume, as our fundamental working hypothesis, that the specificity of each kind of cell depends essentially upon what we call its *organization, i.e., upon the construction of the cell-machine*, in some sense or other—morphological, physical, or chemical" (Wilson 1925, 635; Wilson's emphasis).

Wilson conceded the great gap between the cell and any human-made machine, yet he insisted as a valuable hypothesis that the difference is not one of kind but of degree. The alternative is to insist that the operation and organization of the cell is entirely *sui generis* and unlike anything else of which we have experience or understanding. This was the position of vitalists, such as the embryologist Hans Driesch, and of some so-called holists (Allen 2005). But while it is indisputably true that living cells and organisms are not literally machines, this stance provides no clues to how cells might be understood. By comparing them to machines biologists gained at least some provisional insight into how they might function; for aside from the assumption that living cells must also obey the laws of energy, machines provide familiar and useful examples of systems of distinct parts organized for the fulfillment of some specific function.

Wilson put great emphasis on the idea of cellular organization, citing Brücke's 1861 essay especially.[10] "The fact of importance to the cytologist is that we cannot hope to comprehend the activities of the living cell by analysis merely of its chemical composition, or even of its molecular structure alone; . . . the cell is an *organic system*, and one in which we must recognize the existence of some kind of ordered structure or *organization*" (Wilson 1925, 670; Wilson's emphases).

Of this notion of organization, Wilson admits, "We cannot, it is true, say precisely what organization is, but we can hardly think of it as other than some kind of material configuration of the protoplasmic substance, and one that involves both a differentiation of parts and their integration to form a whole, as Herbert Spencer long since urged. When, therefore, Loeb (to cite still another physiologist) characterizes the living organism as a chemical or colloidal *machine* ('06) he employs a word that implies the existence of such a configuration" (Wilson 1925, 671).

Jacques Loeb was one of the most notably reductionist and materialist biologists of the early twentieth century. Like Cowdry, he had a permanent position at the Rockefeller Institute and was a frequent summer visitor to the Marine Biological Laboratory at Woods Hole.[11] Loeb began the lectures mentioned by Wilson (published as *The Dynamics of Living Matter*) by explaining, "In these lectures we shall consider living organisms as chemical machines. . . . Living organisms may be called chemical machines, inasmuch as the energy for their work and functions is derived from chemical processes, and inasmuch as the material from which the living machines are built must be formed through chemical processes" (Loeb 1906, 1). Loeb's chief motivation for calling the living organism a machine was to counter the vitalist thesis that life involves some unique force or principle distinct from and in addition to those of chemistry and physics, and in the collection of essays published six years later under the title *The Mechanistic Conception of Life*, Loeb was willing to call even a waterfall a machine (Loeb 1912, 117).

But talk of the body or of cells as machines did more than assert a stance against vitalism. Comparing the cell to various sorts of machines and other contrivances provided investigators with working hypotheses about the cell's internal organization or the architecture that permitted it to function as a system. I turn now to several of the chapters in Cowdry's *General Cytology* to illustrate how machine or mechanistic analogies were employed to guide investigators as they grappled with the problem of discerning the sub-visible organization inside the cell.

The first chapter after Wilson's introduction is Albert Mathews's contribution on the general chemistry of cells. Mathews's discussion is founded on the metaphor that the body is a "living machine"—more specifically, "a battery, with a series of resistances and condensers, made up of conductors and dielectrics" (Mathews 1924, 15). Each of its cells is also a battery, he wrote, which generates electricity in a way analogous to the principles of a wet-cell or chemical battery (Mathews 1924, 20).[12] Consistent with this perspective, Mathews declared that "the biochemist has been transformed into an electrical engineer" (Mathews 1924, 15), whose task was, as we would say today, to reverse-engineer the cell's function and design.

In addition to a very speculative discussion about the nature of energy, atoms, mind, and the ether, Mathews was concerned with the origin of the electric current implicated in the cell's irritability and responsiveness to external stimuli. He rejected the view that it is the result of a simple diffusion of a concentration of charged ions in solution across the semipermeable cell membrane, for this would eventually come to an end and life with it. Life is essentially respiration he insisted and respiration is a matter of oxidation. Guided by his battery metaphor, he suggested that, like the flow of electricity between the two compartments of a battery across a metal conducting wire at which ends oxidation takes place at different rates, so there must be in the living cell a conductor "wire," perhaps consisting of a graphite "rod" of carbon atoms, across the cell membrane. Convinced of the plausibility of this explanation, Mathews declared that "the living cell is in fact a battery" (Mathews 1924, 68). Here Mathews is doing more than simply asserting that the living cell is a machine in the sense that it obeys the same laws of physics and chemistry as nonliving matter; he is using the battery metaphor to guide him in creating a hypothesis about the spatial organization of a particular subset of molecular components in the cell's interior. Ultimately, however, this particular hypothesis would prove inferior to those proposing the existence of "pores" or "channels" in the cell membrane which allow for the passage of charged ions in and out of the cell (see Jacobs 1924 for discussion of this idea in early form).

Ralph Lillie, in his chapter on the "Reactivity of the Cell," also compared the transmission of electrical current in protoplasm to the electrical and chemical principles of a battery (Lillie 1924, 189, 193). Lillie noted how in both cases polarization occurs when a current passes across a suitable boundary: in the case of a battery linked to an electrolytic cell, the boundary is represented by the metallic and electrolytic phases of the electrode system; in the case of a living cell, the relevant boundary may lie between the protoplasm and its surrounding medium or between different regions of protoplasm separated by internal films and boundaries within an individual cell. "Protoplasm and the cell media contain salts and are good electrolytic conductors; and the characteristic structure, aqueous phases separated by thin, alterable, semi-permeable membranes, furnishes the conditions for differences of potential, and hence for the production of electric currents" (Lillie 1924, 190).

Lillie also appealed to the transmission of current along a "passive" iron wire immersed in dilute nitric acid as an analog model for the functioning of a nerve (Lillie 1924, 225). This transmission of a chemical influence at a distance by means of an electrical current, he explained, also relies on the oxidation of a film insulating the metal from the chemical solution. Nerves are not metal wires of course, but Lillie's point is that their function is likely to be consistent essentially with known physical and chemical principles and that there may be basic similarities of structural chemistry between a nerve cell and the passage of electrical current through a film-coated wire immersed in solution.

In addition to making mechanical analogies, Lillie also made frequent use of the idea that the cell is a system. In fact the term appears 142 times in his chapter, out of a total of 224 occurrences in the *General Cytology* volume as a whole. Some of the occurrences of the term *system* in *General Cytology* involve long-established uses such as "circulatory system," "respiratory system," and so on, but many are specific to discussions of the cell or to the nature of protoplasm—for example, in speaking of the cell as a "colloidal system" or a "reactive system."

Ralph Lillie wrote, for instance, that "each living cell, e.g., an egg cell or an epithelial cell, as well as a muscle cell or nerve cell, is a reactive system, one whose physiological activity changes in response to changes in its environment" (Lillie 1924, 169). After discussing briefly the chemical processes taking place within the cell by virtue of which it counts as alive and reactive, he states, "The cell is primarily a metabolic system" (Lillie 1924, 170). He then asks what it is that allows the protoplasmic substance of the cell to behave in its peculiar way:

> What are those special features in the composition or constitution of living matter which render its chemical processes so susceptible to influence by changes in the surroundings? Living protoplasm, as we have seen, is a *complex system*, consisting of a large variety of chemical compounds associated in a special type of structure. Both its chemical composition and its structural or morphological constitution (arrangement of parts) are to be regarded as factors in the determination of its special type of activity. Now the purely chemical composition of protoplasm, considered by itself, does not sufficiently explain its special property of reactivity; . . . we must conclude that reactivity depends on other conditions than the mere presence of these various compounds within the space of a single cell. The chief of these conditions is the special *structural constitution or organization* of protoplasm; apparently this structure is responsible for the special peculiarities of its chemical behavior. (Lillie 1924, 171; emphases added)

Like others, Lillie stressed the importance of the structure and internal organization of the cell, essentially noting that the living cell is not simply a bag of chemicals or homogenous lump of protoplasmic jelly. "Broadly speaking," he wrote, "by the term structure as used in biology we mean the permanent spatial distribution and physical state of the essential constituents of the living system; the term organization has a similar significance, referring especially to those permanent features of structure and composition which underlie or determine the specifically vital properties" (Lillie 1924, 171). One chief means by which the protoplasmic system might be organized into a heterogeneous system capable of performing multiple and contemporaneous chemical activities, Lillie suggested, is through the presence of internal films like the outer membrane which would "subdivide the cell into regions which are chemically and structurally dissimilar. A high degree of chemical differentiation within the limits of a single cell thus becomes possible; the basis for a stable and characteristic chemical organization is thus furnished. In correspondence with this structural differentiation a variety of physiological or metabolic activities may proceed side by side in the same cell without interfering with one another (cf. Hofmeister, 1901)" (Lillie 1924, 177).

Here again is the suggestion promoted by Hofmeister and Hopkins that the internal "geography" of the cell could be organized by internal films into heterogeneous environments wherein distinct chemical and physical activities can occur simultaneously, analogous to the way a highly organized machine shop or factory consists of many specialized tools or parts, (i.e., structures) arranged or organized alongside one another in functional compartments or spaces.

Lillie noted that while many forms of living protoplasm appear optically homogenous under the (light) microscope, only under the assumption that it is a "film-pervaded or film-partitioned system" can one explain the variability of the speed with which chemical reactions proceed in the living cell under conditions of irritation or stimulus. "The regions separated by these films and the films themselves," he wrote, "are in many cases not optically distinguishable; but their presence at many protoplasmic boundaries—e.g., the general surface of blood corpuscles and other cells without distinguishable membranes—can be demonstrated by physiological methods" (Lillie 1924, 177). Only in the case of vacuoles, nuclei, or alveoli, where there are differences in color, structure, or refractive index, were such boundaries visible. The chief physiological characteristic of these film boundaries, Lillie believed, was their semipermeability, which prevented the rapid diffusion of dissolved substances. This, he speculated, was also likely involved in the chemical and electrical phenomenon of cellular irritability and the transmission of a stimulus along nerve and muscle cells (see the next section for further discussion).[13] Lillie summed up his views and his chapter by saying, "According to this conception, the essential basis of transmission, as of the other phenomena of reactivity in living protoplasm, is to be found in the polyphasic and film-partitioned character of the protoplasmic system" (Lillie 1924, 229).

Another concept closely associated with the idea of an organized system is that of *mechanism*, a term that appears eighty-six times in the Cowdry volume. Broadly speaking, mechanism-talk refers to any causal or law-like regularity that can be invoked to explain some phenomenon (but see Bechtel, chap. 13 in this volume, for a more precise account). As the name suggests, however, it is derived from the example of a machine, which is frequently an organized system of parts whose function is to cause a particular effect. While the notion of a mechanism has obvious roots in our human experience with building and disassembling machines, the concept eventually became cognitively dissociated from its metaphorical origins (Nicholson 2012).[14]

In his chapter, "The Physical Structure of Protoplasm," Robert Chambers discussed different theories of the nature of protoplasm: whether it be of an essentially solid though contractile substratum with fluid in its interstices or essentially fluid with an alveolar or foam-like structure interspersed throughout, as championed by Bütschli. Chambers describes how the current colloid theory of protoplasm is most consistent with Bütschli's fluid hypothesis, though noting that the alveolar structure can be removed without compromising the viability of the protoplasmic matrix. He then writes, "The microscope thus far has revealed no structure within this matrix—its colloidal nature is indicated not so much by its appearance, as by its behavior. On the other hand, protoplasm is a cellular unit which cannot exist without its nucleus and its cortex and, therefore, must be regarded not as a "stuff" but as a mechanism consisting of visibly differentiated and essentially interrelated parts" (Chambers 1924, 238).

Focus on the cell as a coordinated system of mechanisms is also apparent in Clarence McClung's chapter, "The Chromosome Theory of Heredity." Speaking of the essential role of the nucleus and the chromosomes for normal development, McClung states that

> if the cell is the structural and functional unit which all our studies tell us it is, the operation of its parts must be not only co-ordinated but continuous, and any effects produced must result from the performance of the entire series and not from parts. This, of course, does not mean that there is no differentiation of function, as some assume, or that the "cell as a whole" is anything but the co-ordinated sum of its parts; it does signify that in a mechanism of differentiated, co-operative members they must all work together all the time. (McClung 1924, 676)

Note the social-agential metaphor here ("co-operative members working together"), despite talk of the cell as a "mechanism." In fact, despite the implication of machine imagery, much mechanism talk assumes merely a regular causal arrangement within some system and is quite independent of any assumptions of a specific and concrete machine configuration.

Other contributors to the volume were more cautious about the use of machine and mechanical metaphors and language. Cowdry, for instance, in his chapter, "Cytological Constituents," considered the term Golgi *ap-*

paratus "unfortunate" because it suggested a "mechanism of rather mechanical type" (Cowdry 1924a, 334). Yet, in commenting on what he felt to be an overly narrow approach to the study of the mitochondrion, he himself employed a mechanical analogy: "To take a familiar example, close study of the mainspring of a watch would not tell us very much unless its behavior was carefully considered in connection with all the other parts of the mechanism" (Cowdry 1924a, 332). Similar analogies were actually quite common among cytologists and biochemists of the time when reflecting on the limitations of their investigative approaches and techniques.

Internal Cell Architecture and the Notion of a Cytoskeleton

The rejection of a rigid cell-wall as an essential feature of a cell (the protoplasmic cell concept replacing the original conception of the cell) made the attribution of internal organization in the cell a little more tricky, for the cell is in many instances a shape-shifter, a plastic speck of protoplasmic jelly (think of the changes it undergoes during division and the phenomenon of cell motility). How then to maintain organization and structure in such an environment or system so unlike a machine? Edwin G. Conklin (1924, 562–63) proposed that the cell contained a flexible internal framework he called "spongioplasm" that would orient the centrosomes, spindles, and other components involved in division and organize other cell activity. In 1931 the French embryologist Paul Wintrebert referred to Conklin's spongioplasm as a *cytosquelette*, or "cytoskeleton" (Zampieri, Coen, and Gabbiani 2014). This attempt to invoke some notion of organization internal to the cell was taken up by the biochemist Rudolph Peters in his 1929 Harber Lectures on "coordinative biochemistry," in which he proposed the cell's "architecture" to consist of "an organized network of protein molecules, forming a three-dimensional mosaic extending throughout the cell" that would allow for "independent chemical reactions . . . [to] proceed simultaneously in various parts of the cell" (Peters [1929] 1963, 293). The idea was further elaborated by the physiological embryologist Joseph Needham, a former student of Hopkins, who first used the English term *cytoskeleton* in 1935 to denote this internal cell architecture (Teich 1973; Zampieri, Coen, and Gabbiani 2014).[15]

The term *architecture*, as applied to cells, appears seven times in *General Cytology* (pp. 7, 173, 333, 611, 715), in each case to highlight structure or organization. As noted above, Cowdry, in his preface to *Special Cytology*, described the fundamental focus of *General Cytology* to be the "architecture and activity which cells of different kinds possess in common" (Cowdry

[1928] 1932, ix). In his chapter on the cellular constituents, Cowdry also remarked on the significance of the "almost innumerable" mitochondria as a feature of "cellular architecture," and how they provide an immense "topography" of surface area, separating fluids of chemically distinct density and properties, and promoting a great variety of chemical activity (Cowdry 1924a, 333). Cowdry was no doubt thinking only of the external surface area of the mitochondria and not of the internally folded membrane (cristae) that would only become visible in the 1950s with the use of electron microscopy by researchers like Palade and Sjöstrand (Dröscher 2009).

Metaphor as Heuristic Tool of Discovery and as Explanatory Component

I have tried to argue that, by around the turn of the twentieth century, biologists increasingly regarded the cell as a complex physical-chemical system of diverse parts that was capable of life and its essential properties as a result of its organization; and that because the details of that organization remained invisible at the molecular level, biologists appealed to metaphor and analogical comparison to familiar examples of organized systems and mechanisms to help them understand what the cell's hidden architecture might be like. Unsurprisingly, in addition to the spatial organization of a laboratory or factory, they appealed to examples of technologies and machines familiar to them at the time.

That scientists use metaphor and analogy to create hypotheses and models of the world is well known.[16] Metaphor and analogy are ways of comparing one thing with another, in the case of science typically with the purpose of helping to illuminate one poorly understood system in terms of another more familiar. But they can be used in quite different ways. The philosopher Michael Bradie has proposed that metaphors play three basic roles in science: (1) a rhetorical or communicative role (which would include purposes of pedagogy and talking to non-experts); (2) a heuristic function in the creation of new ideas and hypotheses (as a "tool of discovery"); and (3) a cognitive or theoretical function in the assessment of evidence and the formulation of scientific explanations (Bradie 1999).

As an example of the first, we can turn to T. H. Morgan's contribution to the Cowdry volume, where he made an analogy between a railroad timetable listing the various stations and their stopping times and the distances between genes "mapped" out along chromosomes as determined by crossover events (Morgan 1924, 714). This railroad schedule and map analogy is clearly intended as a pedagogic device with no deeper implications for future research than the suggestion that if one knew the distances between genes on a chromatid one could calculate the likelihood of their being affected by a crossover event. This points to a relatively superficial similarity. The examples discussed earlier in this paper of the cell as a laboratory, a battery, or a machine, were clearly intended to suggest more than a superficial similarity between the two systems for pedagogical purposes. These metaphors and analogies were used as guides to the creation of working hypotheses about the cell's inner state and function. In that sense they might be called heuristic tools of discovery, and need not be completely correct or adequate in order to be useful in the cause of scientific progress. But when a metaphor or analogy does hit at a deep identity of principle or structure between two systems initially considered dissimilar, then we have an example of the third function claimed of metaphor, a cognitive one in which it is used in the construction of scientific explanations and improved understanding.

E. B. Wilson noted how the earlier morphological cytology of the nineteenth century had merged with biophysics and biochemistry to become "a many-sided *cellular biology*" (Wilson 1924, 10). He probably would not have predicted the extent to which cellular biology would become so dominated by the techniques and ideas of molecular biology just two or three decades later. The rise of the molecular revolution, coincident as it was with the creation of electronic computers and the information age, would see scientists' conceptions of cellular architecture and mechanisms habitually described in terms of metaphors borrowed from electronic engineering. The language of the cell and molecular biologist today is replete with talk of genetic programs, signal transduction, switches, circuits, and wiring diagrams.

As explained by William Bechtel in chapter 13 of this volume, early notions of mechanism in "general cytology" were relatively "flat," static, and two-dimensional. But as scientists investigating metabolic pathways recognized that the complex interactions between the components of these pathways required the addition of feedback loops, their notion of mechanisms and the diagrams they drew to represent them became more complicated and less flat. This trend has continued in the investigation of cell-cell communication, where "signaling pathways" have become dynamic "networks," and metaphors of social organization are used to describe how kinases (commonly described as "switches" with the ability to turn on or off other proteins) and other molecular components are said to "recruit" one another and to "cooperate" to form "committees" which make dynamic and flexible "decisions" about cell "fates."[17]

What is one to make of all this metaphorical language? First, we must note that it is more than just a convenient rhetorical device for talking to laypeople and students. These metaphors are what Richard Boyd called "theory-constitutive" (Boyd 1993). Scientists do not translate their hypotheses and experiments into these metaphorical terms only when they are talking with people outside their field; they think about and understand the phenomena they investigate in these very terms. But we might ask, Can metaphors ever be part of a proper scientific explanation? The French cell biologist Claude Kordon expressed a common distrust of metaphor in science when he wrote,

> The metaphors applied to physiology at that time [late nineteenth century] always referred to machines. Some fifty years later, this mechanistic vision was tempered by electrical metaphors, with the brain being conceived as a powerful telephone system. And, of course, the metaphors used in the writings of today's biologists come from computer science . . .
>
> *No metaphor is really explanatory; rather, it reflects the cultural references through which we have been conditioned to decipher reality.* But these cultural references play an important role in the way we look at the world. We have not yet really learned how to teach biology through the cultural references of our time, which perhaps explains why for many of our contemporaries the picture is still somewhat blurred. (Kordon 1993, 95–96; emphasis added)

If one wishes to allow that metaphors can be explanatory or be part of legitimate scientific explanations, then two options seem available: (1) we might insist that the terms of a successful and well-developed science are no longer metaphors—that they are "dead" metaphors, like the familiar physical-chemical terms "force," "bond," or even "cell" in biology, whose meanings are now quite literal; or (2) we might allow that even relatively "fresh" metaphors can be explanatory. The second option raises the issue of what is required for a description or account to provide a proper scientific explanation. If one believes that an explanation should issue in a literally true statement, then metaphor can probably not play a part. But facts or literal truths are typically the things for which we seek an explanation. The *explanandum* is a fact or truth, but is it so clear that the *explanans* must also be restricted to expression in literally true language? If, however, one thinks the object of explanation is to provide understanding or insight into some fact or phenomenon, then an explanation parsed in metaphorical language may be very useful and fruitful for discovering new truths. If we accept that science operates through the construction of models which are used by scientists to represent aspects of reality for various purposes (e.g., prediction, intervention, understanding-explanation) then it need not trouble us necessarily that these models employ metaphorical language. And if the scientists themselves believe these metaphors help them to make progress in understanding the systems they study, then that must count for something. This is not to say, however, that scientists—or those who study them (historians, philosophers, sociologists, etc.)—should not be as critical toward the metaphors and analogies they use as they are of any other hypothesis or technique they employ in the run of their investigations. As people like to say, "The price of metaphor is eternal vigilance" (Ball 2011).

To conclude this section, we can say that, while the sorts of mechanisms and the language used to describe the cell's architecture have changed significantly since the early part of the twentieth century (batteries and machine technologies having been replaced by computer and electronic engineering technology), biologists continue to rely on metaphors and analogies to help them understand the cell's behavior. However, one respect in which this may be changing in modern approaches to cell biology involves the use of computational modeling in the field of systems biology (about which see Bechtel and Gross, chapters 13 and 14 in this volume), for as these techniques are largely mathematical in expression, their reliance on metaphor at least appears to be greatly reduced. That these formally abstract models will receive verbal interpretation and explanation through metaphorical expression, however, remains a possibility.

A Word on Organization and Cellular Biology as an Institution

Both Cowdry and Wilson remarked early in the twentieth century that as cytology transformed into cellular biology, it was becoming more specialized, differentiated and complex, and so required greater integration and organization in order to function as a harmonious whole. The development of new instruments, technologies, and approaches was a major driver of this trend toward specialization in the science of the cell. Alexis Carrel, the tissue and cell culture specialist, commented that due to new techniques, increased specialization, and the need for collaborative work, a new organization was required of the "modern" cytological lab: "The modern conception of cytology and the development of new techniques have profoundly modified the previous requirements for the training of workers and the organization of the laboratories; . . . since he cannot master, during one lifetime, these techniques and those of organic chemistry, physics and physical chemistry as well, he must have the collaboration or the assistance of workers more especially trained in these other sciences. The organization of the laboratory has also become more complex" (Carrel 1932, xviii). In a curious way, then, the organization of cell biology and the locations where it was studied came to be arranged in a way analogous to the architecture of cells themselves: in specialized spaces using distinct techniques and equipment, separated by literal (and conceptual) walls, but with communication between them (think of the description of professional journals as "organs" of communication).

Cytology's historical development has in a sense mirrored the cell's own arrangement. It has moved from individual naturalists like Hooke and Leeuwenhoek, working in relative isolation in their own private chambers or cells, to Wilson (last of the universal cytologists) and to Cowdry and his colleagues, a collaborative team or community working in communal spaces like the Marine Biological Laboratory, the Rockefeller Institute, or a modern university of specialized workers requiring proper organization and communication in order to achieve desired results. The *General Cytology* volume represented an effort to integrate the distinct visions and investigative techniques emanating from the division of scientific labor among the various disciplines (morphology, physiology, biochemistry, biomechanics, and genetics) into a collective and systemic understanding of life's fundamental unit, the cell. To the extent that the project of creating an integrated cellular biology has been successful, the science has become a metaphor for the organization within the cell (or is it the other way around?).

Conclusion

This chapter has laid out some of the history of conceptual shifts involved in thinking about cells as comparatively unstructured specks of protoplasm or bags of enzymes to highly organized and dynamic spaces and systems. The concepts of system and organization, even if interpreted rather differently by scientists like Hopkins or Wilson, working from distinct disciplinary backgrounds, provided a common conceptual space in which diverse scientific approaches and techniques could communicate and collaborate. Biochemists driven by questions of chemical physiology and more morphologically inclined cytologists may have found different sorts of models, analogies, and metaphors useful for working through their particular ideas about cell organization, but they nearly all found it helpful to extend features of organized systems with which they were already familiar to more obscure aspects of the cell. Metaphors like cell "architecture" and "geography" helped to organize biologists' thoughts about the internal state of cells, while metaphorical and analogical comparisons of cells to laboratories, factories, batteries, and wires all emphasized familiar examples of structure and organization and helped to suggest mechanisms worth exploring. These were treated not as established truths but as working hypotheses that provided alternatives to a vitalist conception of protoplasm as a homogenous, undifferentiated substance somehow capable of performing "*en masse*" all vital functions in some unique and unspecified way. Their key effect was to suggest how principles of chemistry and electricity, as understood at the time, might be merged in application to understanding processes of cellular function.

While early twentieth-century comparisons of cells to batteries and machines have been superseded by contemporary talk of signal transduction, genetic circuits, kinase switches, and gene-regulatory networks (complete with wiring and circuit diagrams), scientists continue to use systems and language reflective of their own time and place in history to make sense of the cell.[18] If one is curious about what the cell biology of the future might look like (and its terminology in particular), history suggests it may be prudent to keep a close eye on new developments in technology.

Acknowledgments

I would like to thank Karl Matlin, Jane Maienschein, and Manfred Laubichler for inviting me to participate in the "Updating Cowdry" project and for their very useful feedback on earlier drafts of my chapter. I am also very grateful to the other participants in the two meetings at the Marine Biological Laboratory in Woods Hole who provided stimulating discussions and taught me lots about cell biology past and present.

Notes

1 Some biologists rejected the "cell standpoint" in favor of the "organismal" standpoint. See Reynolds 2010 for the history and nature of criticism of the cell theory.

2 For instance, Oscar Hertwig, who wrote that the cell "is a marvellously complicated organism, a small universe, into the construction of which we can only laboriously penetrate by means of microscopical, chemico-physical and experimental methods of inquiry" (Hertwig 1895, ix).

3 Gray's influential *Textbook of Experimental Cytology*, for example, proclaimed, "The conception of the cell as a colloidal system is probably one of the most important landmarks in the history of cytology" and that "for more than thirty years every student of cell structure has had, of necessity, to follow the rapid march of colloidal chemistry" (Gray 1931, 33). For discussion of colloid chemistry in relation to the cell, see Liu (chap. 10 in this volume).

4 It is not clear when the practice of referring to cells (especially bacterial cells) as "bags of enzymes" originated, though Fruton (1999, 418) suggests it was an expression used by molecular geneticists in the 1950s–60s to refer derisively to biochemists' approach to the cell. But given the significance placed on organization for normal cell physiology, I suspect discussion about this description must have begun earlier, perhaps in the 1930s, when Albert Claude was just beginning his research on the mitochondrion by means of cell fractionation and centrifugation at the Rockefeller Institute. Carol Moberg (2012, 38) quotes Keith Porter as saying "when he [Claude] started tearing cells apart, taking pieces out and examining them, everybody who called himself a decent cytologist or cell biologist was at him, . . . [asking] what was the good of doing that, breaking up that gorgeous structure?"

5 I am grateful to Jutta Schickore for her assistance with the translation of this passage.

6 Teich (1973, 458) suggests that, for Hopkins, "the living cell was not so much a physico-chemical machine as a colloidal system of phases coexisting in a dynamic equilibrium," but the significance of the spatial metaphors associated with machines and apparatus for his understanding of intracellular organization seems undeniable.

7 For discussion of the history of ideas about organization as it specifically concerned the egg and embryonal cells around this time, see Maienschein 1997. Allen (1978, especially chap. 6) provides an important discussion of the history of many of these issues in terms of what he has called mechanistic and holistic materialism. Allen reveals how Otto Warburg's early ideas (ca. 1912) on the reliance of enzyme activity on cellular organization (the *Atmungsferment* theory) paralleled those of Hopkins expressed here.

8 See Shull 1922 for an early occurrence of the "enzyme equipment" metaphor.

9 Accounts of the cell as a system occur in fact as early as 1838, when the botanist Franz Unger (1800–1870, 13), in his *Aphorismen zur Anatomie und Physiologie der Pflanzen*, described the "*Grundsystem*" of plants as being based on "*das Zellsystem*." I am grateful to an anonymous reviewer for drawing my attention to this early usage of system language.

10 Wilson begins chapter 9 ("Some problems of cell-organization," Wilson 1925, 670) with the following quotation from Brücke: "We must therefore ascribe to living cells, beyond the molecular structure of the organic compounds that they contain, still another structure of different type of complication, and it is this which we call by the name of organization."

11 As Maienschein (1981, 39) explains, although Loeb was a regular figure at the MBL, he was in poor health by 1922 and died in 1924, and this may explain why he was not a contributor to the Cowdry volume.

12 Mathews also ran the analogy in the other direction, stating that "every battery has a metabolism" (Mathews 1924, 68), referring to the "consumption," or catabolism, of a metal, typically zinc or copper, in the generation of an electric current.

13 Grote (2010) discusses the importance of films and surfaces in twentieth-century molecular biology in relation to bioelectric currents.

14 Indeed I suspect that many people today would consider the notion of mechanism to be a "dead" metaphor that can be taken quite literally.

15 Needham's 1935 Terry Lectures at Yale University were published the following year under the title *Order and Life*.

16 See Black 1962; Hesse 1966; Bradie 1999; Keller 2002; Brown 2003. Reynolds (forthcoming) provides a specific and detailed study of the role of metaphor in cell biology.
17 See for instance González et al. 2003 or Levy, Landry, and Michnick 2010. Needham (1950, 677) noted quite early the attraction of social analogy in molecular biology. In commenting on the dynamic nature of proteins and their molecular components he wrote, "Hence the protein molecule itself, no less than the cell-structure of which it is a part, is a pattern the components of which are in perpetual motion. In visualising this, it is difficult not to think of it as a prefiguration of the mutual collaboration of social units in maintaining patterns at far higher levels of organisation."
18 In fact, as Alberts (1998) reveals, machine-talk has merely moved down a level in modern cell and molecular biology to that of proteins.

References

Alberts, Bruce. 1998. "The Cell as a Collection of Protein Machines: Preparing the Next Generation of Molecular Biologists." *Cell* 92 (3): 291–94.

Allen, Garland. 1978. *Life Science in the Twentieth Century*. Cambridge: Cambridge University Press.

———. 2005. "Mechanism, Vitalism and Organicism in Late Nineteenth- and Twentieth-Century Biology: The Importance of Historical Context." *Studies in History and Philosophy of Biological and Biomedical Sciences* 36: 261–83.

Anonymous. 1966. *Chambers Etymological English Dictionary*. London: Chambers.

Ball, Philip. 2011. "A Metaphor Too Far." *Nature News*, February 23. Online at http://www.nature.com/news/2011/110223/full/news.2011.115.html.

Black, Max. 1962. *Models and Metaphors: Studies in Language and Philosophy*. Ithaca, NY: Cornell University Press.

Boyd, Richard. 1993. "Metaphor and Theory Change: What Is "Metaphor" a Metaphor for?" In *Metaphor and Thought*, 2nd ed., edited by Andrew Ortony, 481–532. Cambridge: Cambridge University Press.

Bradie, Michael. 1999. "Science and Metaphor." *Biology and Philosophy* 14:159–66.

Brown, Theodore L. 2003. *Making Truth: Metaphor in Science*. Urbana: University of Illinois Press.

Brücke, Ernst. 1861. Die Elementarorganismen. *Sitzungsberichte der Mathematisch-Naturwissenschaftlichen Classe der Kaiserlichen Akademie der Wissenschaften* 44: 381–406.

Bütschli, Otto. 1894. *Investigations on Microscopic Foams and on Protoplasm*. Translated by E. A. Minchin. London: Black.

Carrel, Alexis. 1932. "Introduction." In *Special Cytology*, 2nd ed., edited by Edmund V. Cowdry, xiii–xviii. Chicago: University of Chicago Press.

Chambers, Robert. 1924. "The Physical Structure of Protoplasm as Determined by Microdissection and Injection." In Cowdry 1924b, 235–309.

Conklin, Edwin G. 1924. "Cellular Differentiation." In Cowdry 1924b, 539–607.

Cowdry, Edmund V. 1924a. "Cytological Constituents: Mitochondria, Golgi Apparatus, and Chromidial Substance." In Cowdry 1924b, 311–82.

———. ed. 1924b. *General Cytology: A Textbook of Cellular Structure and Function for Students of Biology and Medicine*. Chicago: University of Chicago Press.

———. 1928. "Preface to the First Edition." In Cowdry (1928) 1932, ix–xi.

———, ed. (1928) 1932. *Special Cytology*. 2nd ed. Chicago: University of Chicago Press.

Dröscher, Ariane. 2009. "History of Cell Biology." In *Encyclopedia of Life Sciences*. Chichester: John Wiley & Sons. Online at http://onlinelibrary.wiley.com/doi/10.1002/9780470015902.a0021786.pub2/abstract.

Fruton, Joseph S. 1999. *Proteins, Enzymes, Genes: The Interplay of Chemistry and Biology*. New Haven, CT: Yale University Press.

González, Pedro Pablo, Maura Cárdenas, David Camacho, Armando Franyuti, Octavio Rosa, and Jaime Lagúnez-Otero. 2003. "Cellulat: An Agent-based Intracellular-Signalling Model." *Biosystems* 68 (2–3): 171–85.

Gray, James. 1931. *A Text-Book of Experimental Cytology*. Cambridge: Cambridge University Press.

Grote, Mathias. 2010. "Surfaces of Action: Cells and Membranes in Electrochemistry and the Life Sciences." *Studies in History and Philosophy of Biological and Biomedical Sciences* 41 (3): 183–93.

Haeckel, Ernst. 1876. *The History of Creation; or, the Development of the Earth and Its Inhabitants by the Action of Natural Causes*. 2 vols. Translated by E. Ray Lankester. London: Henry King. Online at http://www.biodiversitylibrary.org/bibliography/45871#/summary.

Hertwig, Oscar. 1895. *The Cell: Outlines of General Anatomy and Physiology*. Translated by M. Cambell and edited by Henry Johnstone Campbell. London: Sonnenschein.

Hesse, Mary B. 1966. *Models and Analogies in Science*. Notre Dame, IN: University of Notre Dame Press.

Hofmeister, Franz. 1901. *Die chemische Organisation der Zelle*. Braunschweig: Friedrich Vieweg und Sohn.

Hooke, Robert. 1665. *Micrographia: Some physiological descriptions of minute bodies made by magnifying glasses with observations and inquiries thereupon*. London: Jo. Martyn and Ja. Allestry, Printers to the Royal Society.

Hopkins, F. G. 1913. "The Dynamic Side of Biochemistry." In *Report of the British Association for the Advancement of Science*, 652–68. London: John Murray.

———. (1938) 1949a. "Biological Thought and Chemical Thought: A Plea for Unification." Reprinted in *Hopkins & Biochemistry 1861–1947: Papers Concerning Sir Frederick Gowland Hopkins, O.M., P.R.S., with a Selection of his Addresses and a Bibliography of His Publications*, edited by Joseph Needham, 302–18. Cambridge: W. Heffer and Sons.

———. (1932) 1949b. "Some Aspects of Biochemistry: The Organising Capacities of Specific Catalysts." Reprinted in *Hopkins & Biochemistry 1861–1947: Papers Concerning Sir Frederick Gowland Hopkins, O.M., P.R.S., with a Selection of his Addresses and a Bibliography of his Publications*, edited by Joseph Needham, 225–41. Cambridge: W. Heffer and Sons.

Huxley, T. H. (1868) 1968. "On the Physical Basis of Life." In *Collected Essays*, 1:130–65. New York: Greenwood Press.

Jacobs, Merle H. 1924. "Permeability of the Cell to Diffusing Substances." In Cowdry 1924b, 97–163.

Keller, Evelyn Fox. 2002. *Making Sense of Life: Explaining Biological Development with Models, Metaphors, and Machines*. Cambridge, MA: Harvard University Press.

Kordon, Claude. 1993. *The Language of the Cell*. Translated by William J. Gladstone. New York: McGraw-Hill.

Kyne, Ciara, and Peter B. Crowley. 2016. "Grasping the Nature of the Cell Interior: From *Physiological Chemistry* to *Chemical Biology*." *FEBS Journal* 283 (16): 3016–28.
Levy, Emmanuel D., Christian R. Landry, and Stephen W. Michnick. 2010. "Signaling Through Cooperation." *Science* 328:983–84.
Lillie, Ralph S. 1924. "Reactivity of the Cell." In Cowdry 1924b, 165–233.
Loeb, Jacques. 1906. *The Dynamics of Living Matter*. New York: Columbia University Press.
———. 1912. *The Mechanistic Conception of Life: Biological Essays*. Chicago: University of Chicago Press.
Maienschein, Jane. 1981. "Cytology in 1924: Expansion and Collaboration." In *The Expansion of American Biology*, edited by Keith Benson, Jane Maienschein, and Ronald Rainger, 23–51. New Brunswick, NJ: Rutgers University Press.
———. 1997. "Changing Conceptions of Organization and Induction." *American Zoologist* 37:220–28.
Mathews, Albert P. 1924. "Some General Aspects of the Chemistry of Cells." In Cowdry 1924b, 13–95.
May, Raoul. 1926. Review of *General Cytology*. *Isis* 8 (1): 213–15.
McClung, Clarence E. 1924. "The Chromosome Theory of Heredity." In Cowdry 1924b, 609–89.
Moberg, Carol L. 2012. *Entering an Unseen World: A Founding Laboratory and Origins of Modern Cell Biology, 1910–1974*. New York: Rockefeller Press.
Morgan, T. H. 1924. "Mendelian Heredity in Relation to Cytology." In Cowdry 1924b, 691–734.
Needham, Joseph. (1942) 1950. *Biochemistry and Morphogenesis*. Cambridge: Cambridge University Press.
Nicholson, Daniel J. 2012. "The Concept of Mechanism in Biology." *Studies in History and Philosophy of Biological and Biomedical Sciences* 43 (1): 152–63.
Peters, Rudolph A. (1929) 1963. *Harber Lectures, Biochemical Lesions and Lethal Synthesis*. Oxford: Pergamon Press.
Reynolds, Andrew. 2010. "The Redoubtable Cell." *Studies in History and Philosophy of Biological and Biomedical Sciences* 41 (3): 194–201.
———. Forthcoming. *The Third Lens: Metaphor and the Creation of Modern Cell Biology*. Chicago: University of Chicago Press.
Schleiden, Matthias. 1847. *Contributions to Phytogenesis*. Translated by Henry Smith. London: Sydenham Society.
Schultze, Max. 1863. *Das Protoplasma der Rhizopoden und der Pflanzenzellen, ein Beitrag zur Theorie der Zelle*. Leipzig: Wilhelm Engelmann.
Schwann, Theodor. 1847. *Microscopical Researches into the Accordance in the Structure and Growth of Animals and Plants*. Translated by Henry Smith. London: Sydenham Society.
Shull, C. A. 1922. "Respiration of Thermophiles." *Botanical Gazette* 73 (5): 419.
Teich, Mikuláš. 1973. "From 'Enchyme' to 'Cytoskeleton': The Development of Ideas on the Chemical Organization of Living Matter." In *Changing Perspectives in the History of Science*, edited by M. Teich and R. Young, 439–71. London: Heinemann.
Welch, G. Rickey, and James S. Clegg. 2010. "From Protoplasmic Theory to Cellular Systems Biology: A 150 Year Reflection." *American Journal of Physiology. Cell Physiology* 298 (6): C1280–C1290.

Wilson, E. B. 1896. *The Cell in Development and Inheritance*. New York: Macmillan.
———. 1900. *The Cell in Development and Inheritance*. 2nd ed., revised and enlarged. New York: Macmillan.
———. 1924. "Introduction." In Cowdry 1924b, 1–12.
———. 1925. *The Cell in Development and Heredity*. 3rd ed. New York: Macmillan. Orig. pub. as *The Cell in Development and Inheritance* in 1896; 2nd ed., 1900.
Zampieri, Fabio, Matteo Coen, and Giulio Gabbiani. 2014. "The Prehistory of the Cytoskeleton Concept." *Cytoskeleton* 71:464–71.

CHAPTER 4

METHODOLOGICAL REFLECTIONS IN *GENERAL CYTOLOGY* IN HISTORICAL PERSPECTIVE

Jutta Schickore

". . . as many and as diverse methods as possible ought to be employed."
—Merle Jacobs, 1924

In his preface to *General Cytology*, Edmund Cowdry explicitly noted that the plan had been to mention methods of investigation "only when necessary" (Cowdry 1924, v). Several contributors pointed out that detailed discussions of methods were out of place in a textbook. Nevertheless, references to methods and techniques are abundant throughout the book, and several chapters contain a fair amount of critical discussion about methods and sources of error in cytology. Even the introduction to *General Cytology* contains comments on methods and techniques. In it, Edmund B. Wilson pointed out that the advancements in cell theory in the 1870s and 1880s would have been impossible without the introduction of "high-power microscopical lenses" and "methods for the preparation of fixed and stained sections." Notably, at the same time, Wilson emphasized that the rapid development of preparation techniques was in fact a mixed blessing: While these techniques played an "indispensable part in the advance of modern cytology," they nevertheless "involved many possible sources of error." Even skilled observers had "not always found their way safely amid what Michael Foster called the 'pitfalls of carmine and Canada balsam' " (Wilson 1924, 4).

The concerns Wilson raised in his comments about microscopic methods and techniques resonate with the ways in which several contributors to the volume address methodological issues—warnings and cautionary notes are frequent, and several authors indicated that they were not completely certain about the reliability of their results. The epigraph for my chapter is taken from Merle Jacobs's essay on permeability of the cell (Jacobs 1924). It alludes to yet another reality that early twentieth-century microscopists—and not just microscopists—were facing: There were multiple techniques and approaches to choose from, each of them less than ideal and prone to error. We will see that for Cowdry and his collaborators, this situation was both a problem and a blessing in disguise. Wilson himself hinted at this in his introduction. Having detailed the concerns about potentially error-prone cytological methods, he added that at the time he was writing, "a more rational treatment of the whole subject, and a gradual affiliation of cytological methods of the earlier type with those of the physiologist, the physicist, and the biochemist" had grown out of those concerns.

Microscopists have always been acutely aware of the importance of good methods, instruments, and techniques for successful observations and also of the huge practical difficulties of microscopy. But the content of the debates about methods and sources of error in microscopy and the conclusions that were drawn from methodological discussions about proper procedures changed quite dramatically over time.[1] The second half of the nineteenth century was a particularly important period in this history. The methodological statements and comments on techniques and procedures in Cowdry's *General Cytology* can be understood as a result of developments in the debates about methods and sources of error in late nineteenth-century microscopy and in the life sciences more generally.

It is helpful to think about the developments in this period in terms of a transformation of the practitioners' "epistemology of evidence." William Bechtel introduced this term in his book *Discovering Cell Mechanisms* (Bechtel 2006, 120). An epistemology of evidence embodies the ways in which scientists assess their instruments and techniques to determine whether these investigative tools reliably provide information about the phenomena of interest. In this work, Bechtel identifies three main epistemological criteria for the assessment of new research instruments or techniques: first, the new research techniques repeatedly yield a determinate structure; second, the results obtained with the new technology agree with results generated by other established techniques; and, third, the results obtained with the new technology agree with accepted theoretical accounts of the objects and phenomena under investigation (127).

Bechtel's analysis of cell biology in the 1940s and 1950s shows that these three criteria were indeed used by mid-twentieth-century cytologists in their evaluation of the evidence they obtained with new instruments, in particular with the electron microscope and the ultracentrifuge. Like Bechtel, I am interested in criteria for the assessment of research methods and techniques, but I am putting these criteria in historical perspective. In this essay, I show that in the course of the nineteenth century, such assessment criteria underwent quite radical transformations, and how this reconceptualization affected the microscopists' epistemologies of evidence.[2]

The 1830s and early 1840s were a phase of vindication of the microscope as a reliable tool for observation in medicine and natural history (broadly understood). Throughout the mid-nineteenth century and onward, as the microscope was increasingly used in various fields of the life sciences, the questions of what optical properties were desirable in a microscope and of how best to prepare and investigate microscopic objects were matters of intense debate. Microscopists increasingly acknowledged both the diversity and the intrinsic uncertainty of available methods and techniques as well as the insurmountable limitations of the optical technology itself.[3]

In what follows, I outline some trends in the discussion about proper observation, illumination, and the application of chemical agents. I show how the reflections on the possibilities and limits of microscopy ultimately made it plausible for the practitioners to import new methodological strategies into microscopic practice. There is much in the discussions about methods in *General Cytology* that is in tune with the methodological concerns that were raised during the second half of the nineteenth century.[4] Like many other microscopists around 1900, the authors of *General Cytology* noted that available techniques of microscopy were not always certain and not always well understood. They explicitly embraced the diversity of available methods and techniques and argued that in microscopical investigations, a whole range of methods had to be employed to make the process more informative and more secure. Their appreciation of a diversity of investigative tools resonates with broader methodological trends in the experiment- and instrument-based life sciences at the beginning of the twentieth century.

Schleiden's Vindication of the Microscope

According to Wilson, modern cytology began with Schleiden's new theory of the genesis and organization of the plant cell and Theodor Schwann's extension of this approach to animal cells. Schleiden outlined his theory in the *Principles of Scientific Botany* of 1842 (English translation, 1849). The question of whether Wilson's assessment of Schleiden's position as the "founding father" of modern cytology is plausible is beyond the scope of this chapter. Nevertheless, Schleiden's work is an appropriate starting point for a historical survey of methodological aspects of microscopy because it also contains a detailed discussion of methods of microscopy. The book comprised an inductive account of scientific practice as well as concrete advice for the practitioner of the micro-anatomy of plants.

Schleiden's instructions did double duty as vindication of the microscope and as a manual for practitioners. This twofold goal made the overall argument in favor of microscopical observation a little awkward. On the one hand, Schleiden argued that the microscope, like the function of the eye, "is founded . . . upon immutable mathematical laws; that errors consequently are only committed by the erring judgment in all observations, whether instituted with the naked eye or the microscope." Both the eye and the optical instrument, Schleiden concluded, were always right (Schleiden 1969, 583). The microscope, a mechanical thing, could not be in the wrong. Only the incompetent user could fall into error, either from lack of knowledge about the instrument or from immature, superficial, ignorant, misguided judgment. The argument was obviously addressed to the skeptic, whose lingering doubts about the microscope's merits Schleiden hoped to ward off.

At the same time, however, Schleiden pointed to various errors of the microscope. Relating the advancement of microscope production, he drew attention to the fact that "two errors particularly" had only recently been removed from the instrument: "namely, the chromatic and the spherical aberration" (Schleiden 1969, 577). Schleiden's discussion exhibits an intriguing tension. Of course, instruments with spherical and chromatic aberrations or a broken eyepiece are subjected to laws of nature too. Microscopists who want to make reliable observations will not be impressed if they are given a damaged instrument and are told that they merely have to account for the damage in terms of physics, and then they will be able to make correct observations. In other words, if we want to assess the instrument's suitability for answering difficult botanical questions, we should not regard it as a mechanical object. Rather, it must be assessed as a technical artifact, designed to fulfill a certain purpose. If we consider instruments as tools that are made to fulfill certain specific purposes, we can distinguish between well-functioning instruments that meet the assigned purpose and malfunctioning instruments that do not.

This tension can easily be explained if we take into account that Schleiden's readership would have been diverse. Most naturalists and medical men would have been quite happy to use the microscope to do research and were curious about how to do so successfully. But there were others who were not entirely convinced that the microscope was a trustworthy tool. In the methodological chapters of the *Principles*, Schleiden attended to those researchers who might have doubts about the reliability of the instrument. At the same time, he wanted to provide concrete advice to the practitioner of the microscopic anatomy of plants.

Those practitioners needed to learn how to see what Schleiden saw. Schleiden advised on how to distinguish the object of observation from all those phenomena that did not belong to that object, such as dust and bubbles of air; he described Brownian motion and recommended techniques for the handling of opaque and transparent, hard and soft objects, and the preparation of slides. He concluded on an upbeat note: "He who wishes to observe with success, must observe frequently and with the most profound attention: by observing this rule, he may gradually learn to see, for *seeing* is a very difficult art" (Schleiden 1969, 604). Microscopy was difficult but learnable. For Schleiden, microscopy was crucially a method of observation, and he insisted that there was *one correct* method of microscopic observation. He was trying his best to instruct his readers in the correct method of collecting facts.[5] The evidential status of microscopic observations depended crucially on the optical properties of the instrument and the skills of the observers.

Warnings, Disillusionment, and Cautious Optimism

At the time when Schleiden's book was published, microscopy was considered to be a kind of—or an art of—seeing. While not every microscopist might have agreed with Schleiden that there was one "best way" to make successful microscopic observations, advocates for microscopy agreed that the challenges were manageable. In the following decades, as the practitioners conducted their research at the boundaries of instrument-aided vision, more and more requirements for proper microscopical observation were put forward, critically examined, embraced, or dismissed: successful observation required a microscope with certain optical properties, appropriate illumination, suitable preparation fluids, stains, micrometers, test objects, and so forth.

One indication of the growing complexity of conditions and requirements that had to be taken into account is that microscopy manuals became more and more detailed. Heinrich Frey's manual, *The Microscope and Microscopical Technology* of 1872—a translation from the German, originally published in 1863—was almost seven hundred pages long. It comprised chapters on optical theory, apparatus for measuring and drawing, different types of microscopes, testing, examining the object, preparation, chemical reagents, staining, and injection. The second edition of Pieter Harting's standard work *Het mikroskoop, deszelfs gebruik, geschiedenis en tegenwoordige toestand* (*The Microscope: Its Theory, Use, and Current Status*) was published in three volumes with a total of more than a thousand pages.[6] Carl Nägeli and Simon Schwendener's *The Microscope in Theory and Practice* comprises more than six hundred pages; the eighth edition of Jabez Hogg's *The Microscope: Its History, Construction, and Applications* is about seven hundred and fifty pages long, and so forth.[7] In part, the growing attention to methods and techniques was due to the expansion of the community of microscopists and of microscopical research itself. From the 1840s onward, microscopes were increasingly used in natural history, anatomy, and clinical practice, and they were introduced into medical education. Instrument makers in Europe and in the United States offered easy-to-use and affordable microscopes.

The more observers were using the instrument, however, the more varied were the observations and descriptions of microscopic objects. Complaints about this diversity of findings were frequent and expressive. In 1868, Lionel Beale, author of the well-known instruction manual for microscopists *How to Work with the Microscope: A Course of Lectures on Microscopical Manipulation, and the Practical Application of the Microscope to Different Branches of Investigation* warned that

> the process of observing facts is as unsatisfactory and as fallacious as the process of imagining and speculating without observing at all. At this time what a mass of thoroughly conflicting evidence is advanced on almost every question! Three or four views are taught concerning first principles of anatomical and physiological science, each one being quite incompatible with the rest, but nevertheless, supported by an immense amount of what purports to be evidence based upon observation. It is obvious in such a case that many of the statements must be false, and many of the facts advanced must be errors; and yet with what pertinacity are they maintained, and what an amount of work must be done, and what a length of time must elapse before the false facts can be demonstrated to be really false and the true facts proved to be really true! (Beale 1868, 189)

His fellow countryman, the British Rev. Joseph Bancroft Reade lamented,

> Among Nature's *invisibilia* let us select the *Diatom-valve*. A keen eye may indeed see here a minute atom, but no apparent outline; the elegant S-like shape of the *Pleurosigmata* cannot be made out, and a single hemisphere on the surface of a valve is absolutely invisible. How, then, with the aid of the microscope is the Diatom-valve described by the host of observers? So far from there being any uniformity of

> statement we may almost say *Quot homines, tot sententiae*. The "Transactions of the Microscopical Society" contain a curious record of the Protean aspects described by different microscopists, and it is amusing to read of the ingenious modes of playing with the illuminating rays, so that the eye, fortified by a little previous theory, may see at will, in one and the same valve, either elevations or depressions, triangular, quadrangular or hexagonal dots, with rhomboids, pyramids, or spheres. (Reade 1870, 140)

This diversity of observations and the disagreements among different observers not only stimulated technological advancements but also fueled discussions about the best ways to secure or verify observations. One problem was, however, that opinions diverged even on what exactly caused the diversity of observations and how one could best remedy the situation. Beale, for instance, reminded his readers that the observer's skill, acquired during many years spent "in patient investigation," was the necessary condition of successful observation: "It is only by careful and unremitting exercise that he will gradually acquire habits of attentive observation and the power of thoughtful discrimination which can alone render his conclusions reliable." And, somewhat dishearteningly, he added, "Indeed, though he labour hard and earnestly, he will scarcely have properly educated himself ere his powers begin to decay and he become liable to err from the natural deterioration in structure of the Organs upon which the observation of his facts entirely depends" (Beale 1868, 189–90). Reade offered a different diagnosis, however—he did not blame the observers for their lack of expertise and skill; he blamed them for using the wrong kind of illumination. He continued on a more upbeat note because he had a solution to the problem (on which more below).

Debates about proper methods, techniques, and tools of microscopy drew out methodological views and commitments and encouraged more explicit statements of procedure. In the course of these debates, more and more conditions of successful microscopic observations were identified. A number of preparation techniques were tried out; optical powers were evaluated; and various kinds of auxiliary apparatuses, designed to aid microscopic observations, were assessed.[8] Different practitioners promoted different "best" instruments, illuminations, preparation techniques, staining fluids, and so on for different purposes. As a consequence, many practitioners became more vigilant, cognizant of a multitude of ways of doing microscopy, and mindful of all sorts of errors. Eventually, it became clear that microscopes and microscopic techniques would forever be uncertain and less than perfect. Methodologies of microscopy simply had to take this fact into account if they were to be of any practical use. The discussions in the second half of the nineteenth century about illumination, the optical power of microscopes, and various preparation techniques and chemical reagents show how the practitioners' methodological pronouncements shifted from initial optimism to caution to disillusionment—and sometimes back to cautious optimism. In the following sections, I illustrate these dynamics in more detail. I begin with debates about magnifying power; then I consider discussions about appropriate illumination and about chemical agents.

Beware of Higher Powers

In the 1830s, many microscopists emphasized the great merit of the optical instrument: namely, the fact that it made visible what was otherwise invisible to the naked eye. Superior microscopes were those with the "highest powers," those instruments that showed clearly and distinctly the most delicate objects with the finest patterns. In his essay on test objects from 1832, for instance, the microscope maker Andrew Pritchard declared that such an instrument "may be at once pronounced superlative" (Pritchard 1832, 142).

Already in the 1840s, however, several microscopists had become more cautious in their pronouncements. They recommended that microscopists begin every investigation with the weakest magnification and then gradually move to stronger ones.[9] Some even recommended staying away from using the strongest magnification altogether.[10] Some drew attention to the fact that greater magnifying power did not necessarily mean better images—on the contrary, the strongest optical systems did not offer huge gains in image quality (Nägeli and Schwendener 1867, 124–5). Pieter Harting even noted that the visualizing power of the instrument would suffer if too strong magnifications were used (Harting 1866, 1:279).

There are many indications that the practitioners had become rather doubtful about "highest power" as the most desirable feature of a microscope. William Carpenter, the author of a very successful manual of microscopy, put it bluntly: Those microscopes that could resolve the most difficult test objects—those that had "highest powers" according to this measure—were definitely not the most useful for biological research (Carpenter 1883, 146–47). In microscopy books from the second half of the nineteenth century, we typically find the recommendation that microscopic investigations be carried out with a magnification of 300–400x. Even in this respect, however, the agreement was not general, as Lionel Beale's *How to Work with the Microscope* shows. The book included a chapter entitled "Apology for the Use of Very High Magnifying Powers" which was addressed to those "persons [who] still persist in asserting that no advantage is to be gained from powers above 300 diameters" (Beale 1868, 280). Beale carefully described how the microscopist—the *experienced* microscopist—could and should go about viewing specimens with magnifications of 1,000 and above.

It was probably Beale's work that the English physician, ophthalmic surgeon, and microscopist Jabez Hogg had in mind when he noted in his microscopy manual of 1869 that "we have not the amount of confidence in the higher powers that some observers seem to have; we know, indeed, with every increase in this direction, how liable we are to encounter unforeseen errors and exaggerations, and we still prefer the 1/8th or 1/4 inch. of Ross, because we know its precise working and defining power, and that most of the great achievements and discoveries of the microscope have been made with powers in no way superior to the last-mentioned objective" (Hogg 1869, 69–67). The best way of utilizing the microscope's powers was to use the instrument only at medium powers.[11]

In his address to the American Society of Microscopists in 1883, the society's president, Albert McCalla, mused, "One of the most interesting questions we are called to meet to-day, as it seems to me, is whether we can discover any sure and satisfactory method of diagnosis of the real nature of minute structure near the present limits of vision from the images it gives." McCalla reminded his audience that Ernst Abbe's theory of the microscope had greatly advanced the understanding of optical instruments. Yet the time was not ripe for high-power microscopes: "At present we are not ready for such machines" (McCalla 1883, 18).

Increasing Concerns with Illumination

The correct illumination of the object became one of the main concerns for microscopists. Reade's lament about the multitude of descriptions of diatoms, for example, is part of a paper on techniques of illumination. Reade remarked that "it is not too much to say that illumination is the soul of the *complex body* [of the microscope], with all its ingenious mechanism, appliances, and powers" (Reade 1870, 138). Yet even though there was general agreement among microscopists that proper illumination was at least as important for successful microscopical observations as a well-designed optical apparatus, they continued to disagree about what constituted the best illumination. Should the light be strong or faint, direct or oblique? Should the light source be natural or artificial? Should it be reflected by a mirror or prism? Should a condenser be used? And how could one go about establishing the best way of illuminating an object? How could one decide which kind of illumination showed the object best, if nobody knew how the object *really* looked?

In 1855, the prominent histologist John Queckett noted in his *Practical Treatise on the Use of the Microscope* that the "perfect illumination" of an object was "of the utmost importance" (Queckett 1855, 210). He recommended viewing every object with every kind of illumination—but he left it open how one could then decide which kind of illumination was perfect for which object. Other microscopists recommended specific modes of illumination and certain auxiliary apparatuses as "the best." But each of them had a different recommendation. The chapter "Use of the Microscope—Microscopic Examination" in Frey's bulky manual discusses at length how "the best possible illumination" could be achieved, what kind of light source should be used ("dull, white, uniform cloudiness") and how, when the "best illumination" had been found, the object should be placed on the stage (Frey 1872, 83, 84). In contrast, the US army surgeon J. J. Woodward recommended a combination of monochromatic sunlight and high-power objectives specifically for the observation of delicate markings (and assured his readers that monochromatic sunlight would not damage the eye of the observer; see Woodward 1872, 459–60).

Reade had yet another solution to the problem. He described a new kind of condenser with an adjustable aperture, which, he hoped, would answer once and for all the "vexed question of the day,"—namely, the true structure of the Diatom-valve. To the practitioners, settling this vexed question was so important because Diatoms were widely used as test objects.[12] Reade complained that the microscopists had so many different condensers to choose from—those by Wollaston, Brewster, Shadbolt, Wenham, Nobert, Amici, Gillett, Kingsley, Dujardin, Reade, "*cum multis aliis*." They were all different, but they all had fixed apertures, a fault that his own condenser remedied (Reade 1870, 138).[13]

Notably, the paper on the silverfish *Lepisma saccharina* published by the American naturalist G. W. Morehouse in 1873 exemplifies how the use of a variety of illuminations could actually enhance the reliability of certain observations. Morehouse wanted to clear up confusions about the "beads" that many other observers had seen on the surface of the silverfish. He noted that the silverfish had coarser and finer markings, and that the pattern looked different in different illuminations. He had used monochromatic sunlight, "white cloud," lamp, central and oblique beam, mirror, prism, and achromatic condenser with and without central stops, and Wenham's paraboloid,[14] and he carefully described how the appearance of the object changed when he varied the illumination. In the end, he left it open which of these really was "the best" illumination of the silverfish, but this was not his point. His point was about existence not about appearance. He demonstrated that the "beads" were an artifact of microscopic observation because, regardless of which technique he was using, the beads disappeared with larger magnifications.

These examples show that the practitioners' debates about the illumination of microscopic objects brought the diversity of modes of illumination to the fore. At first, various practitioners attempted to figure out which kind of illumination was "the best," but no general agreement could be reached about this point, and different microscopists continued to promote different illuminations and different auxiliary apparatuses for managing and improving it. Morehouse's paper is remarkable because it shows that, for certain questions—particularly for identifying artifacts of the observation procedure—it was actually a good thing to use different modes of illumination. Nonetheless, proper illumination remained a challenge. As late as 1909, the microscopist A. M. Kirsch reminded his readers that "the method of illumination" was crucial for the correct identification of delicate markings (he was also concerned with diatoms). He reported that the delicate lines of the object he was studying were correctly displayed when "the incidence of the direct sunlight took place as nearly as possible parallel to the stage, the microscope being inclined for the purpose, and at right angles to the striae." Kirsch's paper is noteworthy because he stated that he had found this "in experimenting for a long time with various objectives" (Kirsch 1909, 26). For Kirsch, figuring out the best method of making microscopical observations had become an experimental endeavor.

Merits and Demerits of Manipulation

For Schleiden, microscopy was an art of seeing—seeing enhanced with powerful optical aids but still essentially noninvasive *observation*. During the nineteenth century, microscopy was closely associated with practices of intervention and manipulation, and ultimately, microscopy was itself conceived as an experimental practice. This reconceptualization of microscopy is most conspicuous if we track the practitioners' views on the application of chemical agents.

Earlier in the century, the common attitude toward chemicals was one of great caution, and various microscopists warned of the distortions that chemical agents might produce. If we think of the microscope as an aid to *observation*, the early warnings against chemicals make perfect sense. Successful observation required examining specimens in their natural conditions. Of course, it was appreciated that some interventions were necessary just for bringing specimens under the microscope, but the observers tried to utilize preparation techniques that were minimally invasive. Earlier in the century, for instance, microscopists were very much aware that specimens had to be kept moist in order to prevent shriveling and distortion. To do so, they recommended using "pure" and always fresh water so as not to alter the objects (e.g., Schacht 1851, 42). When it became clear that "pure water" actually damaged the finer structures of tissues, microscopists switched to other preparation fluids that they expected to be less destructive.

The author of one of the earliest manuals of microchemistry, Julius Vogel, made a point of addressing the worries his fellow microscopists might have had when he promoted microchemical investigations in 1841. He noted that for many microscopists who sought to understand the form and structure of organic materials, the (micro-)chemical analysis of organic materials was like grinding up a clock to find out how it worked—in other words, a completely futile endeavor (Vogel 1841, 175). Vogel insisted, however, that the chemical analysis of organic materials had something important to contribute to morphological studies of organized parts and wholes.

Within a few decades, many more microscopists became comfortable with and recommended using a wealth of chemicals as aids to microscopical investigations. They increasingly expressed enthusiasm about the information that could be gained from staining and other chemical manipulations. In his 1883 address to the Microscopic Society of America, Albert McCalla even noted that chemical media and reagents, if used wisely and carefully, were the "most efficient means of verification." He declared that "we are now armed, as never before, with means of putting nature to the test and verifying our vision of her most indicate [*sic*] minutiae" (McCalla 1883, 13).

Sometimes, the application of chemicals—especially of stains—was presented in analogy to the enhancement of the optical power of the microscope. Minot, for instance, described the employment of chemical means as "endeavors to render certain characters *more* visible than they are naturally" (Minot 1877, 405) and compared the chemicals to the enhanced letters that enabled blind people to "see" writing with their fingertips.[15] One factor that surely facilitated the general acceptance of staining as a means of microscopic observation was the success in making the "invisible enemy" to human health, the germ, visible. In Robert Koch's laboratory, the cause of tuberculosis became a "tangible parasite" (Koch 1987, 95), whose existence could be revealed through staining and microphotographs. Frank L. James, a physician in St. Louis, enthused, "Of what far-reaching consequence, for instance, was Koch's discovery that the bacilli of tubercle will absorb certain of the aniline colors and cling to them with such tenacity that not even nitric acid will bleach them?" (James 1887, 46).[16]

There is a striking parallel between the discussions about the merits and demerits of staining and those about illumination. We saw above that many microscopists pointed out how crucial the right illumination was for the success of microscopical investigations, and that they continued to disagree about what kind of light and what mode of illumination were actually the right ones for what purposes. While intense discussions were going on about the best way to illuminate an object and about desirable optical properties, numerous novel stains and mounting and fixing agents were developed to enhance the microscope's power. Many microscopists emphasized that the right staining technique was an essential aid to successful observation. But what was the best stain? And how should one apply it? More and more stains, fixing agents, and microchemical reagents for microscopic objects were suggested.

Frey's manual *The Microscope and Microscopical Technology* (1872) illustrates the diversification of preparation techniques. On more than a hundred and fifty pages, Frey's book covers techniques of testing, preparation, staining, injection, and mounting microscopic objects. Among these were the staining fluid carmine and the mounting agent Canada balsam that Wilson mentioned in his introduction to *General Cytology*.[17] Frey described eighteen stains overall, including several different kinds of carmine preparations. Frey's manual was no exception. The chapter on reagents for preserving, hardening, and fixing in Carl Friedländer's manual of microscopical technology for pathological anatomists (1885) is thirty pages long; the thirty pages that follow deal exclusively with staining. There are fifteen categories of stains, and each category contains several items. Friedländer's "Category 15: The Nuclear-Staining, Basic Aniline Dyes," for instance, comprises "Vesuvine (Bismarck-brown), fuchsine, gentian-violet and methyl-violet, methyl-blue; besides methyl-green, dahlia, magdala, etc." (Friedländer 1885, 82). Philip Stöhr's 1903 manual for histologists listed sixty-one chemicals for staining, fixing, and mounting—and those were just the agents that Stöhr thought every beginner should have available (Stöhr 1903).

We saw above that many practitioners developed their own best techniques of illuminating objects. In a similar way, many microscopists developed and promoted particular reagents they liked best. In Frey's chapter on staining, for instance, some of the stains were named after the microscopist who had devised them—there is Thiersch's Carmine Fluid, Beal's Carmine Fluid, and so on. Some individuals wrote papers for the budding journals of microscopy, in which they promoted their own approaches to staining, fixing, and mounting.[18]

Nevertheless, many practitioners were optimistic that staining fluids were "a most invaluable help to exact knowledge," as McCalla put it in his address (McCalla 1883, 13). McCalla belittled those microscopists who regarded staining as purely ornamental and as amusement for the amateur. In fact, there was much more to staining than just improving the visibility of objects. Stains could also serve as aids to identify properties of objects. To McCalla, staining fluids were useful because they could be used to *differentiate* various tissues, which would otherwise remain transparent and invisible. Histologists Alexander A. Böhm and M. von Davidoff even recommended "selective" staining with stains that colored certain tissue elements more than others, whereby preparations could be colored with more than one stain in "simple, double, triple, and multiple staining." Certain parts of tissues took up more stain than others. Such "differential staining," as they called it, "has therefore a value beyond the mere coloring of sections so that they may be seen more clearly" (Böhm and Davidoff 1910, 41).

These passages show how, in the course of the development of staining techniques, microscopy came to be closely associated with processes of manipulation and experimentation. For the microscopists who used stains to identify the kinds of tissues they were examining, microscopy was much more than an "art of seeing." To make the invisible world accessible, the microscopists had to master a diverse set of methods and techniques of intervention.

The association between microscopy and experimentation could take several forms. Böhm and Davidoff compared the staining of sections to "a microchemic color reaction." They explicitly declared that the investigation of fresh tissue was "far from revealing all the finer details of their structure. . . . It is therefore generally necessary to subject tissues or organs to special methods of treatment before they may be studied microscopically with any degree of profit" (Böhm and Davidoff 1910, 22). Microscopy manuals specifically for pathologists and clinicians described in detail the systematic manipulations of preparations such as bacteria cultures that were part of the clinical procedures for the diagnosis of infectious diseases. German physician Hermann Lenhartz's standard manual of clinical microscopy, for instance, demonstrates how elaborate the interventions were that microbiology often required. For instance, in his chapter on "vegetable parasites," which deals with bacteria causing croup, meningitis, and other serious diseases, Lenhartz described in detail the techniques of pure cultivation (including the preparation of the culturing media), various different (!) staining methods for the bacteria, and techniques of observation of both unstained and stained specimens (Lenhartz 1904).

The practice of experimental physiology also had an impact on microscopy. The chapter on blood in Edward A. Sharpey-Schäfer's *Course of Practical Histology*, for instance, shows that the microscope became a tool for the magnification of experiments with body fluids and tissues (Sharpey-Schäfer 1877). Among other things, Sharpey-Schäfer described and depicted intricate set-ups through which the action of electric shocks on blood could be studied. The experiments as they are described in the manual closely resemble macroscopic physiological experiments on the effect of electricity on bodies and body parts.

From Sharpey-Schäfer's text, it is not quite clear what larger research goals these investigations of shocked blood cells were supposed to serve. There was, of course, a long tradition of galvanic experiments in physiology but descriptions of similar experiments under the microscope in other handbooks of microscopy indicate that the larger context for these experiments was the ongoing debate about the physicochemical basis for life. Nägeli and Schwendener described similar experiments; they noted that the conclusions drawn from such experiments were daring and at times completely unjustified, yet they did not doubt that the study of the impact of galvanic currents on cell life could ultimately lead to insights into the nature of organic forces (see, e.g., Nägeli and Schwendener 1867, 475–62).[19] Be that as it may, the point is that books like Sharpey-Schäfer's *Course of Practical Histology* and Lenhartz's *Manual of Clinical Microscopy* were laboratory manuals whose overall purpose it was to introduce students—beginners of microscopy—to a variety of important techniques and skills. These techniques and skills were not aimed at observing specimens in their natural state but at magnifying chemically altered specimens, and the authors even encouraged experiments with microscopic objects. For the students who learned histology from Sharpey-Schäfer's or Lenhartz's manuals, the connection between microscopy and experimentation was there from the outset.

This association of microscopy with experimentation is apparent not just to the historian of microscopy but also to late nineteenth-century microscopists themselves. They sometimes described the very practice of microscopical investigation as an experimental endeavor and insisted that repeated, systematic, controlled interventions could provide the most information about microscopic objects. Pieter Harting reminded his readers that there were two roads to truth in science, observation and experiment, and that both were necessary in microscopy. It was not enough to observe the objects; they had to be intentionally brought under the influence of physical and chemical forces, and the modifications these forces caused had to be studied with the aid of the microscope. The microscope stage had to be a "miniature laboratory"; the microscopic objects had to be manipulated with physical and chemical agents that could in turn help reveal physical and chemical aspects of their nature (Harting 1866, 2:142).

The pathological anatomist Friedländer also made the experimentalist approach to microscopy explicit: "In our microscopical investigations we are, for the most part, engaged, not as observers merely, *but we experiment*, and our results are accordingly compounded of preformed structures, on the one hand, and, on the other, of factors introduced by ourselves. To pass judgment, then, upon the nature of an object at first sight, and without reference to these aids,—for the most part so simple—would be to expose one's self to the greatest errors" (Friedländer, 1885, 30; emphasis added).

The interesting question is, of course, how exactly the experimenter-microscopist should make reference to these aids. Friedländer discussed microscopic investigation from the perspective of chemical experimentation. It was therefore important to him that the microscopists had a good understanding of chemical procedures. He pointed out that "if we know that the particular objects of our quest resist the action of certain reagents and modes of treatment by which the remaining substances are destroyed, we are possessed of a method of examination most serviceable for our definite purposes, even though the structural integrity of the specimen be completely sacrificed thereby. From this it appears, that the use of the microscope cannot always be regarded as something purely mechanical, especially when questions in pathology are concerned; but rather that it frequently demands the exercise of a certain amount of reflection and circumspection, in choosing the course to be pursued" (Friedländer, 1885, 30). Using reagents that destroyed the structural integrity of specimens would have been perceived as a major disadvantage and potential source of error in early nineteenth-century microscopy. For Friedländer, however, chemicals were more than a means to render specimens *more visible*. "Sacrificing the structural integrity" of a microscopic object was no longer unthinkable.

The methodological statements in manuals like Harting's and Friedländer's illustrate an experimental approach to the methodology of microscopy. Importing an experimental perspective into the methodology of microscopy allowed for new strategies for securing empirical evidence or new approaches to "verification," as McCalla called it. One of the factors that facilitated the acceptance of staining surely was the fact that stained bacteria produced recognizable regular patterns and the fact that the observation of the germ agreed so well with one of the main theories of disease causation—the theory that many, if not all, diseases are caused by a tangible pathogen. Both the first and the third of Bechtel's assessment criteria for new research techniques were thus fulfilled. The late nineteenth-century discussions about methodological strategies also suggest that for the late nineteenth-century microscopists, it was crucial to have a precise understanding of how exactly the techniques of intervention affected the microscopic objects.[20]

Notably, while many practitioners agreed on this point, they suggested different ways in which such an understanding could be achieved. As we saw, Friedländer pointed out that it was important to exercise circumspection and reflection so as to establish what exactly the chemical reagents and modes of treatment did to the specimens. In contrast, Vida Latham, curator of the museum at the Woman's Medical School at Northwestern University, took a more pragmatic approach in her paper "A Plea for the Study of Re-Agents in Micro-Work," which was published in the *Proceedings of the American Microscopical Society* in 1894. She suggested that "in order to ascertain the effect and amount of change produced by the treatment necessary in preparations for examination, a wise course would be to take a suite of specimens, put them through a set of hardening solutions, carefully made, then through the various preparations for imbedding and cutting sections, as paraffin, celloidin, freezing, and finally apply the dyes, and on examining the results under the microscope an intelligent opinion could be formed as to the best methods" (Latham 1894, 209).[21] Ultimately, the discussions about chemical agents thus confirm what previous sections of this essay have shown: Late nineteenth-century microscopists agreed that microscopy was error-prone and that it was important not to be overconfident and to be aware of all sorts of dangers and pitfalls. But there were different views about what the possible sources of errors were and how the dangers and pitfalls could best be avoided.

Cowdry's Volume: Uncertainty and Diversity

The methodological statements and discussions that we find in *General Cytology* resonate with the development in late nineteenth-century methodology of microscopy that we traced in earlier sections of this essay. By the 1920s, when Cowdry's volume was published, neither the theories of the cell nor the available technologies were deemed completely reliable. Calibration on a well-established technique was thus not really an option, and the conditions for a "wise" and judicious use of chemicals as Friedländer and McCalla had envisaged them were not entirely attainable. The methodological strategies for the assessment of findings that had emerged indicate the microscopists' acknowledgement that microscopic investigations of cells, cell constituents, and their functions were riddled with pitfalls. It is this general notion of methodological uncertainty that makes *General Cytology* so interesting from the standpoint of the history of methodology of microscopy—in fact the history of methodology more generally. The book documents what strategies the microscopists were envisaging to address and cope with this uncertainty. Many authors contributing to *General Cytology* tried to capitalize on the diversity of available techniques and approaches, just as Latham had suggested in her plea for the pragmatic comparison of reagents.

Several authors of chapters in *General Cytology* stressed the complexity of the issues they were dealing with. Frank Lillie and Ernest Just, authors of the chapter on fertilization, made it clear from the outset that the phenomenon of fertilization was so complex that only a combination of morphological and physical, chemical, and biological approaches could capture it. The different approaches complemented one another, thus providing a more complete understanding of the complex phenomenon.[22] But the issue was even more problematic. Cowdry stated that microscopic structures and phenomena should be studied from a variety of perspectives, because microscopic techniques had certain limitations. Cowdry's discussion of cell components—specifically, of mitochondria and the Golgi apparatus—illustrates quite well how early twentieth-century microscopists handled the methodological challenges posed by uncertain techniques and not entirely convincing theories. Cowdry emphasized that there was a lot of uncertainty about the structure and function of cell constituents. He recommended caution, careful experimentation, and drawing together information from different types of investigations.

Mitochondria had been observed for some time, but researchers were now "entering upon a period of experimentation." Cowdry's review of recent work on mitochondria shows that the investigation of these cell constituents had indeed become an experimental endeavor. He described various kinds of manipulations to clear up the structure and composition, such as the application of dyes or deliberate "experimental injury" by phosphorous poisoning (Cowdry 1924, 329), and he highlighted the importance of controlled experimentation in stable conditions and with standardized techniques (326). He concluded on a rather dispirited note, pointing out that the achievements of the study of mitochondria were "as yet somewhat intangible" and that everyone was "working very much in the dark." There was "a plethora of observations but no new experimental method has brought us noticeably nearer to a solution of the puzzle." In the end, he thought it might be possible that the task was "in reality a synthetic one: we must piece together information from many quarters. . . . We must continually search for new methods of approach, and strive to use to better advantage methods already at hand. Even in experimental animals the physiologic processes are of such complexity as to be very baffling" (331–32).

Cowdry's review of the research on the Golgi apparatus does not sound much more optimistic. He pointed out that microscopists had to "rely upon rather unsatisfactory osmium and silver preparations between which there is little to choose," and that, for the study of the Golgi apparatus in the living cells of vertebrates, a method had yet to be devised. Careful experimentation with the tissue was required, and caution had to be exercised in the interpretation of the findings, because it was very difficult to distinguish between normal changes in the shape and size of the Golgi apparatus and experimentally produced artifacts (Cowdry 1924, 334).

Like Cowdry, Ralph Lillie conceptualized his subject—the reactivity of the cell—as an experimental endeavor. Lillie described the protoplasm as "a chemical-reaction system in which the reactions are controlled by structural conditions" (Lillie 1924, 171). The chapter reviews a number of studies in which this reaction system was intentionally perturbed or destroyed chemically or mechanically in order to gain insight into cell activity. Lillie noted in his discussion of protoplasmic structure in relation to reactivity that the structure of protoplasm was not readily accessible to direct observation, as the structure might appear homogeneous even though its physiological properties were in fact complex. Like Frank Lillie and Just, he saw the solution in combining several approaches. In light of Bechtel's epistemology of evidence, it is striking that in addition he called for a convincing explanation that could accommodate the findings: "The combination of observational and experimental (or physiological) methods of study seems the only possible means of obtaining insight into this problem [the structure of protoplasm]; what is essential is that our conceptions of the structure of protoplasm should be consistent with—or help to explain—its fundamental physiological properties" (Lillie 1924, 176). The explanatory success was taken to enhance the reliability of the findings.

Robert Chambers's contribution on the structure of protoplasm discusses a whole host of methods for studying its physical nature—including crushing experiments, centrifuging, electromagnetic methods, the detection of Brownian movement by dark-field illumination, and micro-dissection and injection (Chambers 1924, 238–39). Some were better, some were worse, but even the best of them—micro-dissection and injection—were open to criticism. Occasionally, however, Chambers pointed to a convergence of results obtained by different methods—for instance, when he reviewed the investigations of viscosity changes in cytoplasm. Even though the centrifuge method was "a rather drastic one," it could show a distinct increase in viscosity after fertilization; and the results of micro-dissection agreed with this finding (249).

The most thorough discussion of methods can be found in Merle Jacobs's essay on the permeability of the cell. He made the reasoning underlying Chambers's conclusions explicit. Jacobs's discussion shows that the appeal to diverse methods and approaches was not confined to microscopy but resonated with other fields in the life sciences around 1900. Having stated that "In a general textbook, a detailed account of methods is out of place," Jacobs proceeded to describe and assess on twelve pages diverse methods of studying cell permeability. Like many nineteenth-century microscopists, Jacobs motivated and justified this discussion of methods with a reference to the diversity of findings. Jacobs noted that

> in the case of cell permeability, the results obtained and the conclusions drawn by different workers are so frequently at variance, and since in so many cases such disagreement appears to be due to differences in procedure rather than to inaccurate observation, it seems wise before attempting to summarize the present state of our knowledge of the penetration of cells by various substances to give a short description of the chief methods that have been employed in this field in the past, and in

particular to point out some of the factors connected with each which are likely to lead to erroneous conclusions. (Jacobs 1924, 102)

Jacobs proceeded to discuss five sets of methods: "(1) Methods depending on visible changes produced within the cell. (2) Chemical methods. (3) Osmotic methods. (4) Electrical conductivity methods. (5) Physiological methods" (Jacobs 1924, 103). The interesting move in Jacobs's argument is not only that he covered such a range of techniques—chemical, physical, and purely observational—but that he gave the methodological discussion about proper procedures a new twist. Explicitly turning away from the notion that there are "best ways" of doing things, he wrote, "It will appear that no single method by itself is entirely reliable, and the obvious conclusion to be drawn is that in any given investigation as many and as diverse methods as possible ought to be employed before an attempt is made to base generalizations on the results obtained" (102).

And it was not only the methods that were uncertain. The theory of the cell and of cell permeability in particular was also "still in uncertainty" (Jacobs 1924, 149). Jacobs thus repeated his call for diversity when he reviewed the available theories of cell permeability, noting that "in general, it may be said that no single theory is entirely satisfactory, as indeed would be expected from the complicated nature of the facts which they attempt to explain. At the same time, there are probably elements of truth in most of them, and if each were regarded by its supporters merely as an attempt to deal with a limited number of the factors concerned in a very complex process rather than as a complete explanation of the behavior of the cell, there would be far less occasion for criticism than actually exists." At the end of his contribution to *General Cytology*, Jacobs reiterated that "what is most needed in the field of cell permeability at the present day is facts. When sufficient accurate quantitative data covering a wide range of material and based upon *a sufficient number of independent methods* have become available, a satisfactory theory will follow as a matter of course. Until that time, speculations should be reduced to a minimum" (156; emphasis added).

Complex Organisms, Invisible Things, Uncertain Techniques

In her 1991 essay on Cowdry's *General Cytology*, Jane Maienschein drew attention to Jacobs's dissatisfaction with available theories of the cell and his call for diverse methods of investigation. She argued that the collaborative approach taken in the volume reflects the expansion of the field of cytology in the early twentieth century. In his famous and influential book *The Cell in*

Development and Inheritance (1896; 1900; 1925) E. B. Wilson had presented a unified account of the cell, "a sustained interpretation of what the cell is, how it arises, and how it works" (Maienschein 1991, 32). Cowdry's book, by contrast, represented a collaborative approach in which "numerous questions, approaches, methods, theories, and data were acceptable" (46–47). Cowdry's volume thus impressively demonstrates that collaboration can be a very productive mode of doing research.

As we have seen, *General Cytology* also exemplifies how the expansion of the field of cytology and the multiplication of approaches and techniques could be turned into a *methodological* strategy. The methodological strategies for assessing evidence that Bechtel identified in his study of mid-twentieth-century cytology were initially only of limited value. Microscopists did seek repeatable patterns, calibrated against known methods, and compared their results with techniques, but the experimental techniques themselves appeared unsatisfying and less than completely reliable. Because of the uncertain theories, instruments, and techniques they had to work with, input and confirmation from diverse lines of investigation was called for. In the 1920s, the plurality of new methods and approaches for the study of ever smaller, sub-visible objects made it possible and plausible to seek confirmation from different lines of research. Because both the methods and the theories of cells and cell components were deemed unsatisfactory, uncertain, and preliminary, the strategy of obtaining multiple confirmations through a diverse set of methods took precedence over other assessment strategies.

When new analytic techniques for examining organic substances and powerful analytic instruments began to shape and transform biological and biochemical research, the strategy of using "as many and as diverse methods as possible" remained the strategy of choice—at least in the long run. For instance, when the ultracentrifuge was established, the results it produced were initially considered decisive by many, but often the excitement did not last. The plurality of methods and approaches gave rise to a new methodological strategy of capitalizing on this diversity, both to obtain a more *complete* picture of an object or process and as a safeguard against the perceived uncertainty of each individual technique.

Acknowledgments

I am very grateful to the participants of the "Updating Cowdry at the MBL" workshop and particularly to the editors of this volume for valuable feedback on previous versions of this essay. I also thank an anonymous reviewer for helpful comments.

Notes

1 I use the term *methodological* in a very broad sense, in the general sense of "related to" or "pertaining to" methods and techniques. In this chapter, I use the term *methodology* instead of *epistemology* because I am focusing on microscopists' accounts of "proper procedure." Ultimately, of course, methodological discussions are part of epistemology, because "proper" techniques and procedures are those that are deemed to be conducive to the production of knowledge.

2 I do not pay particular attention to national differences among communities of microscopists. In some respects, these differences are quite pronounced, but I think it is justified to disregard them for the purposes of a first pass at the longer-term history and dynamics of methodological discussions about microscopy. Communication and technological exchanges across national boundaries were frequent at the time, so it makes sense to treat the British, Continental, and US microscopists as part of one and the same community.

3 In older histories of the microscope and microscopy, the nineteenth century was treated as the period in which microscopy became "scientific." Indeed, many nineteenth-century practitioners of microscopy presented their research in these terms—the use of the microscope became one of the hallmarks of progress and innovation in biomedical research. While it is correct that in the course of the nineteenth century, microscopes became a lot cheaper and easier to use and thus a more frequent ingredient of biomedical research, training, and practice, we should be careful not to follow the received view or the scientists' own histories. More recent work on the history of microscopy has already shown that the eighteenth and early nineteenth centuries were by no means the "dark ages" (see Ratcliff 2009). I suggest that the microscope gave rise to methodological concerns *precisely because* it became more generally accepted.

4 Of course, scientists often do not explicitly address methodological issues, and if they do discuss them, they might not adequately represent the methodologies they did actually use. In this paper, I concentrate on methodological concepts as they were stated, discussed, or justified.

5 The German original reads ". . . Anleitung für die richtige Methode der Sammlung der Thatsachen zu geben" (Schleiden 1845, 189). This sentence is missing from the English translation because it is the introductory sentence to the section on induction. The English editors did not think it necessary to translate this section of Schleiden's *Grundzüge der wissenschaftlichen Botanik*, as the English-speaking scientific world already had two "admirable" treatises on scientific inference, namely Herschel's *Preliminary Discourse on the Study of Natural Philosophy* and Whewell's *Philosophy of the Inductive Sciences* (Schleiden 1969, iii).

6 To my knowledge, the work was never translated into English, but it was widely used in the European context. I am working with the German translation.

7 Seventeenth- and eighteenth-century microscopy books also contained discussions of methods and techniques, and some of them were lengthy. But the emphasis of these works was different. They focused on the relations between the eye and the microscope, regarded as optical instruments, and a larger proportion of the text was devoted to descriptions of natural history objects.

8 In the early twentieth century, discussions had reached a point where the two topics were treated in entirely separate publications. In Philip Stöhr's microscope manual for histologists, for instance, aspects of optics were discussed on just a few pages. The author directed his readers to books specifically on the microscope.

9 See, for example, Schleiden 1969, 580; Vogel 1841; Frey 1872, 89; Nägeli and Schwendener 1867, 257. Nägeli and Schwendener pointed out that the strongest magnification usually shows a lot less than the next weakest.

10 I should add that the practitioners were well aware that there was a difference between magnifying power and what we now call resolving power. But there was also much debate about how these properties could be characterized in physical terms and how the different properties should be called. Microscopists argued about "penetrating power," "defining power," "resolving power," "focal depth," and so forth. I do not address this issue here.

11 Related concerns were raised about large instruments that required complicated adjustments of the optical apparatus (e.g., Minot 1877, 406).

12 On test objects, see Schickore 2009.

13 Graeme Gooday has drawn attention to a striking parallel between physics and microscopy in the second half of the nineteenth century. He shows that the members of both communities were concerned with the effective management of the physical environment in which they practiced their research, seeking to obtain stability of their instruments as well as of the working environment (see Gooday 1997). Gooday demonstrates how varied the solutions were that the practitioners had for the illumination problem. He argues, however, that in "the longer term, both communities were able to enhance these conditions by reengineering their working environments to achieve greater phenomenological orderliness: physicists by acquiring specially purpose-built laboratories, and microscopists by installing auxiliary devices such as the achromatic substage condenser into the very structure of their instruments" (Gooday 1997, 433). Below I argue that the discussions in *General Cytology* suggest a different outcome to the conundrum: namely, the explicit acknowledgement of a diversity of approaches.

14 "White cloud" is a technique of illumination that captures natural light from clouds rather than direct sunlight. Wenham's paraboloid was a reflecting paraboloid, which was used prior to wide-angle condensers.

15 "In both cases, the conditions under which the special sense, whether sight or feeling, has to act are greatly exaggerated, so to speak, thus producing magnified or strengthened perceptions" (Minot 1877, 396).

16 Frank was referring to Koch's "specific" method of staining, whereby the bacilli, once stained, retained their color even after treatment with alcohol or nitric acid.

17 Carmine stain was first introduced by the anatomist Joseph von Gerlach in 1858—largely by accident, as he noted (Gerlach 1858). He utilized the red stain for his observations of nerve cells. Canada balsam is a kind of turpentine made from a fir tree resin. It was used as a fixing agent to make permanent slides.

18 See, for example, Charles Mitchell's paper on hæmatoxylon, in which he introduced to his audience "a new and simple method of preparing a logwood staining fluid, by which a permanent, reliable and satisfactory preparation can be easily made" (Mitchell 1883, 297). See also Duffield 1884. Duffield described various methods that he could "recommend from personal experience" (209).

19 See also Gooday 1997, 421–22.

20 Drawing on Ian Hacking, Bechtel notes that such knowledge often cannot be had and is not required. This is one of the points of Hacking's famous discussion of microscopes in *Representing and Intervening*: Changes in the understanding of the theory of microscopes have little effect on the assessment of the reliability of microscopic

observations (Bechtel 2006, 120; Hacking 1983, 199). I agree with Bechtel's and Hacking's epistemological analysis, but I want to make a different point. My point is that for the practitioners of microscopy in the late nineteenth century, at least for many of them, this knowledge and understanding was indeed deemed an essential part of the "verification" of microscopic observations.

21 Latham's essays on staining are quite interesting: Not only did she survey a large number of stains, she also drew attention to the fact that commercially available stains could differ widely, as well as to the fact that the nomenclature for stains was incredibly confused (see Latham 1891; Latham 1896).

22 As Bechtel notes in his essay on research techniques in cognitive neuroscience, obtaining complementary information about the phenomena under investigation is particularly important in situations in which "any given technique can only provide a very selective and distorted perspective on the phenomenon" (Bechtel 2002, S49). See also Trizio 2012 for a similar instance in current cell biology.

References

Beale, Lionel S. 1868. *How to Work with the Microscope*. 4th ed. London: Harrison.

Bechtel, William. 2002. "Aligning Multiple Research Techniques in Cognitive Neuroscience: Why Is It Important?" *Philosophy of Science* 69, No S3: S48–S58.

———. 2006. *Discovering Cell Mechanisms: The Creation of Modern Cell Biology*. Cambridge: Cambridge University Press.

Böhm, Alexander, and M. Davidoff. 1910. A *Textbook of Histology: Including Microscopic Technic*. 2nd ed. Philadelphia: Saunders.

Carpenter, William Benjamin. 1883. *The Microscope and Its Revelations*. 6th ed. New York: William Wood.

Chambers, Robert. 1924. "The Physical Structure of Protoplasm as Determined by Microdissection and Injection." In Cowdry 1924, 235–309.

Cowdry, Edmund V., ed. 1924. *General Cytology: A Textbook of Cellular Structure and Function for Students of Biology and Medicine*. Chicago: University of Chicago Press.

Duffield, George. 1884. "A Few Hints on Hardening, Imbedding, Cutting, Staining, and Mounting Specimens." *Proceedings of the American Society of Microscopists* 6:209–11.

Frey, Heinrich. 1872. *The Microscope and Microscopical Technology*. New York: William Wood.

Friedländer, Carl. 1885. *Manual of Microscopical Technology for Use in the Investigations of Medicine and Pathological Anatomy*. New York: G. P. Putnam's Sons.

Gerlach, Joseph von. 1858. *Mikroskopische Studien aus dem Gebiete der menschlichen Morphologie*. Erlangen: Ferdinand Enke.

Gooday, Graeme. 1997. "Instrumentation and Interpretation: Managing and Representing the Working Environment of Victorian Experimental Science." In *Victorian Science in Context*, edited by Bernard Lightman, 409–37. Chicago: The University of Chicago Press.

Hacking, Ian. 1983. *Representing and Intervening*. Cambridge: Cambridge University Press.

Harting, Pieter. 1866. *Das Mikroskop. Theorie, Gebrauch, Geschichte und gegenwärtiger Zustand desselben*. 2 vols. 2nd ed. Braunschweig: Vieweg.

Hogg, Jabez. 1869. *The Microscope: Its History, Construction, and Application: Being a Familiar Introduction to the Use of the Instrument and the Study of Microscopical Sciences*. New ed. London: Routledge.

Jacobs, Merle. 1924. "Permeability of the Cell to Diffusing Substances." In Cowdry 1924, 97–164.

James, Frank L. 1887. *Elementary Microscopical Technology, Part I: The Technical History of a Slide from the Crude Materials to the Finished Mount*. St. Louis: St. Louis Medical and Surgical Journal Company.

Kirsch, A. M. 1909. "Simple Method of Easily Resolving Microscopical Test-Objects" *Midland Naturalist* 1 (1): 26–27.

Koch, Robert. 1987. "The Etiology of Tuberculosis." In *Essays of Robert Koch*, edited by K. Codell Carter, 83–96. Westport, CT: Greenwood Press.

Latham, Vida. 1891. "The Use of Stains, Especially with Reference to Their Value for Differential Diagnosis." *Proceedings of the American Society of Microscopists* 13:94–109.

———. 1894. "A Plea for the Study of Re-Agents in Micro Work." *Proceedings of the American Microscopical Society* 15:209–11.

———. 1896. "The Question of Correct Naming and Use of Micro Reagents." *Transactions of the American Microscopical Society* 17:350–58.

Lenhartz, Hermann. 1904. *Manual of Clinical Microscopy and Chemistry*. Philadelphia. F. A. Davis.

Lillie, Frank R., and E. E. Just. 1924. "Fertilization." In Cowdry 1924, 449–536.

Lillie, Ralph. 1924. "Reactivity of the Cell." In Cowdry 1924, 165–233.

Maienschein, Jane. 1991. "Cytology in 1924: Expansion and Collaboration." In *The Expansion of American Biology*, edited by Keith Rodney Benson, Jane Maienschein, and Ronald Rainger, 23–51. New Brunswick, NJ: Rutgers University Press.

McCalla, Albert. 1883. "President's Address: The Verification of Microscopic Observation." *Proceedings of the American Society of Microscopists* 5:1–19.

Minot, Charles Sedgwick. 1877. " The Study of Zoology in Germany." *American Naturalist* 11 (7): 392–406.

Mitchell, Charles L. 1883. "Staining with Hæmatoxylon." *Proceedings of the Academy of Natural Sciences of Philadelphia* 35:297–300.

Morehouse, G. W. 1873. "The Structure of the Scales of *Lepisma saccharina*." *American Naturalist* 7:666–69.

Nägeli, Carl, and Simon Schwendener. 1867. *Das Mikroskop: Theorie und Anwendung desselben*. Leipzig: Engelmann.

Pritchard, Andrew. 1832. *The Microscopic Cabinet of Select Animated Objects; with a Description of the Jewel and Doublet Microscope, Test Objects, &C*. London: Whittaker, Treacher, and Arnot. (Facsimile Edition published by Science Heritage Ltd., Lincolnwood 1987.)

Queckett, John. 1855. *A Practical Treatise on the Use of the Microscope: Including the Different Methods of Preparing and Examining Animal, Vegetable, and Mineral Structures*. 3d ed. with additions. Library of Illustrated Standard Scientific Works, vol. 6. London: H. Bailliere.

Ratcliff, Marc J. 2009. *The Quest for the Invisible: Microscopy in the Enlightenment*. Farnham: Ashgate.

Reade, J. B. 1870. "Microscopic Test Objects under Parallel Light and Corrected Powers." *Popular Science Review* 9 (34): 138–48.

Schacht, Hermann. 1851. *Das Mikroskop und seine Anwendung, insbesondere für Pflanzen-Anatomie und Physiologie*. Berlin: G. W. F. Müller.

Schickore, Jutta. 2009. "Test Objects." *History of Science* 47 (2): 117–45.

Schleiden, Matthias Jacob. 1845. *Grundzüge der wissenschaftlichen Botanik*. 2nd ed. Leipzig: Engelmann.

———. 1969. *Principles of Scientific Botany; or, Botany as an Inductive Science*. New York: Johnson Reprint Corporation.

Sharpey-Schäfer, E. A. Sir. 1877. *A Course of Practical Histology: Being an Introduction to the Use of the Microscope*. Philadelphia: Lea.

Stöhr, Philipp. 1903. *Text-Book of Histology including the Microscopic Technic*. Philadelphia: P. Blakiston's.

Trizio, Emiliano. 2012. "Achieving Robustness to Confirm Controversial Hypotheses: A Case Study in Cell Biology." In *Characterizing the Robustness of Science*, edited by Léna Soler, 105–20. Boston: Springer.

Vogel, Julius. 1841. *Anleitung zum Gebrauch des Mikroskopes zur zoochemischen Analyse und zur mikroskopisch-chemischen Untersuchung überhaupt*. Leipzig: Voss.

Wilson, E. B. 1896. *The Cell in Development and Inheritance*. New York: Macmillan.

———. 1900. *The Cell in Development and Inheritance*. 2nd ed., revised and enlarged. New York: Macmillan.

———. 1924. "Introduction." In Cowdry 1924, 1–11.

———. 1925. *The Cell in Development and Heredity*. 3rd ed. New York: Macmillan. Orig. pub. as *The Cell in Development and Inheritance* in 1896; 2nd ed., 1900.

Woodward, J. J. 1872. "On the Use of Monochromatic Sunlight, as an Aid to High-Power Definition." *American Naturalist* 6 (8): 454–60.

CHAPTER 5

CELLULAR PATHOGENESIS

VIRUS INCLUSIONS AND HISTOCHEMISTRY

William C. Summers

While it was Rudolf Virchow who focused attention on the cell as the unit of pathology, Edmund V. Cowdry provided some of the first understanding of viral pathogenesis at the cellular level. Between 1922 and 1940 Cowdry elaborated his views of viral cytopathology based on the study of cellular "inclusion bodies" and their relations to virus infection and pathology. His classification of inclusion bodies as "Type A" and "Type B" became standard and widely used at least until the 1950s, and in some cases it persists to the present. Although this classification was not particularly clear in its meaning, and its application was subjective, Cowdry's approach combined stain technology and the other methods of histology and histochemistry in the study of the interactions of viruses with cells, providing new knowledge about the biology of these obligate intracellular parasites. His exploitation of the chemistry of compounds that bind to intracellular components as ways to study and differentiate structures and functions contributed both to better understanding of subcellular biology (e.g., of mitochondria) and to the pathogenesis of viral infections.

Cowdry's early interest focused on using the newly developed stains and dyes to study intracellular morphology, especially the cytoplasmic bodies called mitochondria by Carl Benda in 1898, and in the 1920s and 1930s he was to make major contributions to virology by his careful studies of the cytology of virus-infected cells. This was at a time prior to the application of cell culture methods and biochemical characterizations of subcellular fractions to understand the pathogenic effects of virus infection. Cowdry's research pathway from mitochondria to virology was based on similar principles of cytology, microchemistry, and a persistent commitment to the proper way to study cells. That research pathway is the focus of this paper.

Eugene Vincent Cowdry was born in 1888 in Fort Macleod, Alberta, was schooled in England, and received a BA in 1909 from the University of Toronto (Anderson 2009). Although his father, Nathaniel, was a banker by profession, he was also an avid amateur naturalist, which certainly must have influenced his son's interests. Indeed, later in life, Cowdry arranged for his father to join him in his laboratory work. The elder Cowdry published several widely cited papers on cytology, and his natural history collections of both insects and plants were widely known and used (N. H. Cowdry 1917, 1920).

Young Cowdry commenced his graduate work at the University of Chicago under Robert Russell Bensley and received his PhD in 1911. Bensley had investigated the small cytoplasmic bodies called mitochondria, thought by some to be parasitic microorganisms, and handed this work on to Cowdry for his graduate research. They thought mitochondria were "as characteristic of the cytoplasm as chromatin is of the nucleus" (Cowdry 1956). Bensley recognized the significance of Michaelis's observation that mitochondria could be stained in living cells with the redox-sensitive dye, Janus Green B (diethylsafranin; IUPAC: 8-(4-Dimethylaminophenyl) diazenyl-*N*,*N*-diethyl-10-phenylphenazin-10-ium-2-amine chloride; this dye is decolorized by oxygen more rapidly in the cytoplasm than in mitochondria, providing somewhat transient visualization of mitochondria). This tool became Cowdry's way into his dissertation research, which was published in 1913 as "The Relations of Mitochondria and Other Cytoplasmic Constituents in Spinal Ganglia of the Pigeon" (Cowdry 1913).

At the very outset of his research career Cowdry formed the guiding principles of his scientific credo: his dissertation strongly asserts that physiological and even behavioral properties can seek their explanations in cellular biology. Thus, in choosing to study nervous system cells, he pointed out that the "cytoplasm of nerve cells is of absorbing interest to the anatomist, the psychiatrist and to many others. The reason for this is self-evident" (Cowdry 1913, 1). He also committed to a holistic point of view: "Believing that the cell is a harmonious whole and that any attempt to dissociate its constituents is likely to result in error, the general viewpoint of synthesis has been adopted" (Cowdry 1913, 3). This view contrasts to his decades-later assessment of this work in his eulogy to Bensley, where he noted that he, Cowdry, "made little progress, so he [Bensley] returned to the attack 25 years later and in a very original way. . . . He and his student N. L. Hoerr (1934) broke living cells up and separated out the mitochondria by centrifugation" (Cowdry 1956, 973).

Cowdry's basic methodology in his cytochemical investigations was to use the cell as a chemical laboratory in which experiments with dyes and reactions could be carried out and observed under the microscope to obtain what he called "tinctorial evidence" (Cowdry Papers 1924a). His dissertation described this approach as "applying one stain to a cell, fixing it in, adding another and another, or else by staining one component in a specific fashion, and then dissolving out the dye and staining others in the same cell by appropriate methods" (Cowdry 1913). For his dissertation, he examined the spinal ganglion cells of the pigeon without explaining his choice of experimental animal. One reason might be the availability of pigeons in Bensley's lab, which studied mitochondria; pigeon flight muscle, with its high oxidative metabolic rate, was known as a rich source of mitochondria, and the pigeon was an accepted experimental animal.

With his Chicago doctorate in hand, Cowdry joined the Johns Hopkins University in 1913 as an associate in anatomy. Famous for its work on human embryology under Franklin Paine Mall, the Anatomy Department was also the home of Florence R. Sabin, an eminent histologist working on blood cells. Cowdry's work at Hopkins focused on mitochondria in blood cells, contributing to Sabin's program on human blood cells (Cowdry 1921b). As noted above, while in Baltimore, Cowdry also managed to find space for his father to join him in laboratory research. The elder Cowdry, apparently a skilled microscopist as well as an amateur field naturalist, carried out several detailed studies using cytochemical techniques that he learned from his son (N. H. Cowdry 1920). These studies were published by his father in several prestigious journals and were frequently cited.

In December, 1916, Cowdry married Alice Hanford Smith, four years his junior, from Washington, DC. She must have been an adventurous woman, since her new husband was in the process of organizing a career-changing move to China within the first year of their marriage. As early as 1909, at the urging of his adviser, Frederick T. Gates, John D. Rockefeller started work on the "educational, social and religious conditions in the Far East," and in 1917 a new medical school was started in China, supported by the Rockefeller Foundation, located in the capital, Beijing (Peking), and modeled after the Johns Hopkins School of Medicine (Ferguson 1970). The grand old man at Hopkins, William ("Popsy") Welch, had a heavy hand in planning this school, known as Peking Union Medical College (PUMC). The new school, designed to provide China with a world-class Western medical education, had three groups of initial faculty: skilled clinicians, high-quality teachers and researchers, and local Chinese physicians. Cowdry, at age twenty-eight, was offered the chair in anatomy at PUMC, as the first choice of the school's director, Franklin McLean. Cowdry apparently had measured up to the Johns Hopkins standards: Florence Sabin wrote to a confidant that "Dr. Cowdry is going to China with the new Rockefeller Medical School. He is the Canadian who has been assisting me of whom we were talking. It's a fine opportunity for him and we are all glad for him" (Sabin Papers 1917).

When Cowdry took up the professorship at PUMC with the first class of preclinical students in September 1917, he took his ecumenical mission seriously; he made a study of traditional Asian medicine and published several articles on Asian anatomical knowledge and medical traditions in China and Japan (Cowdry 1920, 1921a). Interestingly, in addition to his young wife, Cowdry's father and aunt accompanied him to Beijing, and the elder Cowdry continued his natural history research by making an extensive collection of Chinese marine algae, which became the object of study both by him and others (N. H. Cowdry 1922). Indeed, the second publication from the PUMC Anatomical Laboratory was authored by N. H. Cowdry on work in his son's laboratory at Hopkins and published in the prestigious *Biological Bulletin* (N. H. Cowdry 1918). Cowdry spent much of his first two years at PUMC getting the laboratories constructed, outfitted, and up and running. Even with these burdens of work, the Cowdry's, as seemed customary for many expat professionals in China, managed to return "home" during the hot Beijing summer season. When Alice became pregnant, in 1920, she remained in the United States. With PUMC up and running, it appears that Cowdry, with a young wife and new child, decided that his future would be better in North America, and he was accepted as an associate member of the Rockefeller Institute. Perhaps one of Cowdry's last duties in China was the formal dedication at the opening of PUMC in September 1921, where his former Hopkin's colleague, Florence Sabin was among the international scientific and medical elite who attended. She was one of the few who were invited to present principal addresses. Her title was "The Origin of Blood cells" (Wong and Wu 1936, 680–81).

Cowdry joined the Department of Pathology and Bacteriology as a regular salaried staff scientist at the level of associate member. Only full members, however, had anything approaching a permanent or long-term position. In addition to Simon Flexner, the director of laboratories, other full members of this department were Wade Brown, Karl Landsteiner, Hideo Noguchi, Peyton Rous, and Florence Sabin, who had recently relocated from Hopkins to the Rockefeller Institute. Cowdry's fellow associate members of this department were Jacques Bronfenbrenner, Frederick Gates, Peter Olitsky, and Louise Pearce (Sabin Papers 1925–26). Clearly, he was at the center of the biomedical research world both in terms of talent and material support.

The focus of much of the research in Flexner's department was on infectious diseases, and from Flexner's correspondence with Cowdry, Flexner apparently thought that detailed cytological studies could contribute to the mysteries of certain infectious processes. One such mystery was that of diseases associated with the group of microbes called rickettsia.[1] Experimental and epidemiological data suggested that arthropod vectors could transmit an infectious agent that resisted laboratory cultivation. Cytological examination of ticks that could transmit such disease showed characteristic intracytoplasmic inclusions that were believed to be the causative pathogen.

Such diseases, Flexner believed, needed skilled cytological study. He thought Cowdry was just the person to undertake such studies. From the beginning of his appointment at the Rockefeller Institute, it was clear that Flexner saw Cowdry as some sort of super-technician, with specific, valuable skills, not as a fellow creative scientist. Their correspondence repeatedly seems to disparage Cowdry as a "real" scientist: "I believed your field was as teacher and investigator of anatomy, but at the same time you possessed unusual knowledge and technique of minute histology, which I was eager to have represented for a period at the Rockefeller Institute. On that basis I suggested tentatively your coming to the Institute for year or so, where you could work under, I thought, favorable conditions while waiting also for another opening in anatomy." Then he went on to explain that he was offering a two-year "limited appointment" "without any commitment for the future" (Cowdry Papers 1921). Cowdry apparently fulfilled Flexner's expectations, and during this time he collaborated with the Rockefeller Institute scientist, Peter Olitsky, in a significant microchemical characterization of the staining properties used to distinguish mitochondria from intracellular bacteria (Cowdry and Olitsky 1922).

By the end of his two-year "limited appointment," the Rockefeller Institute decided that Cowdry's expertise with "minute histochemistry" might be useful in its program to investigate agricultural diseases, and arranged for him to spend a year in South Africa to study the scourge known there as heartwater fever.[2] Cowdry apparently took well to extensive research travel, since he later was to volunteer almost whenever possible for exotic overseas postings. For a year he worked in the Laboratory of the Department of Agriculture in Onderstepoort, Union of South Africa, under the directorship of Sir Arnold Theiler (Du Toit and Jackson 1936), and he was able to describe the likely pathogen in heartwater fever. His careful cytochemical studies indicated that the "virus" of heartwater fever was a microbe with all the properties of a rickettsia. He wrote, "Although the association of the microorganisms with heartwater, their morphology, their staining reactions were found to be so definite, it was decided to compare them carefully with normal cellular components and with the products of degeneration and phagocytosis, since, up to the present time, like most *Rickettsiae*, their status as living organisms has not been proved by methods of artificial cultivation" (Cowdry 1925a).

This work with Theiler resulted in two publications on the rickettsia as the cause of heartwater fever in the *Journal of Experimental Medicine*, the flagship publication of the Rockefeller Institute (Cowdry 1925a; 1925b). Indeed, in the margin of Cowdry's informal report to Flexner on his work in South Africa, Flexner penciled, "I desire the paper for the JEM" (Cowdry Papers 1924a). In this work Cowdry used his expertise as a cytologist and histochemist to investigate the differences between "normal" and "diseased" or "infected" material for evidence of changes that might be characteristic of the causative agent or might be the agent itself. As he indicated in his discussion of heartwater, he was rather agnostic about the biological status of the rickettsia as a living being. Indeed, his later work on viruses exhibited this agnosticism as well.

His taste for infectious diseases seems to have been whetted by study of rickettsia and heartwater fever. In September 1924 he wrote to Flexner for permission to expand his work to other diseases:

> The whole problem of Rickettsia-like micro-organisms interests me intensely. . . . I wonder where it would be possible to secure some tissues [from *Tsutsugamushi* disease, i.e., scrub typhus] from Japan fixed in Zenker's fluid and preserved in alcohol, which I might work up upon my return. Perhaps you would let me make another attempt with typhus fever. Dr. Noguchi said that a friend of his in Mexico City, Dr. Pruneda I think it was, might be willing to send some tissues, which should be preserved in the same way. To do this would not interfere with any other plans which you may have. (Cowdry Papers 1924b)

Later, he also expressed interest in a similar approach to study some malady in the colony of rabbits at the Rockefeller Institute: "Acting upon a suggestion by Dr. Theobald Smith that the parasites in the rabbits' brains may represent a relatively recent invasion of our laboratory stock," he wrote to China and Australia (as well as Siam and Brazil) requesting fixed samples of rabbit brains (Cowdry 1924b). This proposal to study rabbit neuropathology reflected his cytochemical approach to the study of infectious disease by which he traced the cellular footprints of the putative infectious agent by tracing its life cycle in both the infected animal and its insect vector, the tick, in the case of rickettsia. The requested samples of distant populations of rabbits would serve as controls to test Smith's hypothesis of "recent invasion." Cowdry's success in deploying what he characterized as "morphological or tinctorial evidence" to investigate pathologic conditions related to infectious disease led him in new directions for more than a decade in the mid-1920s to the mid-1930s. In 1926 he was on his way to Tunis to study anaplasmosis, a hemolytic anemia of ruminants, suspected of being caused by a tick-borne rickettsia. In his proposal to the Rockefeller Foundation for support, he raised concerns that were to be central to his work on virus inclusions: "The anaplasma occur within the red blood cells of sick cattle and opinion is divided as to their nature. Some investigators contend that they are true microorganisms, while others regard them as inclusions produced by the action of a virus which has never thus far been seen" (Cowdry Papers 1926).

All during his tenure at the Rockefeller Institute, Cowdry's relation with Flexner seemed fragile. The correspondence between the two men is full of Flexner's skepticism and Cowdry's deference. Cowdry continually sought Flexner's approval for what seems like simple scientific initiatives. Flexner seemed to go out of his way to remind Cowdry of his nontenured position at the institute, while Cowdry was constantly defending his work to Flexner. Even upon the publication of *General Cytology* in 1924, Flexner was faint in his praise: "First let me thank you for the copy of your 'Cytology,' which looks first rate. I congratulate you on the book. It will serve a very good purpose" (Cowdry Papers 1924c). One is tempted to speculate that Cowdry, in spite of his record of successes and his appointment as inaugural professor at PUMC, was deeply insecure, working first under Florence Sabin, by most accounts a supportive mentor at Hopkins; helped by his father, an amateur scientist, first at Johns Hopkins, then in China, and for several summers at Woods Hole; and later deferring to the dominating figure of Flexner.

By the late 1920s Cowdry had become sufficiently immersed in study of the cellular responses to virus infection to contribute a chapter to Thomas Rivers's influential book, *Filterable Viruses*. Rivers leaned heavily on his Rockefeller colleagues for expertise, and Cowdry was close at hand. Cowdry's chapter was on "Intracellular Pathology in Virus Diseases," and there he presented his system of classification of intracellular inclusions, which persists to the present day (Cowdry 1928).

The main focus of cytopathology was the observation of novel structures seen in the cytoplasm and in the nucleus of infected cells, the so-called inclusion bodies. These changes in cell architecture seemed to be related to virus infection, but their interpretation was hotly debated: Were they viruses themselves? Were they structures produced by the cells in response to viruses? Did they have any relation to virus-induced cell behaviors that were becoming increasingly well-characterized? In other words, what did inclusions have to do with pathogenesis?

Two scientists led the way during this era of virus research, Cowdry and Ernest W. Goodpasture (Long 1965). Cowdry was a pathologist and cell biologist who believed that study of the various inclusion bodies in virus-infected cells would shed light on the pathologic processes of virus infection as well as the nature of virus growth and reproduction (fig. 5.1). While others had noted various types of inclusions that were associated with certain infections—for example, Negri bodies in rabies and Guarnieri bodies in vaccinia and smallpox—Cowdry developed a systematic study of inclusions and proposed a classification of types of inclusions, which he hoped would lead to more clarity in viral pathogenesis. He differentiated cytoplasmic from nuclear inclusions, and was skeptical of some reports of virus-associated inclusions. Some inclusions, he believed, were collections of the viruses themselves, and other inclusions seemed to be cellular products, perhaps made in response to viral infections. Cowdry brought a chemical approach to cytopathology, asking "What is the nature of the material of which the inclusion bodies are composed" and whether this material was present in the cell prior to virus infection. He proposed the key question in viral pathogenesis when he asked, "What alterations in cellular activity are caused by the viruses?" The unsettled nature of such inclusions was reflected in Cowdry's caution: "Many believe that the [inclusion] bodies are neither organisms *sui generis* nor combinations between organisms and cellular components but rather *reaction products* produced by the cell in response to injury caused by infective agencies which are ultramicroscopic" (Cowdry 1928, 115). Others, however, were more certain about some virus-induced inclusions. In their study of herpes infections as early as 1923, Goodpasture and Teague (1923) wrote, "The presence of these intranuclear bodies means the presence of the virus, and that they represent the growth of the virus in the infected nuclei, as claimed by Lipschutz."

Cowdry is famous for his classification of inclusion bodies in virus infections or in tissues thought to be related to virus infections. In his original paper on this classification, he listed eighteen cases in which his "Type A" inclusions were observed, including herpes, yellow fever, chickenpox, whooping cough, kidneys of frogs, louping ill (a tick-borne encephalitis of sheep), and "many species in the absence of disease." His "Type B" inclusions were noted in poliomyelitis, Rift Valley fever, Borna disease, and "many species

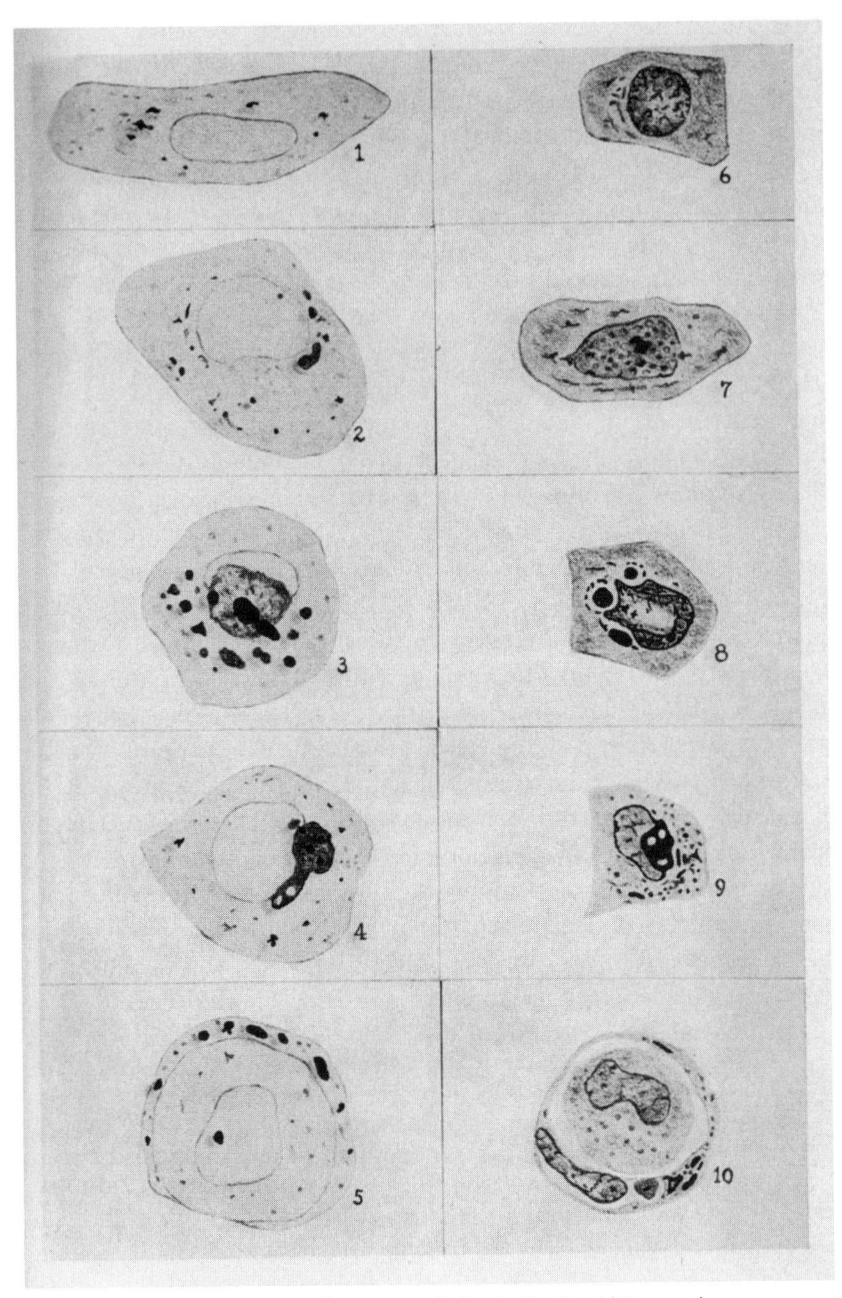

Figure 5.1. Image of vaccinia inclusion bodies in rabbit corneal cells: numbers 1–5, vital stain; 6–10, fixed and stained; 1 and 6 prior to infection with vaccinia. From Cowdry 1928.

unaccompanied by evidence of disease." His criteria for this classification are so vague as to be undecipherable. For example, Type A inclusions "are amorphous or particulate, but may be condensed in rounded masses." With respect to Type B, "The reaction is localized to certain [unspecified] areas of the nucleus, where acidophilic droplets make their appearance" (Cowdry 1934). Surprisingly, Cowdry's cytopathological descriptions of inclusions that may, or may not, have any relationship to virus infection became a central and persistent focus of diagnosis and investigation at least until the 1950s, and even today one finds his terminology still in use. While some cell inclusions are now recognized as manifestations of diverse cell responses to infection, this era of descriptive cytology eventually gave way to more chemical and physiological study of viral pathogenesis at the cellular level.

After this rather exhaustive summary of diverse experimental observations, Cowdry concluded this chapter in a distinctly nihilistic tone: "We do not know even in a single case with any degree of precision the chemical nature of the material of which the inclusion bodies are built." Later, he says, "How the virus enters, if it enters at all, is a mystery." Concluding, he writes, "Until viruses are shown to be living agents, it requires a larger stretch of the imagination to conceive of their initial development apart from living matter than in association with it" (Cowdry 1928, 142–49).

By 1928, when Cowdry had moved to Washington University in St. Louis, he was enthusiastic about collaborating with F. F. Russell of the Rockefeller Institute on cytochemical studies of yellow fever, again tracing the life history of the infection in the suspected vector, the mosquito *Aedes aegypti*.

> I gladly accept the invitation contained in your letter of August 7 to study specimens of *Aedes aegypti* infected with the virus of yellow fever. I have been wondering whom you would get to attack the aspect of the problem, and that you should select me is most gratifying. . . . The task of finding, if possible, traces of the virus appeals to me. . . . Mosquitoes carrying the virus and others of the same age and sex devoid of it should in some way differ. I suppose it is still conceivable that a rickettsia may be involved. (International—Yellow Fever Papers 1928)

In addition to his scientific enthusiasm, however, Cowdry wanted to make sure that, even though he was a full-time employee of Washington University, the Rockefeller Institute would provide him with a life insurance policy in case he should die of a laboratory infection. Apparently he anticipated exposure to the live virus at some point, perhaps in Russell's laboratory, even though Cowdry's methods employed fixed tissue samples.

There were rumors that the Rockefeller Institute had not treated Noguchi's wife well after his death by yellow fever during a research expedition in Africa, and Cowdry wanted to head off such problems "in case of unfortunate outcome of yellow fever research" (International—Yellow Fever Papers 1929).

This exhaustive study, published *in extenso* in a seventy-nine-page paper in the *American Journal of Hygiene* in 1930 did much to establish cytopathology as a tool in the study of the new class of microscopically invisible, filter-passing agents of disease, still called "filterable viruses" (Cowdry and Kitchen 1930). Without the tools of bacteriology—namely, direct staining and observation, together with laboratory culture—virologists were severely hampered until the later advent of cell culture techniques and visualizations in the electron microscope. Cowdry's work, then, provided a window into the biology of viruses with study of the cellular responses in terms of cytochemically visible "inclusions."

His role as a leading researcher in this field was summarized in a Rockefeller Institute memo from 1932: "Doctor E. V. Cowdry has been interested as a cytologist and histologist in the cellular reaction to viruses. Since viruses cannot be cultivated in the absence of living cells, and since to find what happens to cell composition and structure in [*sic*] one of the fundamental ways of learning more about the nature of virus diseases, the cytologist's contribution to the virus problem is at present of recognized value" (Washington University—Virus Research Papers 1932).

In an interview at the same time, the notes taken by his interviewer summarized Cowdry's virus research program as having two principle aims: "(1) To determine the specificity of intra-nuclear inclusions. Has preps for study of yellow fever, chickenpox, herpes, submaxillary disease of guinea pigs, VSV, Virus 3 and an unknown disease of frogs. (2) Study of mineral constituents of cells attacked by viruses. By microincineration . . . shows quite striking shifts in mineral content of affected cells" (Washington University—Virus Research Papers 1932).

The second aim described Cowdry's extension of his cytochemical approaches beyond staining and "tinctorial" evidence to a new technique that he was developing to analyze the mineral constituents of cells by a microincineration technique that relied on microscopical examination and identification of the ash residue after burning away the organic matter in cytological preparations. This new technique involved a collaboration with a physicist who had joined Cowdry's research group at Washington University. He seemed to have an organizational knack and social skill that facilitated collaborations across disciplines and involving diverse organizations to obtain research materials for his broad, comparative approach to cytology.

As an example of his belief in the value of the latter enterprise, in 1933 he wrote to Alan Gregg of the Rockefeller Foundation, the main supporter of his research work, that he was undertaking a comprehensive review of the entire field of intranuclear inclusions:

> Scarcely a year goes by without the discovery of five or six more intranuclear inclusions and it seems desirable to survey the field in order to gain a true conception of their distribution. Arrangements have accordingly been made for the Rectors to utilize the valuable and extensive material collected at the Philadelphia Zoo. (So far studied moles . . . compare to children, guinea pigs, rats)
>
> For years I have been collecting specimens exhibiting intranuclear inclusions as a result of viral action not only from my own material but from investigators far and near. I think I have examples of almost all of the intranuclear inclusions which have been reported.
>
> These [inclusions in nerves caused by stimulation] looked astonishingly like nuclear inclusions caused by a virus. Ranson found somewhat similar appearances could be produced by placing the living nerve cells in hypertonic salt solutions. O'Leary and Jack Lee propose to follow this lead because the intranuclear bodies produced experimentally in nerve cells do look so much like those resulting from virus action that they would certainly lead the unwary astray. I think, however, that the artificial inclusions may be only the acidophilic material which is disturbed. This line of investigation is, I think, of real importance; because, if we can duplicate the response of a cell to a virus, we shall have unearthed the mode of virus action—it is even conceivable that we may have produced some local product which is identical with a virus. You will forgive me for indulging in speculation. I have never been convinced that all [*sic*] viruses are living things; but my mind is open and I want to see the light. (Washington University—Virus Research Papers 1933)

By 1935, however, Cowdry's virus research was coming to an end. The immediate reason was the end of support from the Rockefeller Foundation, but the deeper reasons are less clear. While the economic hard times of the Great Depression and the reduction of Rockefeller income were no doubt important, the new discoveries in virology, such as Stanley's more chemical approaches; the new interest of the Rockefeller Foundation leaders in the biophysical approaches with the ultracentrifuge (Kay 1992) and the promise of the electron microscope; and, especially, Goodpasture's embryonated egg culture methods (Woodruff and Goodpasture 1931) all suggested that Cowdry's microscopic anatomy may have run its course and that new methods might be better ways to advance understanding of filterable viruses. At a time when biochemistry and organic chemistry were showing increasing promise, Cowdry's turn to the inorganic ash left by incinerated cells may, too, have seemed both metaphoric and distinctly esoteric.

Epilogue

A full century after Edmund V. Cowdry started his work on the microchemistry of cell inclusions with only a good microscope and not entirely reliable histologic stains, the two principle objects of his study are now fully established in biological science: mitochondria and intracellular microbes.

Mitochondria, first stabilized as a scientific fact by the discovery that they stain "specifically" with Janus Green B, are recognized as crucial organelles in most cells, responsible for aerobic metabolism, a major evolutionary advance. They are believed to be an early endosymbiont in eukaryotic evolution, allowing cells to exploit oxidative metabolism as a more efficient source of the chemical energy of life. Their function as the locus of many cellular oxidation-reduction reactions, is, of course, the basis of their characteristic reaction with dyes that are sensitive to the oxidation state of the dye molecule, such as Janus Green B.

Cytological "particles," as Cowdry observed, may be specific entities with structure and function, or they may be amorphous, perhaps degraded, material from other cell processes. To distinguish these two possibilities by cytochemical study was Cowdry's central program. In addition to mitochondria, he clarified another group of endosymbionts, the small, obligate intracellular bacteria, often pathogenic, called rickettsia. Even today, because these organisms resist culture outside of cells, immunocytochemistry contributes to their diagnosis and study.

Cowdry's other main group of endosymbionts, the viruses, are still described in Cowdry's classification schemes. As he speculated, his Type B inclusions have not proven to be a useful descriptive category; by contrast, his description of Type A inclusions has been adopted as a key cytological characteristic of herpes virus infections. From about the mid-1950s, Cowdry's name has become eponymous with the cytological appearance of aggregates of nucleocapsids of herpes viruses as "Cowdry Type A inclusion bodies."

While Cowdry's synthetic cellular cytochemical laboratory has given way to the analytical approach of the biochemists and molecular biologists, the insights and fundamental knowledge he established by his meticulous technique and biological insights still inform cell biology a century later.

Notes

1 Rickettsia are endosymbiotic bacteria that have less genetic complexity than most bacteria, and have been impossible to cultivate outside of living cells. After various epidemiological and inoculation experiments with typhus fever, a small microbe was consistently observed in infected animals as well as the lice that could transmit the disease (Ricketts and Wilder 1910). These microbes were only identified by microscopic examination, and their biological status was uncertain until the mid-twentieth century.

2 Heartwater fever is a tick-borne rickettsial disease of ruminants, characterized by accumulation of fluid around the heart and in the lungs of sick animals. This disease is sometimes called "cowdriosis" after the former name of the pathogenic agent, *Cowdria ruminatum*, now known as *Ehrlichia ruminatum*.

References

Anderson, Paul G. 2009. "Edmund Vincent Cowdry (1888–1975)." Online at http://beckerexhibits.wustl.edu/mig/bios/cowdry.html.

Cowdry, Edmund V. 1913. "The Relations of Mitochondria and Other Cytoplasmic Constituents in Spinal Ganglia of the Pigeon." *Internationale Monatsschrift für Anatomie und Physiologie* 30:473–504.

———. 1920. "Anatomy in China." *Anatomical Record* 20:31–60.

———. 1921a. "A Comparison of Ancient Chinese Anatomical Charts with the 'Fünfbilderserie' of Sudhoff." *Anatomical Record* 22:1–25.

———. 1921b. "The Reticular Material of Developing Blood Cells." *Journal of Experimental Medicine* 33: 1–11.

———. 1925a. "Studies on the Etiology of Heartwater: I. Observation of a Rickettsia, *Rickettsia ruminantium* (n. sp.). in the Tissues of Infected Animals." *Journal of Experimental Medicine* 42: 231–52. See especially 244.

———. 1925b. "Studies on the Etiology of Heartwater. II. *Rickettsia ruminantium* (n. sp.) in the Tissues of Ticks Transmitting the Disease." *Journal of Experimental Medicine* 42:253–74.

———. 1928. "Intracellular Pathology in Virus Diseases." In *Filterable Viruses*, edited by Thomas M. Rivers, chap. 4. Baltimore: Williams and Wilkins.

———. 1934. "The Problem of Intranuclear Inclusions in Virus Diseases." *Archives of Pathology* 18:527–42.

———. 1956. "R. R. Bensley, Cytologist." *Science* 124:972–73.

———, and S. F. Kitchen. 1930. "Intranuclear Inclusions in Yellow Fever." *American Journal of Hygiene* 11 (2): 227–99.

———, and Peter K. Olitsky. 1922. "Differences between Mitochondria and Bacteria." *Journal of Experimental Medicine* 36:521–33.

Cowdry, Nathaniel H. 1917. "A Comparison of Mitochondria in Plant and Animal Cells." *Biological Bulletin* 33 (3): 196–228.

———. 1918. "The Cytology of the Myxomycetes with Special Reference to Mitochondria." *Biological Bulletin* 35(2): 71–94.

———. 1920. "Experimental Studies on Mitochondria in Plant Cells." *Biological Bulletin* 39 (3): 188–206.
———. 1922. "Plants Collected at Peitaiho, Chihli." *Journal of the North China Branch of the Royal Asiatic Society* 53: 158–88.
Cowdry Papers. 1921. Simon Flexner to Edmund V. Cowdry, 4 Feb. 1921, box 1, folder 1. Rockefeller Archives Center.
———. 1924a. Edmund V. Cowdry to Simon Flexner, 17 Nov. 1924, box 1, folder 1. Rockefeller Archives Center.
———. 1924b. Edmund V. Cowdry to Simon Flexner, 29 Sept. 1924, box 1, folder 1. Rockefeller Archives Center.
———. 1924c. Simon Flexner to Edmund V. Cowdry, 22 Oct. 1924, box 1, folder 1. Rockefeller Archives Center.
———. 1926. Edmund V. Cowdry, "Objects of Proposed Expedition to Tunis," 11 Dec. 1926, box 1, folder 1. Rockefeller Archives Center.
Du Toit, P. J., and C. Jackson. 1936. "The Life and Work of Sir Arnold Theiler." *Journal of the South African Veterinary Medical Association* 7 (January 1): 135–86.
Ferguson, Mary E. 1970. *China Medical Board and Peking Union Medical College: A Chronicle of Fruitful Collaboration 1914–1951*. New York: China Medical Board of New York.
Goodpasture, Ernest W., and Oscar Teague. 1923. "Experimental Production of Herpetic Lesions in Organs and Tissues of the Rabbit." *Journal of Medical Research* 44 (2): 121–38.
International—Yellow Fever Papers. 1928. Edmund V. Cowdry to F. F. Russell, 9 Aug. 1928, RG1.1 (FA386), series 100; subseries 100.0, Cowdry box 84, folder 778. Rockefeller Archives Center.
———. 1929. F. F. Russell to Edmund V. Cowdry, 3 Jan. 1929, RG1.1 (FA386), series 100; subseries 100.0, Cowdry box 84, folder 778. Rockefeller Archives Center.
Kay, Lily E. 1992. *The Molecular Vision of Life: Caltech, the Rockefeller Foundation, and the Rise of the New Biology*. Oxford: Oxford University Press.
Long, Esmond R. 1965. "Ernest William Goodpasture, October 17, 1886–September 20, 1960." In *Biographical Memoirs of the National Academy of Sciences*, 38:111–44. Washington DC: The National Academies Press.
Ricketts, Howard Taylor, and Russell M. Wilder. 1910. "The Etiology of the Typhus Fever (*tabardillo*) of Mexico City: A Further Preliminary Report." *Journal of the American Medical Association* 54:1373–75.
Sabin Papers. 1917. Florence R. Sabin to Mrs. Denison, 27 Mar. 1917. National Library of Medicine (Profiles in Science).
———. 1925–1926. "Rockefeller Institute Scientific Staff List for 1925–1926." National Library of Medicine (Profiles in Science).
Washington University—Virus Research. 1932. Rockefeller Foundation Memorandum for Officers' Conference, 20 April 1932, box 3, folder 38 RF. Rockefeller Archives Center.
———. 1933. Edmund V. Cowdry to Alan Gregg, 5 July 1933, box 3, folder 38 RF. Rockefeller Archives Center.
Wong, Chi-min, and Lien-Teh Wu. 1936. *History of Chinese Medicine*. 2nd ed. Shanghai: National Quarantine Service.
Woodruff, Alice Miles, and Ernest W. Goodpasture. 1931. "The Susceptibility of the Chorio-Allantoic Membrane of Chick Embryos to Infection with the Fowl-Pox Virus." *American Journal of Pathology* 7 (3): 209–22.

CHAPTER 6

THE AGE OF A CELL

CELL AGING IN COWDRY'S *PROBLEMS OF AGEING* AND BEYOND

Lijing Jiang

The phenomenon of aging is usually seen as an inevitable part of the life course of many multicellular organisms, including humans. Aging is defined as either biological change over time or as decline in health and vigor in later life stages, and has perplexed biologists in its cellular form. When one peeks into a microscope and encounters cells that had been extracted from a healthy individual, but now have condensed chromatin, shrunken vacuoles, and disorganized organelles, should one call these cells damaged or senescent?[1]

If the answer is that these cells display these "symptoms" because of their own aging, then is the timing of such processes predetermined during the cell's individual life course? Or is it responding to a range of external perils—toxins, lack of oxygen or nutrition, wear and tear—that the cells have gone through?

Today many biologists and biomedical researchers still wrestle with the biological meanings of these phenomena associated with cell aging. A number of the processes have been given detailed descriptions and assigned specialized terms, such as *replicative senescence*, *autophagy*, and *necrosis*.[2] Scientists aim to elucidate not only the concepts and terms but also details about their components, pathways, and interrelatedness. Despite differences in interpretation, most researchers today agree that some processes of aging are intrinsic parts of cellular life, evolved to play a protective role for other cells or the whole organism through destroying unwanted, pathological, or cancerous cells. Yet such an understanding is rather recent. In the course of the twentieth century, answers from scientists to the question of whether aging is an intrinsic part of cellular life varied.

Working on the problem of cell aging roughly from the late 1930s to the early 1950s, Cowdry discussed various manifestations of aging of different types of cells in different bodily environments. Regarding what cellular changes could be seen as part of cell aging and what factors might cause these changes, he took an inclusive perspective. For example, to explain different rates of cell aging, he considered different compositions of interstitial fluids, tissue structures, and intercellular interactions. Such a comprehensive view had been shaped by a variety of factors, such as Cowdry's early expertise in cytology, an interdisciplinary approach to gerontology, and his unease in prioritizing one causal factor over another. Cowdry was influenced by Alexis Carrel's early twentieth-century thesis that cells can be intrinsically immortal. In addition, he held a metaphorical understanding of individual cells as members of a cellular society that corresponded with his ideas of the internal structure of cells as organized wholes.

Cowdry's comprehensive consideration of the diverse phenomena of cell aging and their causes, along with the occasional incongruences and contradictions within his interpretations, revealed a changing and uncertain path in cell aging research. The only certainty Cowdry embraced about cell aging was his conviction, inspired by Carrel, that since cells were considered to be intrinsically immortal as long as they were given perfect living conditions, then it would make no sense to search for an inner mechanism that predetermined the lifetime of an individual cell; there would be no such mechanism and no such preset length.

Others held similar views, and the dominant understanding of cell life only changed in the late 1960s, when microbiologist Leonard Hayflick (1928–) demonstrated that human diploid cells maintained in serial culture had a limit to their mitotic life. As a consequence of Hayflick's "limit," between the late 1960s and early 2000s a community of researchers focusing on cell aging largely abandoned Cowdry's earlier comprehensive approach and took on cell culture as the experimental platform for finding one dominant molecular change that caused aging. Since the early 2000s, however, researchers in the field of cell aging have again begun to favor more comprehensive approaches, although most cell biologists have done so without knowledge of Cowdry's earlier work. As Cowdry's earlier research melded cytological and histological traditions with a community understanding of the cell, it served as a counterpoint against the more narrowly focused molecular research programs of the last four decades of the twentieth century. At present, Cowdry's work seems to have gained new relevance with the rise of more holistic and comprehensive approaches to cellular aging, although his work *per se* has been largely forgotten. Yet Cowdry's work not only showed fascination with the roles of the cell in aging processes, but also identified diverse considerations that could be helpful for current research.

What major issues did Cowdry consider about cell aging? How did these reflect the research traditions at the time as well as emerging research programs on aging? How have later developments on cell aging diverged from or converged with Cowdry's earlier visions? This chapter, by first chronologically tracing Cowdry's research and reviews related to cell aging and then comparing his visions to later developments in cell aging research, addresses these questions. Together, they show that Cowdry had already started to consider many of today's meanings of cell aging.

As historian of science Hyung Wook Park has shown, Cowdry's organizational and editorial work was foundational to the formation of gerontology and its infrastructure in the United States, especially in regard to its multidisciplinary setting (Park 2008). In his more recent work, Park regards Cowdry's approaches to cell aging and to gerontology as reflecting an individualistic ideology focused on the aging of individual human beings or individual cells as similar. Park sees Cowdry as having treated the elderly as a homogeneous group, with little space for consideration of differences such as gender or racial minorities (Park 2016, 54–90). American historian Tamara Mann concurs with Park regarding Cowdry's contribution to gerontology, while pointing out further that Cowdry's later move to cancer research should be explained by increased funding for biomedical research, especially for cancer research after World War II (Mann 2014). Focusing on Cowdry's disciplinary changes, both historians have noted how changing demographic trends and funding mechanisms shaped Cowdry's science. Yet another interpretation is that he remained always interested in cell aging, and his research changed primarily because of the many directions that a fuller understanding of cell aging entailed.

This chapter starts with Cowdry's edited collection in *General Cytology* and shows how Cowdry looked at degenerative changes in the cell. This work occurred in the context of Carrel's interpretation of individual cells as immortal, which shaped a generation of biologists' understanding of cell aging. Then we examine Cowdry's editorial work and writings for the first two editions of *Problems of Ageing*, which reflected a comprehensive program envisioned for cell aging research. Finally, with an overview of the research trajectory in cell aging during the latter half of the twentieth century and the beginning of the twenty-first century, I discuss how Cowdry's earlier visions have presciently shown the challenges faced by cell aging research.

Mitochondria, Tissue Culture, and Degenerative Cells in *General Cytology*

The edited textbook *General Cytology*, published in 1924, as Jane Maienschein notes in this volume, took an overall view that the cell is "the fundamental unit that was itself living." In other words, the volume suggested that studies of the morphological structures of cells, including minute changes inside the cell, should be studied with consideration of their physiological functions, such as how cells react to external stimuli.[3] This double focus on both the morphological and physiological was shared by various authors of *General Cytology*, and was one basic assumption of the field known as cell biology. The approach to studies of cells went beyond descriptive cytology and took shape as biologists began to ask a variety of research questions about cellular functions. In subsequent work, this orientation shaped Cowdry's understanding of cell aging. Yet at this stage of his career, Cowdry's primary focus was on the study of mitochondria.

In the chapter "Mitochondria, Golgi Apparatus, and Chromidial Substance," Cowdry described how mitochondria disintegrated as the cell was in certain pathological states or went through fixation treatment before microscopic observation. He pointed out the usefulness of mitochondria in studying pathology because of their relatively high reactivity to external cellular stimuli and injuries. These delicate qualitative responses included how the number of mitochondria in a cell declined and how their filamentous shapes changed into granules, which Cowdry interpreted as part of the cell's "degenerative change" (Cowdry 1924, 326–29, 332).[4]

The vulnerability of mitochondria, however, made Cowdry reflective and cautious in applying methods that might cause cellular injuries. He noted that fixatives that helped reveal nuclear details often damaged or destroyed the structures of more vulnerable cytoplasmic contents, such as the mitochondria. Inconsistent procedures for excising a piece of tissue from the organism and other experimental operations involved in starting and maintaining a tissue culture had similar effects (Cowdry 1924, 332). In addition, Cowdry expressed his hope for a comprehensive approach to studying mitochondria that took account of all related cellular phenomena, noting that "we must piece together information from many quarters, and build up in our mind's eye a dynamic picture of mitochondria in relation to innumerable other cellular constituents" (332). In later years, as Cowdry started to investigate cell aging, a similar emphasis on the proper method and the dynamism of cellular processes continued.

In *General Cytology*, the cell culturists Warren Harmon Lewis and Margaret Reed Lewis discussed degenerative changes in cells. They summarized behaviors of different types of cells maintained in tissue cultures. At the time, using adapted forms of the hanging-drop method introduced by biologist Ross Harrison, they had successfully established cultures of a number of cellular types, such as endothelial, skeletal-muscle, heart, cartilage, and spleen cells. While maintaining these tissue cultures, they closely observed and recorded normal cellular structures, cells in division, and those in differentiation and dedifferentiation, as well as degenerative changes. The authors devoted two sections—"Structure of Degenerating Cells" and "Cell Death"—to describing how some cells became impaired in their intracellular components after some growth, while others disintegrated completely. They noted abnormal accumulation of granules, vacuoles, and broken mitochondria in these degenerative cells and suggested that "lack of proper food, salts, and oxygen, the accumulation of waste products, and changes in the H-ion concentration" were potential causes of such degeneration (Cowdry 1924, 429, 426–28).

Although paying attention to cellular function, their descriptions of degenerative changes in the cell in *General Cytology* reflected the morphological tradition in cytology. Their descriptions could be seen as continuing earlier records of cell death and aging in the nineteenth century. As early as 1842, three years after the recognized formulation of cell theory by botanist Matthias Schleiden and zoologist Theodor Schwann, German zoologist Carl Vogt reported the degenerative changes in cells of the midwife toad that he saw using histological methods of staining (Clarke and Clarke 1996). Later years brought additional records of degenerating cells in life processes such as endochondral ossification, the formation of ovarian follicles, and tissue turnover. In descriptions of these phenomena, the accumulation of vacuoles and the coagulation of various cellular components and nuclei were often mentioned (Majno and Joris 1995). Going beyond the earlier morphological tradition in cytology, however, both the Lewises and Cowdry were also interested in identifying physiological causes of these degenerative changes. They often looked at the cell's internal and external environments for potential causes of change. In this regard, their research direction was significantly shaped by Alexis Carrel's thesis that normal diploid cells were intrinsically immortal.

Dr. Carrel's Immortal Cells and Cell Aging

The notion that individual cells could keep dividing, even forever, if given perfect conditions, had emerged along with the development of serial culture technique since the 1910s. After the Yale biologist Ross Granville Harrison used a hanging-drop method to culture frog embryonic neurons in 1907, the possibility of raising different kinds of cells outside the organismal body using tissue culture techniques attracted a number of researchers. These included Alexis Carrel, a French surgeon then working at the Rockefeller Institute of Medical Research in New York (Maienschein 1978, 54–118; Landecker 2007, 28–67). In 1910, Carrel sent his assistant Montrose Burrows to visit Harrison's laboratory and learn about the technique. Upon Burrows's return, he and Carrel started to culture a number of tissue types, including those from kidney, bone marrow, spleen, thyroid, and human sarcoma (Witkowski 1980).

By inventing a liquid culture and a way to transfer part of the culture to new media, Carrel started to grow cells serially for secondary cultures, thus keeping them living and dividing longer *in vitro*. Amid these experiments, Carrel noticed that ground-up tissue extract obtained from chick embryos, which he called "embryo juice," appeared to be able to reactivate growth and "rejuvenate" the chick heart cells that were being serially cultured. In fact, as one line of the chick heart cells that he and his associate Albert Ebeling kept growing had no sign of demise for months, Carrel began to consider that these cells in culture could have a "permanent life" if the culture medium could be renewed at proper intervals forever (Carrel 1912). Ebeling managed to keep the chick heart tissue culture actively dividing for thirty-four years in the laboratory, which far exceeded the chick's usual life span. By the 1930s, Carrel's "immortal chick heart tissue culture" not only had become a public sensation through media coverage, but also had gained acknowledgments from other influential scientists, such as Harrison, Jacques Loeb, and Raymond Pearl (Landecker 2007, 91–103).

As later experiments in the 1960s showed that normal chick cells actually do have a limited mitotic lifespan, historians of biology have suggested possible errors in Carrel's cell cultures. These suggestions included contamination of existing cells in the "embryo juice," transformation of some cells into a permanent cell line, even intentional addition of extra cells by technicians who did not want to disappoint Carrel (Witkowski 1980). Yet, before these reassessments were proposed, Carrel's immortality thesis achieved great influence and shaped how cytologists and gerontologists understood the nature of cell aging. Among tissue culturists in particular, it had gained such wide acceptance that the following generation often interpreted the demise of cells in their cultures as indicating their own failure.[5]

In *General Cytology*, Warren and Margaret Lewis extensively cited Carrel's work regarding the use of plasma, cultivation methods of blood cells and sarcoma cells, and immortality of embryonic fibroblasts (Lewis and Lewis 1924, 389–90, 408, 418–20). Cowdry himself had maintained a good relationship with Carrel, as they both worked at the Rockefeller Institute in the late 1910s and corresponded afterwards. In his discussion about mitochondria, Cowdry mentioned the usefulness of Carrel's tissue culture for observing mitochondria accurately and monitoring them quantitatively in living cells (Cowdry 1924, 332). Carrel's notion about the immortal nature of individual cells, however, had a more far-reaching influence on Cowdry's understanding of the nature of cell aging and the extent to which tissue culture technique could be useful for research.

One implication of Carrel's supposition was that cells only age or die because of lack of nutrition or accumulation of injuries from toxins or waste products in their environment. The implication, as Hannah Landecker puts it, was that "senility and death of tissues are not a necessary fate but an accidental one" (Landecker 2007, 74). This suggestion had a double-edged influence regarding the studies of cell aging. On one hand, the suggestion seemed to offer an opportunity for using tissue culture to study the mitotic lifespan. One could potentially isolate an aged cell, and put it into a culture medium with perfect composition to measure the true length of the cellular life span by measuring how long the cells kept dividing. On the other hand, healthy cells that age *in vitro* were seen as artificial, because components of their culture medium were different from the normal environment within the body.

In later years, Cowdry held a general skepticism regarding the usefulness of the tissue culture for studying cell aging. Yet he did consider the possibility of measuring accurate length of cellular life using tissue culture, although he still concluded that the technique was not yet ripe for extensive studies because many types of cells were difficult or impossible to culture at the time (Cowdry 1942b, 630). In the next section, by reviewing the content and context of Cowdry's research on cell aging, I reveal some subtle aspects of Carrel's influence on Cowdry's attention to cellular environment.

Gerontology and Cellular Environment in *Problems of Ageing*, 1939

When Cowdry started to study cell aging, he had been engaged with building the discipline of gerontology in the United States. As issues of human aging were complex both scientifically and in their social aspects, Cowdry was quite reflective about the potential isomorphism between the social and biological, which not only shaped his multidisciplinary approach to aging in general, but also further precipitated his attention to the cellular environment, community, and diversity while studying cell aging.

In the late 1930s, the Great Depression and then World War II hurt the economic conditions for the elderly in the United States and elsewhere. At the time, Cowdry noted that while many of the aged were being neglected, unemployed, and left to become ill and die during the Great Depression, this was not the case in China, where he had worked in the 1920s and where the old were "highly venerated." For Cowdry, there was a lesson to be learned from the lives of cells in the body that could guide social policies applied to the elderly population (Cowdry 1936). He felt that the "body anatomic" provided a lesson for the "body politic" of many modern nations concerning the treatment of senior citizens. And for both cells and individual humans to enjoy adequate and long life, Cowdry emphasized "the importance of stability of environment" (220). In the paper he wrote for *Scientific Monthly*, "Body Anatomic and Body Politic," Cowdry thus noted, "Many aged and dead cells are not only utilized but are given positions of great importance. Firmly bound together in a dense layer on the surface of the skin, dead epidermal cells act as a shield and protect the living cells within" (224). With knowledge about cellular aging and death that manifests in more diverse ways than those of the skin cell, Cowdry highlighted skin cells particularly to make his point that senior citizens need to be employed for the good of society and that their health ought to be maintained and improved through science.

At the time, the Josiah Macy Jr. Foundation became interested in funding aging research, and Cowdry's new concerns aligned well with the foundation's vision. Dedicated to multidisciplinary research, Kate Ladd, the foundation's founder and youngest daughter of Josiah Macy Jr., had hoped to help relieve human suffering from chronic diseases through a more integrated medical approach. This approach diverged from treatment of patients by focusing on parts of the body isolated from the whole person, a way that she thought modern medicine often functioned. Having funded Cowdry's research on arteriosclerosis for some years, in the late 1930s the foundation expanded its disease-based research to a multidisciplinary research program on aging (Mann 2014). In aging research as in cytology, Cowdry's organizational skill helped to cultivate a sense of community and to attract funding. The "Club for Research on Ageing," an informal discussion group that Cowdry helped organize as a result of these endeavors, continued throughout World War II.

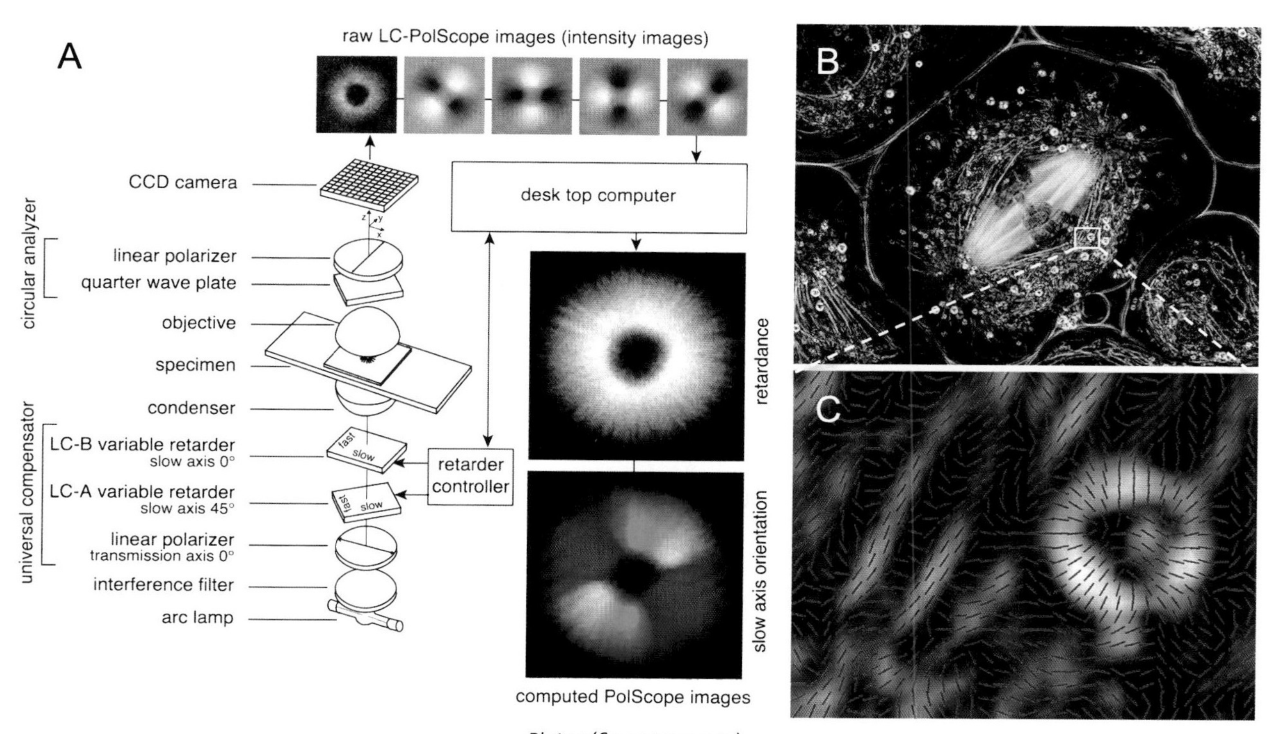

Plate 1 (fig. 12.10, p. 292)

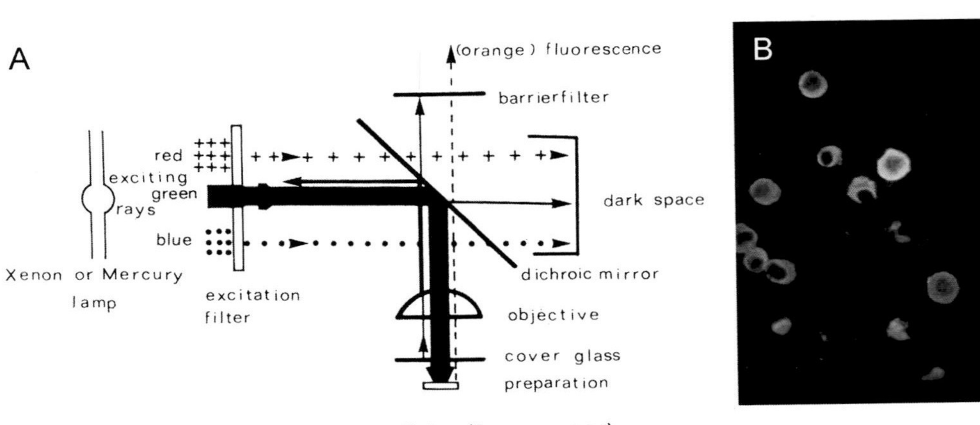

Plate 2 (fig. 12.11, p. 294)

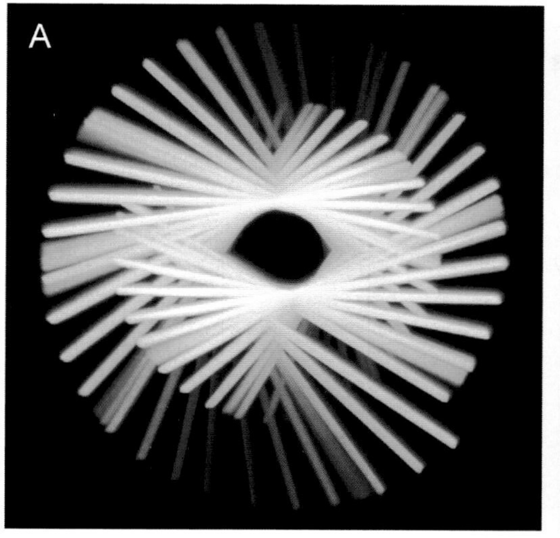

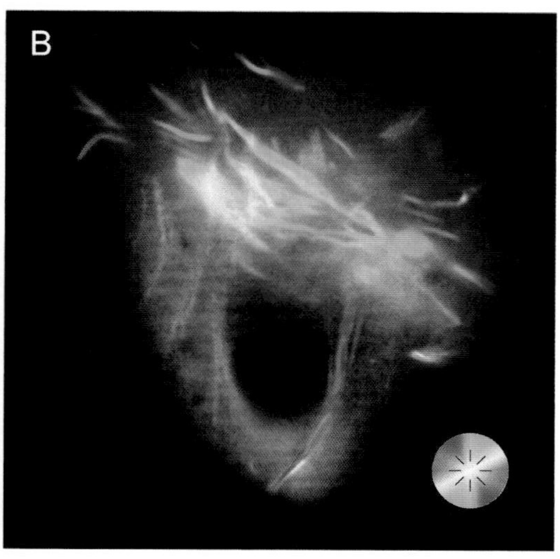

Plate 3 (fig. 12.12, p. 295)

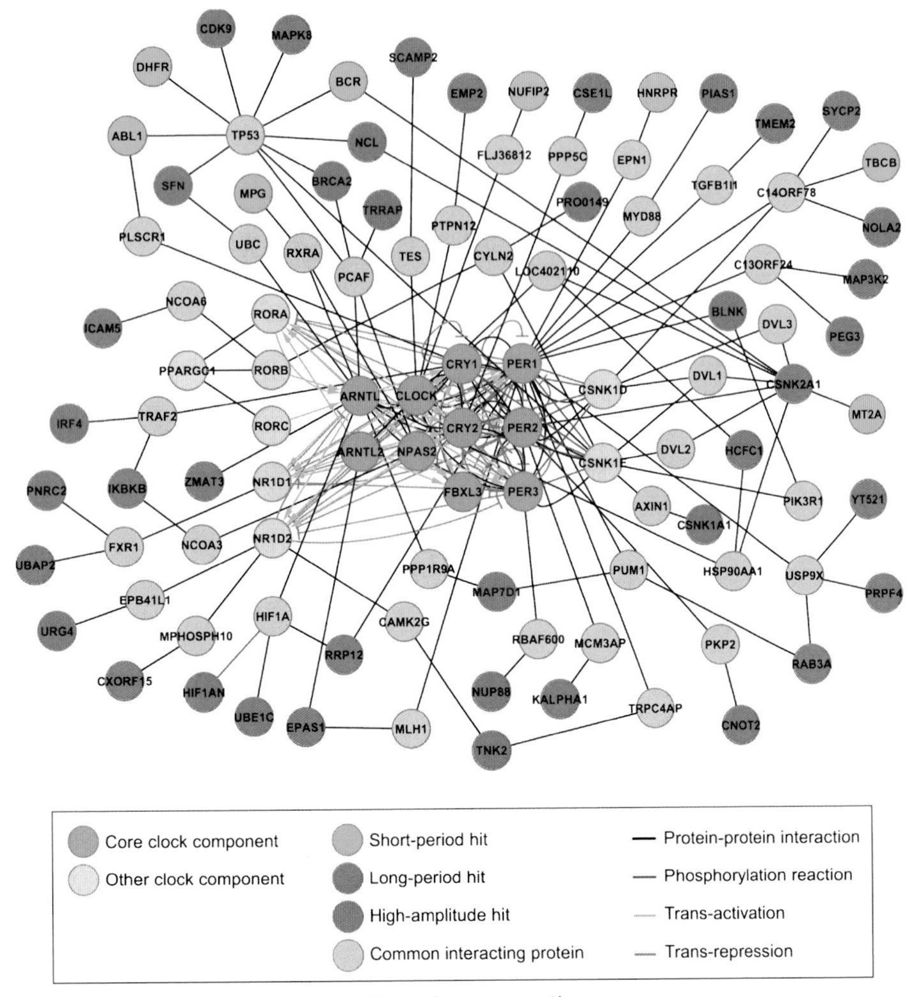

Plate 4 (fig. 13.3, p. 318)

In 1937, Cowdry turned to investigating the problems of aging by organizing another group of scientists and scholars to tackle the problem through a collaborative, multidisciplinary conference. Again, the group met at Woods Hole and was reminiscent of the one that led to *General Cytology*. The resulting edited book, the first edition of *Problems of Ageing*, was published in 1939 and then extensively supplemented and revised for the second and third editions, all turning into disciplinary classics in the new field of aging research. Because Cowdry's primary concern was to lay out the most important aspects of aging for future studies, individual chapters usually had a wide scope and did not primarily concern cell aging *per se*. Hoping to establish a science of aging to benefit human welfare, Cowdry included contributions from a significant number of medical researchers who specialized in the diseases and aging of particular organs and tissues, as well as several philosophers, psychiatrists, and mathematicians.

In the first edition of *Problems of Ageing*, no chapter was dedicated to the aging of individual cells. Perhaps due to the influence of Carrel's conviction about the intrinsic immortality of the cell, and thus the predominant importance of the cell's environment in the rate of aging as a corollary, Cowdry did not write directly about individual cell aging, even if it was his primary expertise. Instead, he contributed a general chapter entitled "Ageing of Tissue Fluids" that gave a structural and relational analysis of how various kinds of interstitial fluids, including those found in epithelial, vascular, bone marrow, and muscular tissues, interacted with nearby blood streams and cells (Cowdry 1939a). Discussions about aging in different types of cells were dispersed in chapters focusing on the aging of plants, protozoa, insects, and vertebrates contributed by established biologists, such as Herbert Spencer Jennings, as well as chapters concerning aging in tissues and biological systems, such as the cardiovascular system and blood, the urinary system, and skin (Cowdry 1939b).

Cowdry's chapter starts with a general statement that the "balancing mechanisms" of tissue fluids lose their original stability over their lifetimes, and these disturbances cause biological maladaptation in older individuals. Thus, Cowdry gave himself the task of describing quite detailed variations in the composition of various tissue fluids and the modes of interaction among cells and these fluids in different types of tissues. Particularly, he discussed thirteen types of tissues, including epithelial, vascular, splenic, articular, and muscular tissues (Cowdry 1939a, 649). He chose to focus on tissues because he was convinced not only that the deterioration of the balance in body fluids caused the deterioration of cells, but also that the composition and structure of different tissues shaped the processes that gave specific morphology and function to each cell embedded within them, thus affecting the aging process of the cell as well. The whole chapter, in the end, was highly derivative of previous literature in histology and reads more like a textbook introduction to different tissue structures and dynamics than a description of their aging. He regarded such an approach as a starting point for future studies of tissue changes over time.

Cowdry painstakingly detailed the diverse components of different tissues. When he paid attention to the morphology and distribution of cells in these tissues, he usually treated them as reactors responding to their surroundings. Regarding such a treatment, prominent physiologist Walter Cannon expressed disapproval. Cannon not only thought that the cells could maintain their individual functions and thus have a general rate of aging without much influence from the cellular environment, he also did not think that tissue fluids surrounding different cells were that different. As Cannon was a physiologist who expanded on Claude Bernard's concept of homeostasis, and he himself paid particular attention to interactions between cells and their environment, the objection constituted a serious blow. This resulted in a protracted debate between Cowdry and Cannon regarding the extent to which the different compositions of tissue fluids affected the rates of aging in individual cells (Park 2008, 557–59). Cowdry cited Carrel's immortality thesis in his rebuttal of Cannon, which demonstrated Carrel's influence on Cowdry. Nevertheless, the occurrence of the debate itself also showed the limitation of Carrel's influence among physiologists.

Overall, the volume under Cowdry's editorship committed to showing the multiplicity of aging processes. As noted in the foreword by Lawrence K. Frank of the Macy Foundation, "In the search for a uniform, general process of ageing there is a risk of neglecting the various factors, influences or conditions that enter into this differential rate of ageing in each individual at different periods of his life" (Cowdry 1939b, xv).

Taking on Individual Cells and Cancer in *Problems of Ageing*, 1942

The first edition of Cowdry's *Problems of Ageing*, upon publication, met with a favorable reception. By readers' demand, it was reprinted soon after its publication. This was also the time when further funding for disciplinary development of aging research became possible. Since research on aging and its publicity quickly expanded with a series of conferences on the topic, it seemed that a second edition became desirable for academic reasons as well as for an eager readership (Cowdry 1942b, iv).

In the second edition, published in 1942, nine chapters were added, including Clive Maine McCay's report on effects of diet in aging and several chapters on psychological aspects and clinical treatments related to aging patients. While these newly added chapters usually dealt with more practical and social concerns, Cowdry himself contributed a new chapter on specific cytological issues, "Ageing of Individual Cells" (Cowdry 1942a). In this review of various processes of aging in different types of cells, Cowdry's emphasis on the diversity of causes persisted. The overall orientation of the chapter expressly denied the use of tissue culture as a primary experimental method for studying aging, noting that many processes of aging were interdependent, "so that we must think in multifunctional relationships" and "can not isolate a single aging process in 'pure culture' and study it alone" (Cowdry 1940, 52).

Here, we can see the extended influence of Carrel's research on Cowdry's conception of cell aging and the best method to study it. The belief that cells could live forever and divide indefinitely under ideal conditions meant that to study cell aging would involve investigating the loads of cellular injuries induced by metabolites, toxins, and other harmful substances existing in the cell's surroundings. Because the interstitial fluids within the body and the culture medium had different compositions, it also implied that the aging process of cells inside the body might well differ from those outside the body. Such differences between aging in tissue culture and in bodily context made many researchers of aging, including Cowdry, suspicious about the effectiveness of using tissue cultures to study aging in general.

Another subtle influence from Carrel's work with tissue cultures was reflected in Cowdry's consideration of different modes of aging, depending on whether the cells kept an actively mitotic life or not. While Carrel's research on tissue cultures showed how actively dividing cells age outside the body, it did not provide the most relevant clues about how those more dormant types age over time. Since cells that kept dividing and cells that did not might well age in different ways, the beginning of Cowdry's new chapter grouped different cells according to their capabilities in cell division. Some cells, such as basal cells of the epidermis, primordial blood cells, and spermatogonia, sustain continuing mitosis (intermitotic); in contrast, others go through long periods of quiescence before resuming division, such as some endothelial and epithelial cells or smooth muscle cells (reverting postmitotic). Still others, such as nerve cells and skeletal muscle cells, were considered to have reached an end-point in their mitotic life and were thus called fixed postmitotics (Cowdry 1942a, 626–29). Cowdry noted that these groupings of cells more or less determined how they would eventually perish—whether their lives were renewed in mitosis or ended in cell death. He then considered the ways in which cells in relatively stable, nonmitotic states underwent the aging process, such as nerve cells, skeletal muscle cells, epidermal cells, and neutrophilic leucocytes. The factors he considered were diverse, including changes in these cells' nucleocytoplasmic ratios, decreases in active protoplasm, decreases in water content, and aging of the colloids (Cowdry 1942b, 638–49).

With much care about details and specifics, Cowdry nevertheless regarded the ultimate goal of research on cell aging as explaining the whole life course of a cell, including potential malignancy. He noted that "it would be lamentably lacking in orientation and perspective if they were considered simply as cells only at one period in the age of the body and not as changing members of a changing cellular community of great complexity" (Cowdry 1942a, 650). Consequently, in the section "Changes in Cell Life with Aging," he described the overall dynamics of cells over a lifetime, including the changing populations of cells from the formation of the embryo, in which intermitotic cells form the majority, to the unfolding of hereditary traits and to the waning of special functions in later life. In the end, Cowdry discussed "the onset of malignancy," raising the state of malignancy as a stage that some but not all cells could reach. To understand malignancy along with the aging of cells, Cowdry suggested that one ultimately needed to sequentially arrange all types of cells that made up the body at different ages, and identify the cause of malignancy by comparing the malignant tendencies of these cells (657).

Eventually, the broad conception of the goal of aging research catalyzed Cowdry's research interest in cancer. In the early 1940s, Cowdry became the research director of the Barnard Free Skin and Cancer Hospital in St. Louis, and obtained access to human skin samples. He thus took this chance to delve into studies of aging and cancer in the skin. He and his colleagues compared the cellular morphology and tissue structures of skins from human subjects of different ages and measured how various stressors such as sunlight or physical pressure affected such differences (Evans, Cowdry, and Nielson 1943; Cowdry, Cooper, and Smith 1947).

In the 1950s, Cowdry's research interest migrated further into studies of cancer etiology, for which the skin tissue again provided major research material. His later interest in cancer, although related to aging, might also have diverted some research energy from more focused studies of aging. The third edition of *Cowdry's Problems of Ageing: Biological and Medical*

Aspects, for the first time included Cowdry's name in the book title. Yet this edition was actually edited by Albert I. Lansing, who had done postdoctoral work with Cowdry at St. Louis. In the 1960s and 1970s, although Cowdry continued to publish on the topic of aging, including an edited volume entitled *Care of the Geriatric Patient* (1971) and a singly authored book *Aging Better* (1972), these works focused on clinical treatments and education about aging instead of experimental research.

Cowdry's Conception of Cell Aging In Retrospect

Having distanced himself from researching aging through tissue culture and from proposing a simple explanation for cell aging, Cowdry's vision of how aging should be studied on the level of the cell were generally shared by his contemporaries. Beginning in the late 1950s, however, as tissue culture techniques matured to a level that single-cell plating and mass production of cell cultures became possible, others worked out new hypotheses about cell aging and new uses of cell culture. Leonard Hayflick suggested that normal human cells have a limit to their mitotic life and worked out a new way of using cell culture to study mitotic aging.

By the 1950s, partly through promotional courses in cell culture held by the Tissue Culture Association (TCA), the practice of mass multiplication of cell culture had spread widely and attracted a number of microbiologists (Landecker 2007, 134–39; Puck 1972). Hayflick, then a microbiologist-turned-cell-biologist working at the Wistar Institute as the chief of the cell culture facility, isolated a number of human cells from fetal tissues (Hayflick 1998; Hall 2003, 21–22; Wadman 2013). By 1961, after Hayflick had amassed twenty-five strains of cells from human fetal tissues, including those of the lung, skin, muscle, kidney, heart, liver, thymus, and thyroid, and kept them growing, he eventually realized that they all stopped dividing after continued growth for about half a year (Hayflick and Moorhead 1961). Hayflick started to consider the possibility that the observed cessation of cell division *in vitro* was a manifestation of a normal cellular aging process possibly responsible for aging of the organisms as a whole. He eventually made this suggestion, first in a paper published in 1961 coauthored with the cytogeneticist Paul Moorhead, and then a single-authored paper in 1965. In these papers, Hayflick divided normal cellular life *in vitro* into three periods. While the first two were characterized by prosperous growth and proliferation (Phases I and II), visible deterioration occurred in the third period (Phase III). Through this analysis, he defined the mitotic limit as an intrinsic property of normal diploid cells. In 1965, Hayflick called for more research about "a general cellular theory of aging" that explained why normal diploid cells had a limit to their maximal cell doubling, while cancer cells seemed to by-pass this limit *in vitro* (Hayflick 1965).

Hayflick's suggestion that the human diploid normal cells have an intrinsic limitation in their capacity for continued mitosis contradicted Carrel's earlier immortality thesis that had held sway. Not surprisingly, his suggestion initially met with resistance and rejections. The *Journal of Experimental Medicine*, where Hayflick and Moorhead hoped to publish their coauthored manuscript, rejected it on the ground that "the inference that death of the cells in some of the uninfected cultures is due to 'senescence at the cellular level' seems notably rash. The largest fact to have come out from tissue culture in the last fifty years is that cells inherently capable of multiplying will do so indefinitely if supplied with the right milieu *in vitro*."[6] Yet, the rising interest in the biological investigation of aging from a new generation of biologists who searched for general mechanisms and effective tools toward this end soon brought cell culture into use for aging research.

In the 1960s, the biology of aging was a small yet expanding field. As the number of physicians within the research community on aging grew, a postwar generation of biologists who wanted to explain aging with fundamental biological principles started to criticize the existing gerontological discipline as overly conservative. The fresh notion that cells can age *in vitro* and thus that aging can be studied *in vitro* thus attracted a few biogerontologists. This was a time when physicists and chemists joined the study of biological problems and helped form what we know today as molecular biology. The mishmash of methods and results from different species and human tissues that spoke incoherently about what caused aging, which Cowdry had carefully taken into account, was not these newly converted biogerontologists' cup of tea. They preferred cleaner data that could reveal regularities or even a universal mechanism of aging. Hayflick's cell culture model of aging seemed to offer an experimental platform for searching for such a mechanism.

After all, tracing cellular changes inside higher organisms was difficult and time-consuming, if not downright impossible. Cell culture, in contrast, was a much more feasible platform for experimentation. Most notably, in the early 1970s, the British molecular biologist Robin Holliday adopted Hayflick's diploid cell culture model to study whether mutations in the DNA or the errors in the protein and protein synthesis caused aging. Working as the director of the Genetics Division of the National Institute for Medical Research in Mill Hill, United Kingdom, Holliday trained a few cell biologists, biochemists, and biomathematicians who eventually devoted their careers to the problem of cell aging (Jiang 2014). Besides Holliday, those who established lasting and influential projects focusing on cellular aging *in vitro* include Vincent J. Cristofalo at the Wistar Institute, Samuel Goldstein at the University of Arkansas, George M. Martin at the University of Washington, James Smith at the Baylor College of Medicine in Houston, and Álvaro Macieira-Coelho and Woodring Wright, who both studied with Hayflick in the 1960s. They offered various interpretations of cell aging *in vitro*, which usually framed cell aging as an integral part of cellular life, a suggestion that went against Carrel's thesis. By the mid-1970s, the conception that cell division in human diploid normal cells has an intrinsic limit, along with Hayflick's cell culture model of aging, were widely accepted. With the new focus on a universal mechanism of cell aging, the expanding studies on cell aging nevertheless abandoned Cowdry's earlier conception of cell aging as well as a comprehensive, holistic view.

Yet, as more researchers investigated the mechanism through which cells age *in vitro*, others questioned some initially convincing hypotheses and experimental results, often with new findings that contradicted earlier conclusions. As the question of why normal cells age (why they stopped dividing *in vitro*) plagued the community, it also encouraged further investigations and discussions about the complexity of the aging process (Cristofalo 1972). By the late 1970s, as researchers had already started to explore alternative mechanisms of cell aging, one could argue that at this point, Cowdry's earlier emphasis on the diversity of causes of aging seemed to be well worth revisiting. Yet one road to reconsidering diverse causes of aging within and without the cell was tortuous—only after the telomere-shortening hypothesis of aging, which suggested a molecular counting mechanism, became highly successful in the 1990s, did more influential criticism of the focus on a single explanation of cell aging begin to be taken seriously (e.g., Sozou and Kirkwood 2001).

The study of the telomere began in the early 1970s, when molecular biologist Elizabeth Blackburn, as a PhD student in Fred Sanger's lab at Cambridge University, worked on the end structure of the chromosomes of the protozoa *Tetrahymena*, which originally was seen as a rather obscure study of an unimportant problem. The situation changed, however, after Blackburn moved to the University of California, Berkeley, and Blackburn's graduate student Carol Greider realized while collaborating with a researcher on cell aging that the telomere's length was associated with the number of times a cell divided in culture (Brady 2007).

Aging of the human normal cells and the Hayflick limit began to be explained by the stepwise shortening of the telomere, the end structures of the chromosomes, with each cell division. The hypothesis stated that once the whole chromosomal structure was shortened to a threshold and became unsustainable, senescent changes (Phase III) would become observable in the cell. Telomere research soon boomed. In the early 1990s, many cell aging researchers started to study telomeres, which helped to further boost the influence and publicity of the field. In 2009, for their work on telomere biology, Blackburn, Greider, and molecular biologist Jack Szostack were awarded the Nobel Prize in Physiology or Medicine.

As the telomere hypothesis of aging entered the scientific limelight, however, many veterans of cell aging research and their students became unsatisfied with the elegant yet simple thesis that telomere shortening served as an explanation of cell aging—it seemed too simple. For example, Thomas Kirkwood, working at the University of Newcastle, tried to connect telomere shortening with other mechanisms proposed by cell aging researchers, such as changes caused by oxidative stress and somatic mutations (Kirkwood 2008). Kirkwood's program eventually evolved into the Centre for Integrated Systems Biology of Ageing and Nutrition in 2004, in which cell aging was one of the most important problems his programs tackled.

Others, such as Judith Campisi, have asked questions about how individual cell aging could affect other groups of cells by exerting stress signals and hormones. She showed that although cell aging may help to prevent cancer, aging cells also secrete stress hormones and signals that harm other cells. The word *integration* often appeared in her publication list. In Campisi's depictions of cell aging, neighboring cells, body fluids, and communities of cells came back into view (Campisi 2005). These more comprehensive, integrative approaches remind us of Cowdry's earlier attention to diverse cells, cellular communities, and cellular environments, as well as his cautions in using the cultured cell for aging research.

Conclusion

From *General Cytology* to various editions of *Problems of Ageing*, Cowdry's work on degenerative cells evolved from the studies of changing mitochondria to experimenting with and reviewing various cellular degenerative changes in relation to their diverse environments. Working at a time when neither the concept of cell aging nor the method for studying it was mature, his research was characterized by continued attempts to present a comprehensive picture of how cells age through piecing together specific cellular changes in diverse biological contexts. Although Carrel's cell immortality thesis had drawn Cowdry's attention from a proper understanding of the nature of aging in dividing cells and that of the legitimacy of using cell culture for aging research, one could argue that, even if Cowdry had had all the knowledge of the Hayflick limit and the cell culture model of aging, his considerations regarding cell aging would still have been as varied.

Cowdry's focus on the diversity of biological conditions, in retrospect, was what biological research on cell aging lacked between the 1960s and the early 2000s. Although a few researchers pointed to the complexity involved in aging with sometimes divergent evidence, these four decades of research had focused on looking for a single, universal explanation for the Hayflick limit. At present, when scientists tackle the problem of aging through more recent approaches labeled as "omics science," big-data biology, and systems biology, it seems that many observations and suggestions that Cowdry made more than half a century ago, and the exhaustive ways he described them, could still inspire renewed insights as we assign more value to alternative and holistic research programs.

Acknowledgments
The author thanks Jane Maienschein and Karl Matlin for helpful comments in improving this chapter.

Notes

1 At present, shrunken chromatin is usually associated with the process of apoptosis. Before the term was coined in 1972, however, it had been seen as part of the degenerative changes associated with cell death or aging (Kerr, Wyllie, and Currie 1972).

2 Certainly, these phenomena were not only related to cell aging, but also to a range of other physiological and pathological processes, such as immune response, development, and wound healing.

3 This vision was summarized by Ralph Lillie in his chapter "Reactivity of the Cell": "All living systems, including single cells as well as complete organisms of all kinds, react to changes occurring in their immediate environment, or to changes in their relations to the environment, by exhibiting characteristic alterations in their own special activity" (Cowdry 1924, 167).

4 Also, because of the relatively high sensitivity of mitochondria to the external environment, Cowdry cautioned about the potentially wrong information brought about by harmful, artificial techniques that were designed to reveal the nuclei instead of the mitochondria, noting that mitochondria would show degenerative changes under the influence of many such fixatives and dyes (Cowdry 1924, 326). See also Jutta Schickore, chapter 4 in this volume.

5 For example, unbeknownst to most cell culturists, two scientists working at Western Reserve University, H. Earle Swim and Robert F. Parker, carefully documented the patterns of proliferative cessations in human diploid cell cultures and published the

results in 1957. Their conclusion was framed in an indecisive and strictly technical way, as stated in the end of their report: "Normal human fibroblasts will not proliferate indefinitely in the media used but may nevertheless yield permanent lines of cells as a result of infrequent alterations in their nutritional requirements" (Swim and Parker 1957, 242).

6 Peyton Rous. Correspondence from Peyton Rous to Hilary Koprowski, 24 April 1961. Personal Letter Collection of Leonard Hayflick.

References

Brady, Catherine. 2007. *Elizabeth Blackburn and the Story of Telomeres: Deciphering the Ends of DNA*. Cambridge, MA: MIT Press.

Campisi, Judith. 2005. "Senescent Cells, Tumor Suppression, and Organismal Aging: Good Citizens, Bad Neighbors." *Cell* 120 (4): 513–22.

Carrel, Alexis. 1912. "On the Permanent Life of Tissues outside of the Organism," *Journal of Experimental Medicine* 15:516–28.

Clarke, P. G. H., and S. Clarke. 1996. "Nineteenth-Century Research on Naturally Occuring Cell Death and Related Phenomena," *Anatomy and Embryology* 193 (2): 81–99.

Cowdry, Edmund V., ed. 1924. *General Cytology: A Textbook of Cellular Structure and Function for Students of Biology and Medicine*. Chicago: University of Chicago Press.

———. 1936. "Body Anatomic and Body Politic." *Scientific Monthly* 42 (3): 222–29.

———. 1939a. "Ageing of Tissue Fluids." In *Problems of Ageing: Biological and Medical Aspects*, edited by Edmund V. Cowdry, 642–94. Baltimore: Williams & Wilkins.

———, ed. 1939b. *Problems of Ageing: Biological and Medical Aspects*. Baltimore: Williams & Wilkins.

———. 1940. "We Grow Old." *Scientific Monthly* 50 (1): 51–58.

———. 1942a. "Ageing of Individual Cells." In *Problems of Ageing: Biological and Medical Aspects*, edited by Edmund V. Cowdry, 626–63. 2nd ed. Baltimore: Williams & Wilkins.

———, ed. 1942b. *Problems of Ageing: Biological and Medical Aspects*. 2nd ed. Baltimore: Williams & Wilkins.

———, Zola Cooper, and Warren Smith. 1947. "Program of Research on Aging of the Skin." *Journal of Gerontology* 2 (4): 31–44.

Cristofalo, Vincent J. 1972. "Animal Cell Cultures as a Model System for Aging." *Advances in Gerontological Research* 4:45–79.

Evans, Robert, Edmund V. Cowdry, and Paul E. Nielson. 1943. "Ageing of Human Skin: I. Influence of Dermal Shrinkage on Appearance of the Epidermis in Young and Old Fixed Tissues." *Anatomical Record* 86 (4): 545–65.

Goldstein, Samuel. 1990. "Replicative Senescence: The Human Fibroblast Comes of Age." *Science*, n.s. 249:1129–33.

Hall, Stephen S. 2003. *Merchants of Immortality: Chasing the Dream of Human Life Extension*. Boston: Houghton Mifflin.

Hayflick, Leonard. 1965. "The Limited *in Vitro* Lifetime of Human Diploid Cell Strains." *Experimental Cell Research* 37:614–36.

———. 1998. "A Brief History of the Mortality and Immortality of Cultured Cells." *Keio Journal of Medicine* 47 (3): 174–82.

———, and Paul S. Moorhead. 1961. "The Serial Cultivation of Human Diploid Cell Strains." *Experimental Cell Research* 25:585–621.

Jiang, Lijing. 2014. "Causes of Aging Are Likely to be Many: Robin Holliday and Changing Molecular Approaches to Cell Aging, 1963–1988," *Journal of the History of Biology* 47 (4): 547–84.

Kerr, John F. R., Andrew H. Wyllie, and Alastair R. Currie. 1972. "Apoptosis: A Basic Biological Phenomenon with Wide-Ranging Implications in Tissue Kinetics." *British Journal of Cancer* 26 (4): 239–57.

Kirkwood, Thomas B. L. 2008. "A Systematic Look at an Old Problem," *Nature* 451 (7): 644–47.

Landecker, Hannah. 2007. *Culturing Life: How Cells Became Technologies*. Cambridge, MA: Harvard University Press.

Lewis, Warren H., and Margaret R. Lewis. 1924. "Behavior of Cells in Tissue Cultures." In Cowdry 1924, 383–448.

Maienschein, Jane Ann. 1978. "Ross Harrison's Crucial Experiment as a Foundation for Modern American Experimental Embryology." PhD thesis, Indiana University.

Majno, G., and I. Joris. 1995. "Apoptosis, Oncosis, and Necrosis: An Overview of Cell Death." *American Journal of Pathology* 146 (1): 3–15.

Mann, Tamara. 2014. "Old Cells, Aging Bodies, and New Money: Scientific Solutions to the Problem of Old Age in the United States, 1945–1955." *Journal of World History* 24 (4): 797–822.

Park, Hyung Wook. 2008. "Edmund Vincent Cowdry and the Making of Gerontology as a Multidisciplinary Scientific Field in the United States." *Journal of the History of Biology* 41 (3): 529–72.

———. 2016. *Old Age, New Science: Gerontologists and Their Biosocial Visions, 1900–1960*. Pittsburgh, PA: University of Pittsburgh Press.

Puck, Theodore T. 1972. *The Mammalian Cell as a Microorganism: Genetic and Biochemical Studies in Vitro*. San Francisco: Holden-Day.

Swim, H. Earle, and Robert F. Parker. 1957. "Culture Characteristics of Human Fibroblasts Propagated Serially." *American Journal of Epidemiology* 66 (2): 235–43.

Sozou, Peter D., and Thomas B. Kirkwood. 2001. "A Stochastic Model of Cell Replicative Senescence Based on Telomere Shortening, Oxidative Stress, and Somatic Mutations in Nuclear and Mitochondrial DNA." *Journal of Theoretical Biology* 213 (4): 573–86.

Wadman, Meredith. 2013. "Cell Division," *Nature* 498:422–26.

Witkowski, J. A. 1980. "Dr. Carrel's Immortal Cells." *Medical History* 24 (2): 129–42.

CHAPTER 7

VISUALIZING THE CELL

PICTORIAL STYLES AND THEIR EPISTEMIC GOALS IN *GENERAL CYTOLOGY*

Beatrice Steinert and Kate MacCord

In 1925, the American Library Association listed the edited volume *General Cytology* as one of the forty best American books of the year (University of Chicago Press Archives 1925). This "textbook of cellular structure and function for students of biology and medicine" (Cowdry 1924a) spanned the gamut in its coverage of cytological topics—from chemistry to subcellular structures, from processes in which cells engage (e.g., fertilization, differentiation) to their relationship with heredity. As noted throughout this volume, the making of *General Cytology* required the attention of preeminent specialists of the American school of biology, and nearly all of them were long-time Marine Biological Laboratory inhabitants during the summers.[1]

This text, constructed of contributed chapters, was meant as a steppingstone for students and researchers—a means of brushing up on adjacent specialties and a resource for finding recent publications. As one reviewer of the text noted, "The eminent men who have contributed to this great book are all specialists on the chapter which they take up, and this fact is enough to insure that *General Cytology* is not a 'text-book,' but a series of original contributions" (May 1926, 214). These original contributions from multiple perspectives amounted to a view of the cell as a dynamic whole (May 1926). This breadth of subject, in combination with the expertise with which the contributors addressed their topics, rendered the book extremely useful for students and researchers alike, who sought the most comprehensive understanding of cell biology possible at the time.

In this chapter, our focus is on the illustrations produced for *General Cytology*. Among the 754 pages of this text, there are 181 illustrations. However, these illustrations are not evenly distributed. The first three chapters, by Albert Mathews, Merle Jacobs, and Ralph Lillie, contain very few to no illustrations. While it is certainly interesting to posit why these chapters on cell chemistry, permeability, and reactivity include such sparse visual information, we do not discuss them here. We focus instead on the other seven chapters, which incorporate numerous illustrations of various kinds.

Illustrations in biology—especially the biology of cells and development—are crucial for conveying knowledge (Hopwood 2015). Illustrations within scientific texts range from photographs to line drawings, schematics, and beyond, and they play a number of different roles within a text (Lynch 2006). These roles range from being an integral part of the scientist's argumentation and heuristic strategy (Cambrosio, Jacobi, and Keating 2005) to more or less direct translations of the text, serving as little more than ornamentation (Lynch 1991). These different types of illustrations serve different epistemic functions for a pedagogical text, highlighting a variety of preparatory methods, theories, and knowledge bases.

There are massive corpora of literature, spread across fields like visual culture studies and the German *Bildwissenschaft* (see, e.g., Pauwels 2006; Bredekamp, Dünkel, and Schneider 2015; Anderson and Dietrich 2012) that address illustrations in the sciences. While these fields offer fruitful insights into scientific imagery, our interpretation aligns more closely with the tradition, well-established within the history and philosophy of science literature (see Griesemer 2007 and Maienschein 1991 for excellent examples), of understanding scientific illustrations in terms of their epistemic goals. That is, we ask, What kinds of knowledge, theories, or methods are the images included in the chapters of *General Cytology* meant to convey? And, how are these epistemic goals met within the images?

Throughout the chapters in *General Cytology*, there is a clear correlation between the pictorial style used to construct the images in each chapter, the way in which that image is used, and the kind of information it is intended to convey to the viewer. This relationship between images and their epistemic goals can also be clearly traced throughout Edmund Beecher Wilson's *An Atlas of Fertilization and Karyokinesis of the Ovum* and his multiple editions of *The Cell*, a widely used cell biology textbook of the same period as *General Cytology*. In *An Atlas*, Wilson writes, "It is extremely difficult for even the most skillful draughtsman to represent the exact appearance of protoplasm and of the delicate and complicated apparatus of the cell." Thus, Wilson primarily employs photographs in this text to illustrate cellular structure and processes in a way that "gives an absolutely unbiased representation of what appears under the microscope" (Wilson 1895, v).

However, in his multiple editions of *The Cell*, the first of which was published in 1896 and the last in 1925, Wilson transitioned away from using detail-rich photographs to using increasingly abstract drawings and diagrams of cells and their components. This trend toward employing simplified diagrams was correlated with an increased confidence and certainty in what was being depicted. Unlike photographs, abstract diagrams also provide more specificity of detail, calling attention to particular aspects of the structure or phenomenon being described (Maienschein 1991). Similarly, it is a combination of confidence in the visual information being presented, specificity of the topic communicated, and visibility of the phenomena discussed that determines what kind of image each author uses in *General Cytology* and the kind of knowledge each seeks to convey. While some aim to present more general information, others aim to communicate a specific piece or aspect of a structure or a particular procedure or process.

Pictorial Styles in *General Cytology*

The illustrations in *General Cytology* can be grouped into three distinct pictorial styles: realistic illustrations, experimental diagrams, and theory diagrams. The term *style* has a long and contentious history (see Bredekamp, Dünkel, and Schneider 2015 for an introduction to this history; also see Jones and Galison 1998). Here we employ a simple definition—a style "designates recognizably shared traits of created forms that transcend the individual producer" (Bredekamp, Dünkel, and Schneider 2015, 18). To this definition we add the contention that styles are deployed to different ends and represent different epistemic goals. Simply put, different ways of illustrating serve different roles within scientific texts. Thus, each of the pictorial styles that we outline in this chapter is joined by a shared form and addresses a shared epistemic goal.

Usually a choice of style is regarded primarily as an artistic decision. Often certain visual elements are worked into images for aesthetic reasons or to evoke certain emotions or parts of the imagination. In this discussion, however, we show that decisions regarding the kinds of marks made and where to place or exclude them have significant epistemic implications; the choice of pictorial style here is not an artistic decision but a scientific one (Bruhn 2011; Tversky 2011). These scientific decisions determine the way in which the viewer engages with the image, the kind of information presented, and the way visual arguments are constructed.

Each of the pictorial styles we address in the following discussion comprises three main components: an epistemic goal, visual or pictorial elements, and the method by which the illustrations were made. While each style is distinct, there is a certain degree of overlap among the components of each style. For example, all the styles we discuss contain an element of abstraction and make use of simplified marks. However, they do so in various ways and to different ends. The illustrations that make up each style use many of the same visual elements to convey different types of knowledge to the reader and serve different purposes within the text. Thus, while these three styles are not meant as hard and fast pictorial types, and many of the chapters employ more than one style, grouping the illustrations in *General Cytology* along these lines serves as a useful tool for analysis.

Realistic illustrations aim to highlight gross processes and make anatomical features visible. Realistic, in this context, refers to attempts by the authors to present their audience with as literal as possible a translation of what they see through the microscope's eyepiece, largely through the inclusion of a wealth of detail and information. This can be achieved either through meticulous and carefully considered drawing or through the medium of photography, which allows for the quick and easy capture of multiple cellular structures at once.

Of course, certain decisions must be made or methods used when creating a realistic illustration, or any kind of representation, that limit its ability to be an exact copy of what it illustrates. These limitations become especially apparent when it comes to microscopic, often translucent, entities such as cells, which themselves must be manipulated and taken out of their normal context in order to be illustrated. The degree to which these illustrations achieve realism is certainly interesting, but it is not our main focus here. Rather, we unpack how the visual elements of the illustrations in the realistic style achieve their shared epistemic goal, which is to invite readers to inspect them closely and to compare them with their own observations. In providing a wealth of visual information, these images allow for "virtual witnessing" of the cellular structures presented (Shapin and Schaffer 1985, 60). This permits readers to make their own observations of the specimens and to verify their existence. In this way, readers are being asked to engage with the images and formulate their own interpretations regarding the reality of the materials. Finally, realistic illustrations also serve to familiarize the reader with relatively unknown cellular structures or behaviors. This familiarization is achieved through the inclusion of multiple illustrations of a specific cellular structure or groups of cells, which are often arranged as a tableau on the page.

The rest of the illustrations in *General Cytology* fall into one of two styles, or categories: experimental diagrams and theory diagrams. While all diagrams are usually characterized by visual simplicity and the use of abstract marks to clearly illustrate certain structures or ideas, here we contend that they can serve different purposes and communicate different kinds of information (see Tversky et al. 2000). Not only can diagrams illustrate concepts or theories that are not directly observable and must be abstracted from multiple observations, but they can also serve to elucidate and demonstrate concrete things such as structures or processes (Bredekamp, Dünkel, and Schneider 2015, 152). The diagrams in *General Cytology* serve all of these purposes.

The group of illustrations we call experimental diagrams mainly accompany the chapters that turn to experimental approaches to answer questions about cellular structure and behavior. These diagrams are used to explain experimental techniques clearly and employ abstract and simplified drawn marks to do so. This visual specificity focuses the reader's attention on exactly *what* manipulation is being done *where*. Experimental diagrams also demonstrate the outcome of experimental manipulations by unambiguously highlighting particular alterations of cellular structure or behavior. While such diagrams certainly contain realistic elements, in that they are meant to refer to and look like the actual cells they depict, their shared epistemic goal lies in demonstrating experiments. Rather than asking viewers to inspect and compare them to their own observations, these diagrams serve as a means by which the experimenter can explicitly communicate particular techniques or results.

The final group of images, which we call theory diagrams, illustrate theories that are either derived from an amalgamation of observations from various specimens or abstracted from other empirical evidence. Unlike the other two styles, this illustration style displays phenomena that are, in several ways, not directly observable in a single cell. This is because many are used to present general theories, mainly general theories of heredity that span multiple specimens, individuals, and generations. These theories are formulated by the researcher based on various forms of evidence and thus require clear communication of their components. This epistemic goal is achieved through use of simplified marks to represent components of the theory, such as the behavior of particular genes or chromosomes, and the arrangement of these marks on the page indicates the relationships between them.

Realistic Illustrations

The first group of images in *General Cytology* that we discuss are realistic illustrations, which are those that seek to present the reader with a detailed depiction of gross processes and anatomical features of cells. Realistic illustrations are found primarily in the chapters by Cowdry, Warren and Margaret Lewis, and Frank Lillie and E. E. Just. With a few exceptions, all of the illustrations in these chapters are done in the realistic pictorial style.

In the beginning of his chapter on cytological constituents, or organelles, Cowdry starts the sections on mitochondria, Golgi apparatus, and chromidial substance by laying out the current state of research on each. When dealing with subcellular structure, it is the task of the researcher to develop methods of making otherwise invisible phenomena visible and accessible to the human eye. "Thus," Cowdry writes in discussing research on mitochondria, "it has only been during the past thirty-five or forty years, slowly, with the gradual improvements in technique, that the distinctive characteristics of mitochondria have come to light. *We are now entering upon a period of experimentation*" (Cowdry 1924b, 313; emphasis added). Cowdry is clearly situating the status of research on organelles as "preexperimental." Thus the aim of this chapter is to present what is known about the structure of these three organelles primarily based on microscopical observation, not experimental manipulation.

But here observation is not as simple as sitting down and looking at something. As Jutta Schickore discusses in chapter 4 of this volume, approaches to microscopical methods increasingly incorporated experimentation as the nineteenth century progressed. Microscopists slowly began to realize that tissues could be manipulated; stained, and dyed in such a way that would make them more visible and that would also make otherwise invisible structures detectable. A lot of experimenting is inherent in developing the techniques for visualizing subcellular structures. This is certainly the case for subcellular structures such as organelles. In discussing the methods of preparation necessary to study organelles such as the mitochondria, Cowdry writes, "The technique is not really difficult, but those who have had no experience in cytology cannot expect to be immediately successful. A little experimentation is necessary" (Cowdry 1924b, 314).

In situating the status of organelle research in this way, Cowdry not only calls attention to its preexperimental nature, but he also more generally calls attention to the fact that it is still in its infancy and that methods for just making organelles visible and observable are still in an experimental stage. Because of this, Cowdry cautions readers about the interpretation of the results and asks them to be aware that methods of preparing tissue could artificially alter the structure of the organelles. In discussing methods for studying the Golgi apparatus, Cowdry writes, "Great caution has to be exercised in interpreting the findings, particularly in respect to changes in the shape and size of the Golgi apparatus occurring normally and induced experimentally" (Cowdry 1924b, 334). This question of interpretation is likely one of the main motivations for presenting as much visual information about the organelles as possible—so that readers can observe them for themselves and verify what might be going on.

The main images Cowdry uses in his chapter are realistic, almost photographic-looking drawings that depict organelles of interest, mitochondria, the Golgi apparatus, and chromidial substance, in cells from different organisms. For each organelle examined, there is at least one group of images arranged as a tableau, some of which fill an entire page (fig. 7.1; also see Cowdry 1924b, figs. 18–26, 28, 32, and 35). These tableaus largely consist of illustrations borrowed from various sources and authors, a common practice of that time for works on cell biology (see Wilson 1925 for several more examples). The illustrations in each group depict individual cells from various organisms stained in such a way that the organelle of interest becomes visible. This organization emphasizes and encourages both close inspection of each image and comparisons among them.

While calling attention to the differences in the structure and arrangement of each organelle in various cells, these tableau arrangements even more importantly make a visual argument for their being fundamental components of the cell (Bredekamp, Dünkel, and Schneider 2015). The display of the individual images in this way allows the reader to grasp in one glance that the same structure is present both in diverse cell types across various organisms, albeit with some modifications to structure (fig. 7.1), and in the same cell type through time (see Cowdry 1924b, figs. 18–22, 37). In this way, these drawings of cells and their organelles serve as evidence for the existence of these structures in diatoms, fungi cells, insect intestinal epithelial cells, and human spinal ganglion cells (images numbered 2, 3, 9, and 14 in fig. 7.1, respectively), establishing them as basic structures in the vast majority of cell types.

As not much was known at the time about these structures, a realistic pictorial style was important for presenting this information in a way that allowed readers to make their own observations. Here the drawn lines and marks are rendered so as to stand in for the organelles they depict, inviting readers to observe them as they would the organelles themselves under the microscope. For example, Cowdry admits that for both mitochondria and the Golgi apparatus, "it is difficult to say when we are dealing with a true [organelle] or with other materials which may be in part its products" (Cowdry 1924b, 335). Thus, while these collections of images work together to demonstrate the existence of and roughly characterize a given organelle, they only establish an "elastic and very tentative type of homology" for each organelle, and are thus open to interpretation (335).

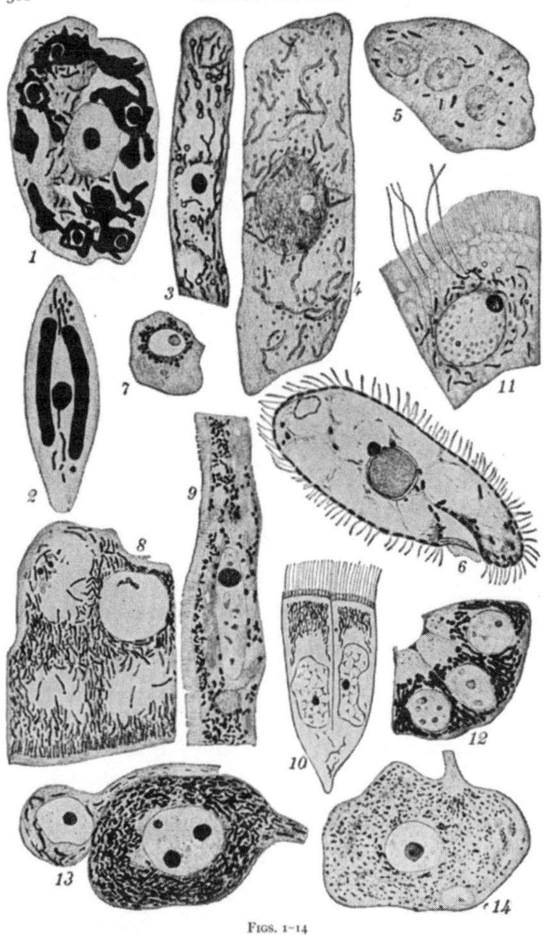

FIGS. 1–14: FIG. 1.—Filament of *Spirogyra maxima*, after Guilliermond (1921*j*) containing typical rodlike and filamentous mitochondria. FIG. 2.—A diatom, after Guilliermond (1921*j*) containing similar mitochondria. FIG. 3.—A fungus, *Pustularia vesiculosa*, after Guilliermond (1915*b*). FIG. 4.—A spermatophyte, *Narcissus poeticus*, after Guilliermond (1919*c*). FIG. 5.—A myxomycete, *Arcyria denudata*, after N. H. Cowdry (1918). FIG. 6.—A protozoön, *Glaucoma piriformis*, after Fauré-Fremiet (1909). FIG. 7.—A coelenterate, *Aurelia aurita*, ovarian egg after Tsukaguchi (1914). Note perinuclear accumulation of mitochondria. FIG. 8.—An arachnid, *Amblyomma americana*, Malpighian tubule. FIG. 9.—An insect, *Cimex lectularius*, intestinal epithelium. FIG. 10.—An amphibian, *Rana esculenta*, pharyngeal epithelial cells after Saguchi (1917), showing distal condensation. FIG. 11.—A selachian, *Scyllium canicula*, cell from choroid plexus after Grynfeltt and Euzière (1913*a*), illustrating perinuclear clumping of mitochondria. FIG. 12.—Kidney cells of a white mouse with mitochondria in proximal cytoplasm. FIG. 13.—Small cell of *locus coeruleus* and large cell of mesencephalic nucleus of the fifth nerve of a white mouse to indicate difference in amount of mitochondria. FIG. 14.—Human spinal ganglion cell.

Figure 7.1. Figures 1–14 from the chapter by Cowdry (1924b, 316) in Cowdry's *General Cytology*. These illustrations show mitochondria in various cell types.

The tableau arrangement of the figures reflects the reference-guide structure of the chapter on the whole. The chapter is divided into three parts, one for each organelle, and each part is split into numbered subsections. The type of illustration used and the way information is presented visually reflect the kind of work the text achieves, which is to characterize each organelle and lay out the details of what is known about the structure of each. Thus the images are also meant to familiarize readers with what these cells and organelles look like under the microscope. In order to achieve this, they must be rendered in such a way as to contain as much visual information as possible. In other words, the images must be as faithful as possible to what would be presented to a researcher looking at these structures through a microscope. This is also likely one of the main reasons why, despite being drawings, these illustrations look markedly photographic.

Like the methods for detection and study of organelle structure, *in vitro* tissue culture was a method just beginning to become standard in cytological investigations at this time. Unlike older, more established techniques that involved fixing and staining specimens, tissue culture "revealed new facts concerning the behavior, structure, and physiology of the living body cells" (Lewis and Lewis 1924, 385). These methods and the new kinds of information about the characteristics and behaviors of living cells are the focus of Warren and Margaret Lewis's chapter in *General Cytology*.

Similar to Cowdry's chapter, the Lewises' text is accompanied by extensive visual information documenting various attempts to culture different cell types *in vitro*. Also like Cowdry, these authors seek to show cellular structure and arrangement in a realistic way, so that readers can see for themselves the phenomena being discussed. Another important epistemic goal of these images is to familiarize the reader with the characteristics and behaviors of cells that tissue culture can reveal, which were relatively novel to the field at the time.

Unlike Cowdry, however, the Lewises employ photography to describe cellular phenomena, a medium with even stronger historical connections to the ideals of realism than any form of scientific image-making that directly involves the human hand (Breidbach 2002; Tucker 2006, 117; Wilson 1895). Another virtue of the use of photography to illustrate cells and tissues is that it allows for the quick and easy capture of multiple structures in a single image. As these authors aim to show the morphological outgrowth and patterning of various cultured cells, the medium of photography allowed them easily to capture this visual complexity and to display the details of multiple cells at once (fig. 7.2; also see Lewis and Lewis 1924,

PLATE I

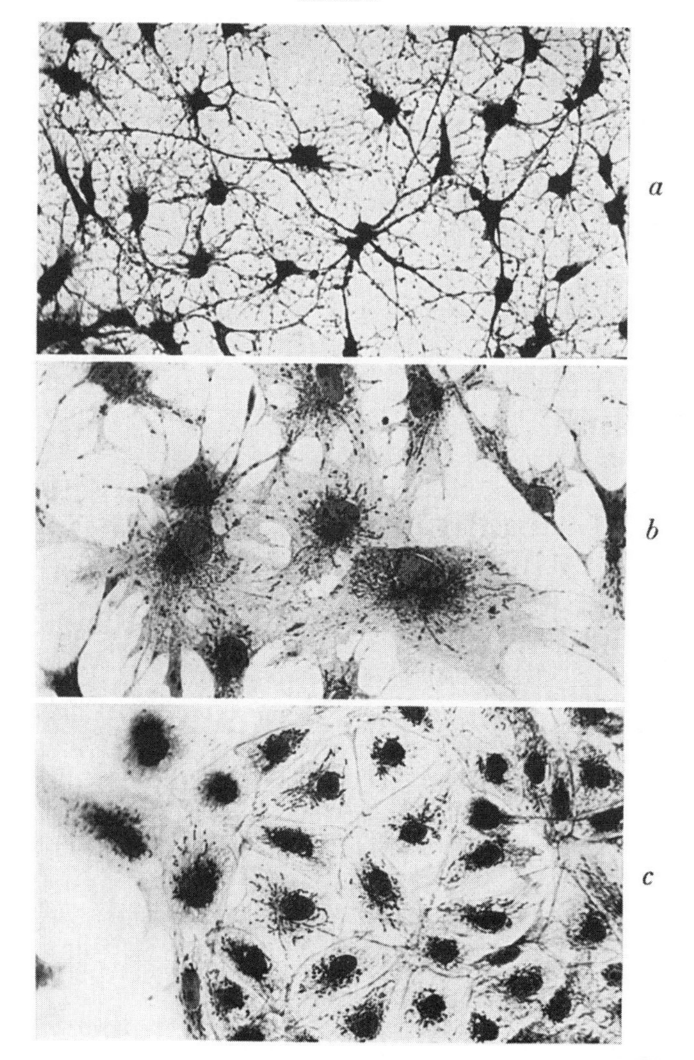

PLATE I

a. Subcutaneous mesenchyme, two-day culture, eight-day chick embryo; Locke-bouillon-dextrose; Janus green, iodine. ×480.

b. Subcutaneous mesenchyme, three-day culture, eight-day chick embryo; Locke-bouillon-dextrose; Janus green, iodine. ×480.

c. Mesothelium, three-day culture from stomach, six-day chick embryo; 0.9 per cent NaCl; Janus green, iodine. ×480.

Figure 7.2. Plate I from the chapter by Lewis and Lewis (1924) in Cowdry's *General Cytology*. Note the numerous cells in each image as well as the use of the vital stains to make cellular structures visible.

plates II–IX). The wealth of visual information in these images invites readers to make their own observations and formulate their own interpretations.

To create these photographs, Lewis and Lewis transferred small samples of various cell types into "hanging drops" suspended from microscope slides, sealed off the slides with a coverslip and Vaseline or paraffin wax, and then incubated them to observe the outgrowth and behaviors of the cells (Lewis and Lewis 1924). Although they do not explicitly state what happened next, in most cases it is likely that they stained the cells with vital stains before photographing them. Each photograph is accompanied by a caption explaining the tissue type, the time of culture, and the stains used to highlight various cellular structures.

The third chapter containing images that primarily fit into this group is the one by Frank R. Lillie and E. E. Just on fertilization. With a few more diagrammatic exceptions (see Lillie and Just 1924, figs. 2, 4, 7), the majority of the thirteen illustrations in this chapter look remarkably similar to those in Cowdry's chapter and are drawn in the same style (see Lillie and Just 1924, figs. 1, 3, 5, 6, 8–10). While these authors discuss both the morphology and physiology of fertilization, all but one of the images are included in the half of the chapter on morphology. By using a realistic pictorial style, they aim to familiarize the reader with the cellular morphology of fertilization and to encourage close inspection of the illustrations.

Experimental Diagrams

The next group of images in *General Cytology* we call experimental diagrams. The chapters that primarily contain illustrations in this style are those by Robert Chambers and Edwin Grant Conklin. With the exception of a few in Conklin's chapter, all of the diagrams in these chapters depict experimental procedures on cells and their outcomes. Unlike the realistic illustrations, these diagrams leave little interpretation to the reader and instead impart the experiential knowledge or observations of the experimenter. Additionally, they serve as a reenactment of a technical procedure and highlight particular features so that a student would know what to expect when using the technique. These diagrams achieve these epistemic goals through the high level of visual simplification of complex structures and the use of special marks, both of which draw the reader's attention to particular features relevant to the experimental outcome being communicated or the manipulation technique being demonstrated.

The purposes of the experiments discussed in chapters such as those by Chambers and Conklin were to reveal the characteristics of cell structure

FIG. 6.—Ciliated cell from sea-urchin ovary showing effect of injury with a needle.

Figure 7.3. Diagram from the chapter by Chambers (1924, 258, fig. 6) in Cowdry's *General Cytology*.

and behavior that are unobservable in either fixed and stained or unmanipulated living tissue. For example, Chambers writes that conceptions about the structure of cellular protoplasm were skewed by the use of fixing agents, which tended to coagulate it and present deceptive visual information, such as the appearance of filamentous structures. "Many, however," he writes, "who devoted themselves to experimental studies on living protoplasm maintained its essentially fluid nature" (Chambers 1924, 237). While experimental methods such as micro-dissection and micro-injection received criticism for altering the natural condition of the object under study and depending in large part on individual interpretations (Chambers 1924), these approaches revealed otherwise hidden phenomena.

In Chambers's chapter on experimentation on cellular protoplasm, the demonstration of precise techniques necessitated diagrams that could clearly demonstrate to the viewer exactly the procedure being carried out. It needed to be clear exactly where a needle was inserted, a cut was made, or a cellular structure disturbed or excised. Most of the diagrams include the manipulation needle and consist of series of simple drawings that together demonstrate exactly how the needle interacts with the tissue and what its effect is (fig. 7.3; also see Chambers 1924, figs. 3, 7, 11–12, 15–16). These diagrams draw the viewer's attention to the action of the needle and highlight the pertinent effects of the manipulation.

Another useful aspect of diagrams is that special marks can be used to denote a particular effect or outcome of an experimental manipulation. Rather than displaying them in a realistic way, the author can use a special mark to represent or stand in for a certain cellular structure. Often these marks have no direct relationship to the structure or phenomenon they are meant to represent, so that the author must make the relationship explicitly

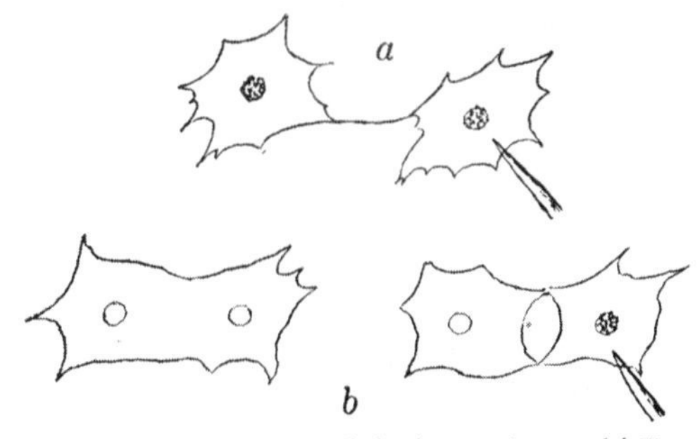

FIG. 2.—Mesenchyme cells in tissue culture. (*a*) Two daughter-cells still connected by a bridge which transmits injury from one cell to the other. (*b*) Two contiguous cells, one of which only is injured by pricking.

Figure 7.4. Diagram from the chapter by Chambers (1924, 242, fig. 2) in Cowdry's *General Cytology*. Injured cells are denoted with speckled nuclei, and uninjured cells with unfilled nuclei.

clear (see Chambers 1924, fig. 45). For example, Figure 7.4 demonstrates the effect of pricking one of two attached chick mesenchyme cells. Diagram 7.4a shows how when one of two cells that are connected by a protoplasmic bridge is injured that injury is transferred to the connected cell. Here Chambers denotes an injured cell as one with a speckled nucleus, indicating that it has coagulated. Unlike in 7.4a, diagram 7.4b shows how when one of two mesenchyme cells that are only contiguous and not attached by a bridge is pricked, the injury is not passed on to the other cell. One cell with a speckled nucleus and one with a nucleus that is not filled in represent this experimental result. Not only does this kind of notation make the outcome of the manipulation immediately clear to the reader, but it also indicates that these changes are the most important parts of the diagrams.

While the purpose of experimentation was to reveal new, otherwise unobservable features of cellular structure and behavior, the experiments described in *General Cytology* were performed on well-studied and characterized cellular structures. Recall Cowdry's comment about research on mitochondria only then entering an experimental period, after roughly thirty-five years of establishing basic structure and morphology. Many researchers of this period held that invasive experimental manipulation should only commence once as much knowledge as possible could be gleaned from direct observation alone (Conklin 1897; Cowdry 1924a). Of course, there was a great deal of experimentation involved in developing methods of seeing and observation, but in working out these protocols, great pains were taken to preserve "normal" conditions as much as possible. Thus most of the experiments undertaken in cytology, such as the ones demonstrated in figures 7.3 and 7.4, were performed on already widely studied cells, such as those found in sea urchins and chick embryos. Because it was generally known what these cells look like, it was not necessary to represent them in a realistic manner here. Doing so would only detract from the aim of the diagrams.

The second main chapter in *General Cytology* with experimental diagrams is Edwin Grant Conklin's chapter on cellular differentiation. All of the images Conklin uses in this chapter are diagrams of eggs and embryos of various species, although not all of them depict the results of experimental manipulation. Those that do not, however, show well characterized structures and are included primarily for comparison with the experimental diagrams. The two main organisms Conklin discusses and includes images of are *Crepidula* and *Styela*. For both, Conklin includes diagrams both of experimental results and of the tissues under normal conditions for comparison.

The main experimental technique Conklin employs here is centrifugation to establish the effect of displacing cellular components from their normal locations. His central aim in doing this is to investigate the relationship between nuclear and cytoplasmic placement within cells and the axis of cleavage and differentiation.

In figure 7.5, Conklin demonstrates the effect of centrifuging the eggs and early stage embryos of *Crepidula*, a species he studied extensively throughout the 1890s. In these diagrams he shows that the size and growth of the nucleus depend on the volume of cytoplasm that surrounds it. By centrifuging *Crepidula* eggs to change the location of cytoplasm and yolk as the egg undergoes division, he is able to affect the growth of the nuclei (Conklin 1924). In parts C and D of figure 7.5, Conklin depicts two embryos, one two-cell stage and one four-cell stage, respectively, that were centrifuged at the single-cell stage to disrupt the normal positions of cytoplasm and yolk. As Conklin shows in these diagrams, in the subsequent cell divisions, the cytoplasm, which is represented by speckle-filled areas, is unevenly distributed to only half of the daughter cells. In these cells the nuclei are clearly much larger than in those cells with no cytoplasm and only yolk, which is represented by the unmarked spaces bounded by the cell outlines.

Although they are not realistic, these line diagrams effectively convey the result of Conklin's experimental manipulation—namely, the movement of the cytoplasm and the resulting size of the nuclei in the daughter cells.

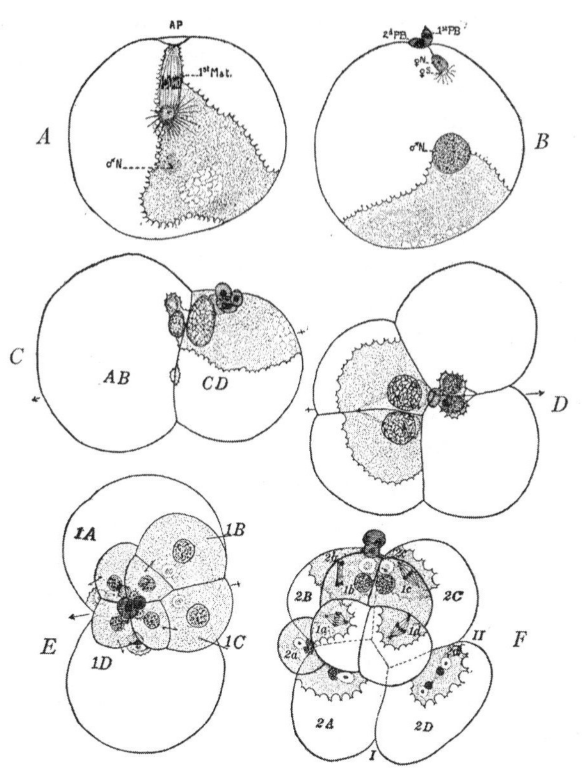

Fig. 4.—Eggs of *Crepidula* centrifuged in the 1- or 2-cell stages. *A*, Cytoplasm centrifuged away from the animal pole (AP) leaving first maturation spindle attached to that pole. *B*, Cytoplasm centrifuged away from animal pole leaving egg nucleus (♀N) and egg sphere (♀S) in an area of yolk, and sperm nucleus (♂N) in an area of cytoplasm. The latter is accordingly much larger than the former. *C*, Egg centrifuged during first cleavage, most of the yolk going into the cell AB, and of the cytoplasm into CD; daughter nuclei are proportional in size to the field of cytoplasm in which they lie. *D*, Four-cell stage of egg centrifuged during first cleavage; nuclei proportional to volume of cytoplasm in which they lie. *E*, Centrifuged during second cleavage separating purely protoplasmic macromeres (1B, 1C) from yolk-containing ones (1A, 1B). The four unequal macromeres have produced four micromeres of approximately equal size. *F*, Egg centrifuged at close of first cleavage so as to cause the second cleavage to be equatorial in position; first group of micromeres (*1a–1d*) not at animal pole, second quartet (*2a–2d*) forming in abnormal positions; evident attempt of cytoplasm to return as near as possible to animal pole.

Figure 7.5. Diagram from the chapter by Conklin (1924, 554, fig. 4) in Cowdry's *General Cytology*. Cellular cytoplasm is represented by speckle-filled areas, and yolk by the unmarked spaces bounded by the cell outlines.

Compared to some of Conklin's earlier, more realistic renderings of *Crepidula* eggs, these diagrams look markedly more simplified (see Conklin 1897). Significantly, within the images of this chapter, Conklin overemphasizes the outline and visibility of the cytoplasmic component and nucleus of each cell. He does this both by rendering them darker than they actually appear and by eliminating all other unnecessary details, such as the details of the yolk, which would distract from what he is trying to communicate (fig. 7.5; also see Conklin 1924, figs. 2, 12). This stylistic change reflects a shift in purpose. In his earlier works, Conklin was presenting for the first time the cell divisions in early *Crepidula* embryos in order to allow readers to observe the developmental process for themselves; here he is trying to make a specific point about cell division and differentiation following experimental manipulation.

Theory Diagrams

The third and final group we call theory diagrams, which are used primarily in the last two chapters of *General Cytology* by Clarence E. McClung and Thomas H. Morgan. These chapters are united by the presentation of general theories of chromosomes and heredity. Unlike the other two pictorial styles, theory diagrams depict phenomena that are not directly observable, because they present theories or concepts that arose out of multiple observations, specimens, individuals, and generations. Across these two chapters, only thirteen of the fifty-three illustrations are theory diagrams. The rest are primarily realistic illustrations that are included to support the theories laid out in diagrams. In several cases, these realistic illustrations depict the directly observable cellular phenomena on which the theories are based.

Theory diagrams can be further refined within these chapters into two subcategories: aggregates and abstractions. Aggregates are used to order visual evidence for phenomena by bringing together observations derived from several specimens. In the case of McClung's chromosome theory, aggregates are often bloblike illustrations of chromosomes, acquired from a number of cells and across generations (see McClung 1924, figs. 5–8, 15, 24, 26–27, 30–32). These aggregates serve as evidence for McClung's theory. Abstractions, on the other hand, depict phenomena that exist beyond the realm of observation but whose reality is inferred from empirical observations. In the case of Morgan, abstractions serve as translations of his verbal arguments, highlighting in a visual format the same process that he outlines within the text.

McClung's chapter on the chromosome theory of heredity employs diagrams to describe the structure and behavior of chromosomes in various cell types. The majority of diagrams in this chapter are aggregates (although he also uses abstractions; see McClung 1924, figs. 3, 29) that display the formation and movement of chromosomes over multiple stages of cell division or multiple generations. The chromosomes are almost exclusively depicted as simplified, solid black shapes with arms bent at angles of varying degrees. McClung writes that "[heredity] also is continuous and a function primarily of the group and not the individual. . . . The value of the chromosome theory of heredity is due to the fact that it furnishes clear evidence of this group continuity—a continuity which manifests itself, not in a series of fixed forms, but through the repetition of a determinate series of cyclical changes" (McClung 1924, 614).

As McClung states here, the chromosome theory of heredity deals with phenomena that take place over extended periods of time and that exist among groups of individuals. Both of these conditions render this theory not directly observable in a single cell or group of cells; rather, it must be abstracted from a large group of individuals over time. It is for this reason that the diagrams shown in McClung's chapter are even less realistic than those depicting experimental manipulations and that they contain abstract shapes and other markings.

Similar to McClung's chapter, Morgan's chapter discusses heredity, although with more of a focus on Mendelian heredity and its relationship to observations of chromosome behavior in cells. Many of the genetic theories Morgan presents are beyond the realm of the visible and are thus illustrated with abstractions. The two phenomena that constitute the majority of the chapter are the linkage of genetic traits, or genes, and crossing-over. Morgan explains that traits are linked when they go into a genetic cross together, meaning that they are both present in the same individual and tend to remain so in later generations. This linkage between two traits "is expressed by the percentage of cases in which they are found to remain together" (Morgan 1924, 698). These links can be broken, however, through a process called crossing-over, whereby an exchange takes place between genes on the maternal chromosome and the corresponding genes on the homologous paternal chromosome (fig. 7.6; also see Morgan 1924, figs. 7–9).

Unlike the realistic illustrations and the experimental diagrams in *General Cytology*, the abstractions that describe crossing-over are not based on direct observation of chromosomes or genes. However, Morgan claims that this does not make them any less credible. In a discussion of studies on crossing-over he writes, "In this respect, genetics has proved a more refined instrument . . . than direct observation of the germ cells themselves, and while this advance may appear more theoretical than the conclusions based on observations of the cell, this need not mean that it is less reliable" (Morgan 1924, 693). Thus, the diagrams illustrating crossing-over are almost entirely constructed based on observations of the outcomes of genetic

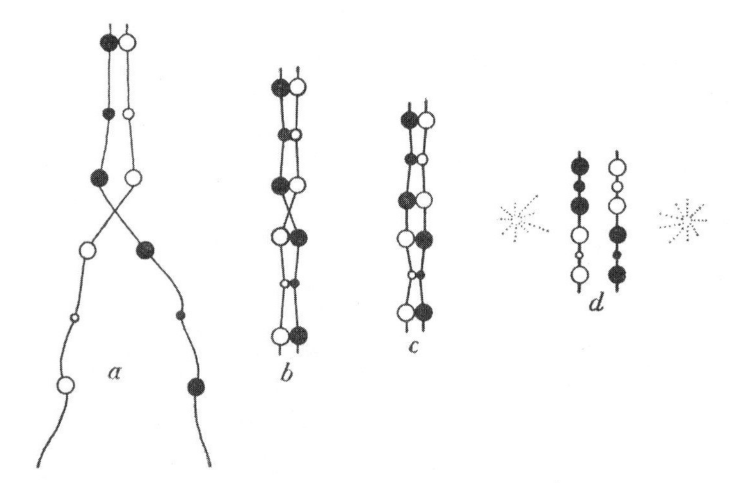

Figure 7.6. Diagram from the chapter by Morgan (1924, 707, fig. 5) in Cowdry's *General Cytology*. The filled-in circles on each chromosome represent genes from one parent, and the open circles represent genes from the other parent.

crosses, mainly observations about the inheritance of certain traits by the offspring, not on direct observation of the cellular structures themselves.

As with the experimental diagrams, special marks are used here to represent particular features of the phenomena being discussed that the author has deemed important. These marks serve as representations of certain entities within the cell that defy direct observation and make immediately clear to the reader the theory being described. In figure 7.6, for example, each circle represents a different gene on the chromosome. The filled-in circles represent genes from one parent and the open circles represent genes from the other. However, unlike the experimental diagrams, these illustrate a theory formulated from observations of several organisms over time, not particular individual cells. Instead of standing in for experimental outcomes that the experimenter wants the reader to pay particular attention to, these marks represent theorized structures and phenomena based on inference.

Conclusion

Throughout *General Cytology*, scientists used images for a variety of purposes by deploying a range of pictorial styles. In this chapter, we have attempted to parse the ways in which the authors used their images to demonstrate a range of knowledge, theories, and methods—that is, the epistemic goals of the images. Here we have posited that at least three pictorial styles exist within the volume: realistic illustrations, experimental diagrams, and theory diagrams.

Realistic illustrations offer the highest degree of reader engagement by providing a wealth of visual detail and information. This style is used to highlight gross processes and make anatomical features visible in a way that invites readers to inspect them closely and even interpret their meaning themselves. This style is used extensively by Cowdry, Lewis and Lewis, and Lillie and Just.

Experimental diagrams are used to display experimental techniques and tools that require visual specificity regarding exactly *what* manipulation was being done *where*. This style draws the reader's attention to specific cellular structures and the outcomes of different interventions in cellular processes. Little interpretation is left to the reader; instead, the goal is to impart the experiential knowledge and observations of the expert experimenter, who serves as the primary interpreter of cellular phenomena. Thus, experimental diagrams serve as reenactments of technical procedures and highlight particular features so that a student would know what to expect when using the technique. These diagrams achieve both of these goals through visual simplification of complex structures and special marks, both of which draw the reader's attention to particular features relevant to the argument being made or the technique being demonstrated. This style is deployed mainly within chapters, such as those of Chambers and Conklin, that turn to experimental approaches to answer questions about cellular structure and behavior.

Finally, theory diagrams display phenomena that are unobservable because they present theories or concepts that arose out of multiple observations, specimens, individuals, and generations. Within this style, we have further distinguished two substyles: aggregates and abstractions. Aggregates bring together observations from across specimens and order them in such a way as to provide visual evidence for otherwise unobservable phenomena. Abstractions depict phenomena that cannot be observed, such as crossing-over, but for which there is a great deal of empirical evidence.

While the illustrations in *General Cytology* can be categorized into these three styles, the boundaries between them are, as has been touched on briefly throughout this discussion, somewhat blurry. For example, the same kinds of marks are used both in the experimental diagrams and the theory diagrams, although to different ends. The use of simplified circles and line segments in both of these styles (compare fig. 7.2 with fig. 7.6) does not negate the fact that each has distinct epistemic goals. The shared characteristics between realistic illustrations, experimental diagrams, and theory diagrams stem in large part from the fact that they are all two-dimensional translations of complex, four-dimensional entities: living cells and their myriad components and behaviors (Rheinberger 2003, 624). Thus the visual elements of all three styles are confined to what kinds of marks and spatial relationships can be rendered on paper, whether by a human hand or photographic developing chemicals.

Such a translation of four-dimensional entities into two dimensions also necessitates certain decisions about what aspects of living cells to focus on and what to depict, a process that always results in some abstraction. The question, therefore, is to what degree this abstraction is implemented and to what end. In the realistic illustrations discussed here, the authors chose a method of depiction that included as much detail, and thus as little abstraction, as possible to allow the reader to observe and inspect the structures. In both the experimental and theory diagrams, however, a high degree of abstraction is employed to unambiguously direct the reader's attention to the experimental manipulation or theory being presented by the author.

Our motivation for closely examining the illustrations included in *General Cytology* was multifold. First, we thought it necessary that in a volume reflecting on this influential book the central role of images and visual argumentation not be overlooked. In a text entirely devoted to biological phenomena that are in many ways beyond the reach of the naked human senses, presenting and receiving knowledge in a visual manner is paramount. Given this fundamental role of visual information, we also sought to lay out a framework with which to unpack the illustrations of cells within this text. Just like the written word, images can convey a variety of meanings depending on the way they are constructed and the different elements they weave together. Thus they can and should be seen and analyzed as such. Recognizing the epistemic goals of illustrations and the ways those goals are met through different kinds of marks, spatial arrangements, and degrees of abstraction is not just important for an understanding of the illustrations themselves; it is crucial for thoroughly grasping the science they display.

Notes

1 In the Preface, the editor Edmund Cowdry remarked: "The volume, as it stands, is to be considered, to some extent at least, as a contribution from the Marine Biological Laboratory" (Cowdry 1924a).

References

Anderson, Nancy, and Michael Dietrich, eds. 2012. *The Educated Eye: Visual Culture and Pedagogy in the Life Sciences*. Hanover, NH: Dartmouth College Press.

Bredekamp, Horst, Vera Dünkel, and Birgit Schneider, eds. 2015. *The Technical Image: A History of Styles in Scientific Imagery*. Chicago: University of Chicago Press.

Breidbach, Olaf. 2002. "Representations of the Microcosm: The Claim for Objectivity in 19th Century Scientific Microphotography." *Journal of the History of Biology* 35 (2): 221–50.

Bruhn, Matthias. 2011. "Life Lines: An Art History of Biological Research Around 1800." *Studies in History and Philosophy of Biological and Biomedical Sciences* 42: 368–80.

Cambrosio, Alberto, Daniel Jacobi, and Peter Keating. 2005. "Arguing with Images: Pauling's Theory of Antibody Formation." *Representations* 89 (1): 94–130.

Chambers, Robert. 1924. "The Physical Structure of Protoplasm as Determined by Micro-Dissection and Injection." In Cowdry 1924a, 235–309.

Conklin, Edwin Grant. 1897. "The Embryology of Crepidula: A Contribution to the Cell Lineage and Early Development of Some Marine Gasteropods." *Journal of Morphology* 13 (1): 1–226.

———. 1924. "Cellular Differentiation." In Cowdry 1924a, 537–607.

Cowdry, Edmund V., ed. 1924a. *General Cytology: A Textbook of Cellular Structure and Function for Students of Biology and Medicine*. Chicago: University of Chicago Press.

———. 1924b. "Mitochondria, Golgi Apparatus, and Chromidial Substance." In Cowdry 1924a, 311–82.

Griesemer, James. 2007. "Tracking Organic Processes: Representations and Research Styles in Classical Embryology and Genetics." In *From Embryology to Evo-Devo*, edited by J. Maienschein, and M. Laubichler, 375–433. Cambridge, MA: MIT Press.

Hopwood, Nick. 2015. *Haeckel's Embryos: Images, Evolution & Fraud*. Chicago: University of Chicago Press.

Jones, Caroline A., and Peter Galison, eds. 1998. *Picturing Science, Producing Art*. New York: Routledge.

Lewis, Warren H., and Margaret R. Lewis. 1924. "Behavior of Cells in Tissue Culture." In Cowdry 1924a, 383–447.

Lillie, Frank R., and E. E. Just. 1924. "Fertilization." In Cowdry 1924a, 449–536.

Lynch, Michael. 1991. "Pictures of Nothing? Visual Construals in Social Theory." *Sociological Theory* 9 (1): 1–21.

———. 2006. "The Production of Scientific Images: Vision and Re-Vision in the History, Philosophy, and Sociology of Science." In *Visual Cultures of Science: Rethinking Representational Practices in Knowledge Building and Science Communication*, edited by Luc Pauwels, 26–40. Hanover, NH: Dartmouth College Press.

Maienschein, Jane. 1991. "From Presentation to Representation in E. B. Wilson's *The Cell*." *Biology and Philosophy* 6 (2): 227–54.

May, Raoul M. 1926. "Book Review: *General Cytology* E. V. Cowdry; *The Cell in Development and Heredity* Edmund B. Wilson." *Isis* 8 (1): 213–15.

McClung, Clarence E. 1924. "The Chromosome Theory of Heredity." In Cowdry 1924a, 609–89.

Morgan, Thomas H. 1924. "Mendelian Heredity in Relation to Cytology." In Cowdry 1924a, 691–734.

Pauwels, Luc, ed. 2006. *Visual Cultures of Science: Rethinking Representational Practices in Knowledge Building and Science Communication*. Hanover, NH: Dartmouth College Press.

Rheinberger, Hans-Jörg. 2003. "Scrips and Scribbles." *MLN* 118 (3): 622–36.

Shapin, Steven, and Simon Schaffer. 1985. *Leviathan and the Air-Pump: Hobbes, Boyle and the Experimental Life*. Princeton, NJ: Princeton University Press.

Tucker, Jennifer. 2006. "The Historian, the Picture, and the Archive." *Isis* 97 (1): 111–20.

Tversky, Barbara, Jeff Zacks, Paul Lee, and Julie Heiser. 2000. "Lines, Blobs, Crosses and Arrows: Diagrammatic Communication with Schematic Figures." In *Theory and Application of Diagrams*, edited by Michael Anderson, Peter Cheng, and Volker Haarslev, 221–230. Berlin: Springer.

Tversky, Barbara. 2011. "Visualizing Thought." *Topics in Cognitive Science* 3 (3): 499–535.

University of Chicago Press Archives. 1925. Box 127, folder 1 on Cowdry's *General Cytology*. The Press archives are held in Special Collections at the University of Chicago Library. Thanks to Karl Matlin for finding and sharing this information.

Wilson, E. B. 1895. *An Atlas of Fertilization and Karyokinesis of the Ovum*. New York: Macmillan. Available online at http://www.biodiversitylibrary.org/bibliography/6244#/summary.

———. 1925. *The Cell in Development and Heredity*. 3rd ed. New York: Macmillan. Orig. pub. as *The Cell in Development and Inheritance* in 1896; 2nd ed., 1900.

CHAPTER 8

THOMAS HUNT MORGAN AND THE ROLE OF CHROMOSOMES IN HEREDITY

Garland E. Allen

The final two chapters in E. V. Cowdry's massive volume, *General Cytology* (1924) are the only ones devoted to the role of chromosomes in the life of the cell. Thus, of the total 734 pages, only 123 (around 17%) focus on chromosomes. From our perspective today, this may seem surprising. For one thing, extensive and detailed studies of chromosomes had consumed much time and effort among late nineteenth- and early twentieth-century cytologists as they attempted to determine the role these structures and their elaborate movements played in the cell cycle, particularly their presumed relationship to heredity. For another, compared to most other cell structures known in 1924 (mitochondria, Golgi apparatus, chloroplasts and plastids in general, and the various cell membranes), chromosomes and their behavior, including their pair-wise alignment during meiosis, were larger and considerably easier to study in detail with the light microscopes of the day. By contrast, Golgi bodies, mitochondria, and cell membranes showed relatively little structural detail and consequently their roles in the cell were even harder to determine. And third, after a decade of rapid growth in the Mendelian-chromosome paradigm of heredity,[1] the mapping of genes on chromosomes and their association with such central biological issues as the inheritance of sex, chromosomes had come to occupy a central position in cell biology.

There are several possible reasons why chromosomes may not have received as much attention in the organization of the Cowdry volume as we today might have expected. (1) Their role in heredity and the process of mapping had been so clearly worked out and was so well known as not to require any extensive review at the time. (2) As a corollary of (1), Cowdry and the other organizers may have felt that the book should concentrate on the less understood features of cell biology, that is, the areas in which future research should be directed. (3) Given the movement of biology in the 1920s toward a more explicitly experimental and functional orientation, it may have seemed appropriate to focus on dynamic and physiological, as opposed to structural, aspects of cell life. Examining the table of contents indicates that seven of the ten chapters (excluding only E. B. Wilson's overview in chapter 1) are devoted to such topics as general cell chemistry, reactivity of cells to stimuli, the selective permeability of membranes, the physical composition of protoplasm, behavior of cells in tissue culture, and fertilization and cellular differentiation. Morgan and many of the other authors (Wilson, Mathews, Ralph and F. R. Lillie, the Lewises) were strong proponents of taking biology toward a more functional and away from the more descriptive set of concerns that had characterized much of the field in the later nineteenth century.

While no doubt all three reasons were at play, I suggest that, despite the general acceptance of the Mendelian-chromosome paradigm of heredity (MCPH hereafter) by the early 1920s, there was still formidable opposition to the work of the Morgan school. This will account, I believe, for the fact that Morgan devoted most of his chapter not only to reviewing but also to defending the MCPH as the simplest and best interpretation of the combined facts from cytology and breeding experiments. Both reasons (2) and (3) appear to have been consciously discussed in the planning stages during 1922 and 1923 at MBL. In the book's preface Cowdry claims that the volume was aimed to show "the close rapprochement between physico-chemical and morphological points of view" and to assign chapters to those authors working directly in the laboratory at the forefront of their specialties, thus emphasizing current work and unsolved problems. While the volume was meant as a textbook (for advanced students and investigators), it was not merely a compendium of "received wisdom," but also pointed to directions for future research.

In this paper, I take advantage of Morgan's presentations about chromosomes in the final chapter to examine three issues of importance to the history of genetics in this period: (1) The serious objections that still lingered regarding the MCPH within the biological community. (2) Morgan's style of presentation—his "persuasive practice," as historian Kevin Amidon has termed it—in making a convincing case for the MCPH to cytologists and geneticists alike. Although over the years I have read virtually all of Morgan's writings, this chapter provides a new insight into his rhetorical skills and his method of presentation that balances arguments for his own views along with all the possible objections that had been or were being raised. (3) An interesting third issue becomes apparent if we consider the penultimate chapter by Clarence E. McClung in conjunction with Morgan's. Since McClung was trained as a classical cytologist, and Morgan as an embryologist-turned-geneticist, these two chapters provide a glimpse into the budding field of cytogenetics in the years just before the field emerged dramatically in the late 1920s and 1930s and ended nearly all controversy about the MCPH.

Background: The Chromosome Theory of Heredity

What was known as the "chromosome theory of heredity" had a long history stretching back at least to the 1860s. Chromosomes were observed to condense from an amorphous mass of "nuclein" material as long, rod-shaped bodies in cells preparing to divide. In many cases chromosomes seemed to have a regular structure, so that within a given cell they could be recognized by size and shape. The number of chromosomes in cell preparations was sometimes difficult to determine, and it was not clear, even by the turn of the twentieth century, whether the chromosome number was constant for all members of a given species, or even for different tissues within a single individual. While the hypothesis had been put forward that differentiation might involve selective loss of individual chromosomes during embryogenesis, there was little evidence to support it. And by 1924 it had become clear that, in virtually all cases examined to date, all cells in the body of an individual contained the full complement of chromosomes, and that this was constant (in most cases) for the species as a whole.[2] A particularly puzzling observation, however, was that when cells were not dividing, the chromosomes seemed to lose their rod-shaped structure and revert to the thread-like mass.

Chromosomes in any dividing cell were observed to exist in morphologically similar pairs. The complex movements of chromosomes in both mitosis (where the chromosomes are replicated prior to division of the cell into two identical daughter cells) and meiosis (in gamete formation where, through two separate divisions, the number of chromosomes is reduced in half) were well described in both plants and animals by the 1880s, though their significance remained controversial. By 1883 German zoologist August Weismann (1834–1914) reasoned that meiosis was a logical necessity to insure that the number of chromosomes remained constant in every generation. The highly regular movements of the chromosomes (they were sometimes referred to metaphorically as "dancing" through their mitotic and meiotic stages) suggested they might have something to do with heredity. But, it was only after the work of cytologist-embryologist Theodor Boveri (1862–1915) had shown that fertilized egg fragments lacking certain chromosomes developed abnormally that the association with heredity became well established (Laubichler and Davidson 2008; Baltzer 1967, 83–89).

The state of knowledge about the chromosome theory in the early 1920s was summarized for the Cowdry volume by Clarence Erwin McClung (1870–1946) in chapter 10. A distinguished cytologist at the University of Pennsylvania from 1912 onward, McClung was one of the early advocates of the view that the accessory or supernumerary chromosome (now called the X chromosome) might have something to do with sex determination (McClung 1902). At the exact same time (also in 1902), one of E. B. Wilson's students, Walter S. Sutton (1877–1916) pointed out the striking parallel between the cytological observations of separation of chromosomes at anaphase of meiosis I and Mendel's hypothesized segregation of factors in gamete formation (Sutton 1902). In some quarters by the very early 1900s, there was a growing realization that, although they were then separate fields, Mendelism and cytology might have more to say to each other than had previously been thought. It was into this environment that Morgan stumbled when he found his white-eyed male *Drosophila* in 1910.

McClung's chapter ("The Chromosome Theory of Heredity") in the Cowdry volume presents a summary of the cytological work on chromosomes from the 1870s onward. McClung notes that the term *heredity* has been very loosely and poorly defined by biologists over the ages. It has been referred to, he states, "as a 'law,' 'rule,' 'force,' 'material contribution,' 'act,' 'relation,' 'process,' 'fact,' 'principle,' 'link' and 'organization.' " "Little wonder," he notes, "that discussions of the subject are so lacking in clearness and precision when the central conception is so poorly defined" (McClung 1924, 613). After offering his own definitions, McClung devotes much of the rest of the chapter to the various topics regarding chromosome structure and organization. Among the most important of these topics are the details of meiosis and its differences from mitosis, the significance of variations in chromosome numbers between and within species (for example polyploidy), the physical structure of chromosomes, and the relationship between nucleus and cytoplasm.

There was, however, much more to learn about the physical structure and chemical nature of the chromosomes, as well as about their functions and behavior. For example, What was the structural relationship between the chromatin (highly-staining part of the chromosome) and the other structural elements? Were the latter largely a scaffolding to which the chromatin (hereditary material?) was attached or was the chromosome as a whole the actual hereditary material? What was the chemical composition of the chromosome as a whole, and the chromatin in particular? How did the chromatin condense and then dissolve between mitotic (or meiotic) cycles? How were chromosomes attached to the spindle fibers, and did they move of their own accord during anaphase or were they pulled by the spindle fibers? How did they line up so perfectly as paired homologs during prophase of meiosis? Particularly important, what was the functional relationship between the chromosomes and the cytoplasm, and how did chromosomes direct and participate in development and differentiation during embryogenesis? Yet despite the gaps in current knowledge, McClung could conclude his chapter by claiming, "The chromosome theory as it stands is logical, consistent, and generally applicable to both plants and animals. Admittedly incomplete, it yet stands as one of the highest achievements in biology and offers the most promising guide to further advances" (McClung 1924, 681). To elaborate on this fundamental concept and relate it to the observations on plant and animal breeding was the purpose of T. H. Morgan's chapter, which follows McClung's and concludes the Cowdry volume.

T. H. Morgan: Mendelian Heredity in Relation to Cytology

Thomas Hunt Morgan grew up in Lexington, Kentucky, receiving his bachelor's degree from the State University of Kentucky (now the University of Kentucky) in 1886 and his PhD from the Johns Hopkins University in 1891, working under William Keith Brooks (1848–1908). Although trained as a classical morphologist, attempting to deduce the phylogenetic origin of the Pycnogonids (sea spiders), Morgan early on fell under the spell of the Naples Zoological Station and the work of Wilhelm Roux and Hans Driesch in experimental embryology. By the mid-1890s he had become a vociferous exponent of experimentation in biology as a counter to unbridled, and untestable, speculation (especially the kind of phylogenetic hypotheses so common in morphological work at the time). After a dozen years on the faculty of Bryn Mawr College (1891–1904), he was hired by his friend and colleague Edmund Beecher Wilson to join the Zoology Department at Columbia University, where he remained for twenty-four years. In 1928 he was persuaded to become the first director of the newly founded Division of Biology at the California Institute of Technology in Pasadena, a position he held until his retirement and death in 1945. It was during his first few years at Columbia that Morgan began his work with *Drosophila melanogaster*, the vinegar, or fruit fly, which converted him from skeptic to ardent advocate of the Mendelian-chromosome paradigm.

In the introduction to his chapter for the Cowdry volume, Morgan reflects on the changes since those early days of fruit fly studies and explains that

"The extraordinary advance in our knowledge relating to the germ cells that took place during the last quarter of the last century prepared the way for a cytological interpretation of Mendel's principles immediately after their recognition in 1900. Within two years after the rediscovery of Mendel's paper in 1900 the application of the results of cytology to Mendel's laws was postulated (Sutton 1902). Since that time the significance of this relationship has become more and more apparent with every new advance, both in genetics and in cytology."

He goes on to say, "The behavior of the chromosomes in the maturation of the egg and spermatozoon not only furnishes an exact parallel to the genetic behavior of the postulated Mendelian factors, but crucial situations have also been met with (such as nondisjunction, the elimination or addition of specific chromosomes), that have furnished very strong confirmation of the view that the chromosomes are the bearers of the genetic elements. Other genetic phenomena such as linkage and crossing-over, that were unknown to Mendel, have also been brought into the relationship with chromosome behavior" (Morgan 1924, 693).

The structure of Morgan's paper is, in fact, aimed at presenting the evidence from both breeding results and cytology as "one long argument" (in the same way that Darwin saw his own work) supporting his interpretation that Mendelian genes are physical entities arranged in a linear pattern on the chromosomes. Morgan's approach is a superb example of what has been called "persuasive practice" in scientific writing. Historian Kevin Amidon has examined the works of a number of German biologists in the late nineteenth and early twentieth centuries (e.g., Ernst Haeckel, Rudolf Virchow, Adolph Meyer-Abich) to analyze the ways they use scientific writing skills to *persuade* their readers of the merits of their various arguments (Amidon 2008). I think this approach helps inform our understanding of what Morgan was undertaking in his article in the Cowdry volume (and now that I look at it, in many of his other more general and public writings as well).

In contrast to several of the figures Amidon examines, however, Morgan's approach is overtly nonpolemical, presenting evidence indicating the weak points or problems with his views. His approach, very much like Darwin's, as I suggest later, is to stress repeatedly how each kind of observation, though sometimes problematic in its own right, *could make sense only by admitting the validity of the central claim: the linear arrangement of genes on the chromosomes.* He uses a kind of *consilience* argument, by showing that putting all the varied lines of evidence together makes the conclusion almost inescapable.

Before examining Morgan's argument more closely, it is necessary to ask a more contextual question: What was the level of acceptance of the MCPH idea in 1924? Was Morgan overly sensitive, or were there serious doubts still lingering about the reliability, or even reality, of the whole cytological interpretation (including mapping) enterprise? In a nutshell, the answer appears to be that while many biologists, especially in the United States, had accepted the MCPH by 1920 (Brush 2002), there were still a number of formidable critics whose arguments, if accepted, would be devastating to the whole edifice of the MCPH, including claims about the linear array of genes on the chromosome and the mapping procedure. I want to suggest that Morgan was using the vantage point of the Cowdry volume to offer a focused, persuasive account of all the evidence at the time that supported the Mendelian-chromosome theory, and to counter arguments that were still being raised against it. In a 1917 review Morgan had already summarized some of those arguments as follows:

> It has been said, for instance, that the factorial [Mendelian] interpretation is not physiological but only "static," whereas all really scientific interpretations are dynamic. It has been said that since the hypothesis does not deal with known chemical substances, it has no future before it, that it is merely a kind of symbolism. It has been said that it is not a real scientific hypothesis, for it merely restates its facts as factors, and then by juggling with numbers pretends that it has explained something. It has been said that the organism is a whole, and to treat it as made up of little pieces, is to miss the entire problem of "organization." It has been seriously argued that Mendelian phenomena are "unnatural," and that they have nothing to do with the normal process of heredity and evolution as exhibited by the bones of defunct mammals. It has been said that the hypothesis rests on discontinuous variation of characters, which does not exist. It is objected that the hypothesis assumes that genetic factors are fixed and stable in the same sense that atoms are stable and that even a slight familiarity with living things shows that no such hard and fast lines exist in the organic world. (Morgan 1917, 513–14)

Ironically, a number of those objections are ones that Morgan himself had advanced about both the Mendelian and chromosome theories in the period before 1910! But he had come to embrace both theories shortly after initiating his work on inheritance in *Drosophila melanogaster* in 1910–1911. By 1924 other objections had emerged as well, so that these needed to be addressed in order to establish the MCPH as the major (and only) concept of inheritance that was consistent with all the evidence at hand.

Critics of the MCPH, 1910–1924

Who were the most influential critics/skeptics by the early 1920s, and what were their objections to the MCPH as advanced by the Morgan group? Among the most skeptical and influential critics of the MCPH at the time were William Bateson (1861–1926), Wilhelm Johannsen (1857–1927), William Ernest Castle (1867–1962), and Richard Goldschmidt (1878–1958). Since the details of the objections launched by each of these individuals have been discussed in depth elsewhere, they can be quickly summarized here (citations for individual critics are given below).

Bateson was of course, the person most responsible for initially making Mendel's work known to the English-speaking world, and one of the staunchest supporters of the new science (even coining the term *genetics* in 1906). His specific criticism of the MCPH was that it was a "complex web of theory . . . so exceedingly elastic that it can be fitted to any facts" (Bateson 1916, cited in Brush 2002, 34). Bateson's long-standing suspicion of assigning a material reality to Mendelian genes goes back to his early skepticism about the quasi-materialist theories of heredity put forward in the late nineteenth century by Ernst Haeckel (1834–1919), August Weismann (1834–1914), and many others. These theories were materialist only in the sense that they tried to ground the hereditary process in hypothetical, but supposedly real, material particles, which Weismann, an expert cytologist, assumed were contained in the chromosomes (Churchill 2015, 422–23). Bateson found such speculations philosophically unacceptable; it was crude materialism and offended his British empiricist predilection. Historian William Coleman has also attributed Bateson's skepticism about chromosomes to a conservative, intuitionist penchant for philosophical idealism that also made him skeptical of other hard-core materialist theories in science (Coleman 1970). Despite his major public concession after visiting the Morgan lab in December, 1921, Bateson still held reservations. He died, however, in 1926 without ever coming to firm grips with the MCPH.

Wilhelm Johannsen in Denmark was a pharmacist's apprentice when he took a job at the Carlsberg (brewery) Laboratories and began to learn about plant breeding (varieties of cereals and hops were used in brewing). Later, at the Copenhagen Agricultural college and finally at the University of Copenhagen, he emerged as one of the most significant biomathematical students of heredity and selection at the time. His famous "pure-line" experiments in 1900–1901, and his genotype-phenotype distinction in 1911 were among the most important theoretical contributions to the early development of genetics. Although he was a Mendelian (he thought Mendel's system was notational, as in chemical notation), he was initially skeptical, like Bateson, of tying Mendel's "factors" to chromosomes. From a mathematical perspective, he appreciated the power of Mendel's hypothesis to predict, but thought it unnecessary, even scientifically unsound, to provide a direct mechanism that seemed to have no basis in reality (Churchill 1974; Roll-Hansen 1978).

Indeed, Johannsen argued that the genotype functioned "as a whole": "I am unable to see any reason for localizing the 'factors of heredity' . . . in the nuclei. The organism is in its totality penetrated and stamped by its genotype constitution" (Johannsen 1911, 154; quoted in Falk 2009, 72). But after his compatriot Otto Winge (1886–1964) invited Norwegian Otto Louis Mohr (1886–1967), one of Morgan's first European postdoctoral students (1919–1920) to Copenhagen to discuss Morgan's work, Johannsen did eventually come around to accepting the basic tenets of the MCPH (Allen 1978, 280). But that was not until the third edition of his *Elemente der exacten Erblichtslehre* in 1926, still several years in the future when Morgan was composing the Cowdry chapter.

William Earnest Castle was a distinguished mammalian geneticist at Harvard (both Sewall Wright and Leslie C. Dunn were his students) who had a proclivity for getting involved in controversies from which he ultimately had to capitulate. Two such controversies claimed his attention in the period just prior to Morgan's writing the Cowdry chapter. The first was his claim that selection in a given direction could alter the gene for a trait in the direction of selection, thus attacking the very basis of the Mendelian principle of "purity of the gametes" (i.e., the claim that genes remain constant as they are shuffled from generation to generation, and are not "contaminated" or changed by other alleles with which they come to reside). Both Sturtevant and Muller took Castle to task on this point, arguing that selection only accumulated a greater or lesser number of "modifying factors"—that is, subsidiary genes that altered the expression of a major gene (for example a color gene), increasing or decreasing the intensity of expression in a quantitative way. Castle eventually capitulated and accepted the concept of "modifiers," which by the 1920s had become a major component of the MCPH.

No sooner had Castle extricated himself from this controversy than he took aim at the most central feature of the breeding work of the Morgan group: the linear arrangement of genes on the chromosomes and the very method of mapping itself. Although he accepted the association of genes and chromosomes, as well as the recombination frequencies published by the Morgan group, he argued that the claim that "the arrangement of the genes within a linkage group is strictly linear seems for a variety of reasons doubtful. It is doubtful, for example, whether an elaborate organic molecule ever has a simple string-like form" (Castle 1919b, 501). Citing evidence that the distances calculated for genes further apart are less than the summation of short distances between them, Castle argued that a three-dimensional structure of the chromosome gives results more consistent with the breeding data. According to Castle, "To account for this discrepancy Morgan has adopted certain subsidiary hypotheses, of 'interference,' 'double crossing over,' etc." (501). These seemed to be ad hoc hypotheses with no independent lines of evidence supporting them. Castle then introduced his own solution, a three-dimensional model of the chromosome, which he actually constructed with wires: the length of each wire representing the crossover value between two genes (see fig. 8.1*A*; compare to the Morgan group's construction in fig. 8.1*B*).

The complexity of this model quickly earned it the name of the "rat-trap model" by the Morgan group. Muller again took on Castle's claim, arguing, among other things, that double crossovers would be expected as much as single crossovers for genes lying at some distance from each other, and there should be no reason to assume otherwise. That the double crossover values, when calculated, turned out to give the precise distances that would be expected, indicated this was no ad hoc hypothesis, but a clear and reasonable inference to the best explanation from the data.

Marion Vorms has argued that Castle's argument was based on his commitment to a more physiological than a structural explanation, and to a strong preference to stay as close to the data as possible (Vorms 2013). In that light, his objections do not seem quite so idiosyncratic as previous authors, beginning with Muller and Sturtevant, have claimed. However, it became clear that Castle misunderstood some aspects of the linear model, and that his own model was in some ways more complex than Morgan's. Eventually, after trying to fit data from another lab for eight genes linked on chromosome III, his "rat-trap" model had to be flattened out to the point that it resembled the linear maps of the Morgan group. As Castle had to admit by 1920, "Obviously the arrangement approaches the linear. . . . But if we grant that the arrangement is in any sense linear, then it must be granted also that double and triple crossing-over are likely to occur"

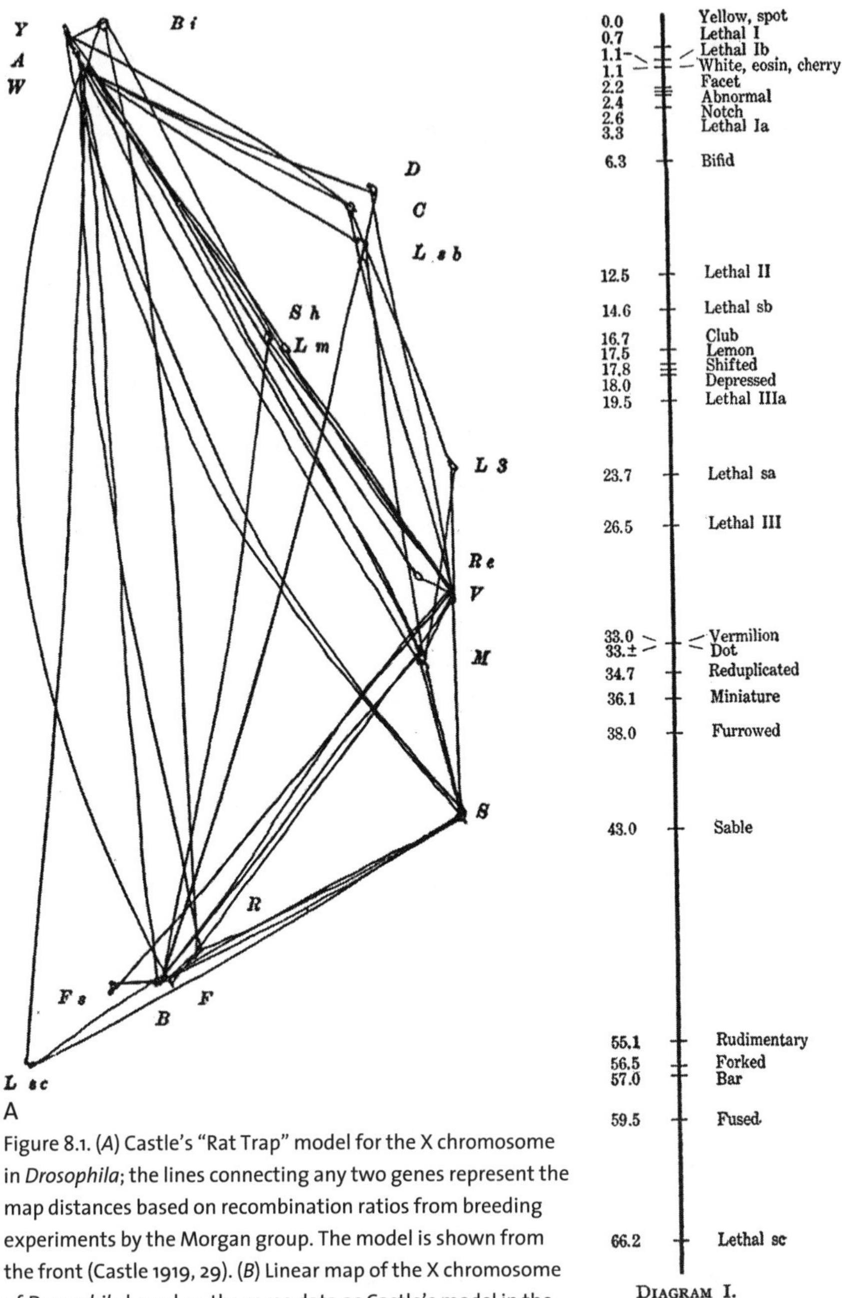

Figure 8.1. (*A*) Castle's "Rat Trap" model for the X chromosome in *Drosophila*; the lines connecting any two genes represent the map distances based on recombination ratios from breeding experiments by the Morgan group. The model is shown from the front (Castle 1919, 29). (*B*) Linear map of the X chromosome of *Drosophila* based on the same data as Castle's model in the previous figure. The Morgan group's procedure involved the assumption of double or triple-cross-overs to calculate their distances, which Castle avoided by mapping only two genes at a time. From Morgan and Bridges 1916, 27.

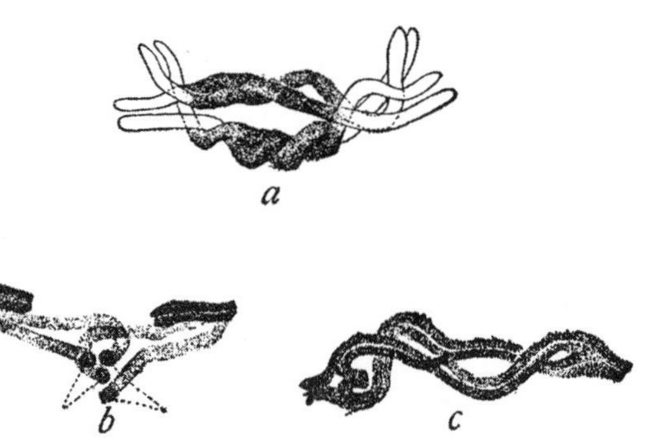

Figure 8.2. F. A. Janssens's drawings of chiasmata, showing intertwining of homologous chromatids and possible crossing-over during prophase of meiosis I. From Morgan 1924, fig. 4, 707; after Janssens.

(Castle 1920). Although Castle backed down, the controversy was fresh in the minds of geneticists and cytologists in the period in which Morgan was writing his chapter for the Cowdry volume.

A final critic whose views had to be taken seriously was Richard Goldschmidt, who, in 1917, had just been appointed to the prestigious position as head of the Genetics Department at the newly organized Kaiser-Wilhelm Institute for Biology in Berlin (Dahlem). Soon after he had assumed his new post, Goldschmidt launched one of his many attacks on the chromosome theory. He called into question a fundamental assumption of the mapping procedure: the theory of "chiasmatype," or crossing-over and exchange of parts between homologous chromatids,[3] as a means for generating maps based on recombination frequencies among offspring in breeding experiments. First, he argued that there was no independent evidence that breakage and rejoining of homologous chromatids occurred during meiosis, other than that it was consistent with Janssens's drawings of chiasmata, as shown in figure 8.2 (Richmond and Dietrich 2002). Like Castle, Goldschmidt thought this was just another ad hoc hypothesis.

He further questioned the prevailing assumption that chromosomes retained their same linear structure after they appeared to "dissolve" (i.e., became threadlike and visually indistinct) during interphase of the cell cycle. Goldschmidt argued that when the genes reassembled in the next cell cycle, they might be brought back into their proper "places" (loci) by a special cohesive force, and thus account for linkage without having to postulate any physical connection (Goldschmidt 1917). However, the order could sometimes be altered simply by chance (like typographical errors) and thus give the impression of crossing-over. Since visually there was no way to tell if the chromatin retained its physical integrity when it recondensed into rod-shaped chromosomes, Goldschmidt thought his "force" theory was just as reasonable as the chiasmatype theory to explain both linkage and recombination without what he considered the dubious assumption of breakage and recombination. This claim was also consistent with Goldschmidt's view that the chromosome as a whole was the unit of heredity, and not individual, discrete "genes" (Dietrich 1995; Allen 1974).

We can summarize briefly the major criticisms of the Morgan group's use of recombinant data for characters in breeding experiments with cytological data about chromosome behavior to construct genetic maps. (1) The method requires too many ad hoc hypotheses, such as double crossing-over, breakage of homologous chromatids, and exchange of parts. (2) Linearity of the chromosome is assumed from the start when other geometrical arrangements can account for the distance estimates between genes. (3) There is no cytological (visible), independent evidence that crossing-over actually occurs other than Janssen's drawings. Thus, the whole attempt to correlate breeding data and cytology is only an assumed relationship, the very point the experimental work is supposed to demonstrate.

Morgan's Review: The Evidence Summarized

It seems to have been with these (and other) critiques in mind that Morgan begins his summary of the evidence supporting the MCPH. His treatment is in many ways similar to the approach Darwin used in *On the Origin of Species*: Whewellian consilience, presenting numerous, often totally independent lines of evidence, all of which, taken both individually and collectively, support the interpretation being presented (Reznick 2010, 14–16; Browne 1995, 437–39; Mayr 1991, 9–10; Ruse 1975). The lines of evidence Morgan used fall into two large and independent categories: breeding results and cytological observations—what Lindley Darden and Nancy Maull long ago referred to as an "interfield theory" (Darden and Maull 1977). Along the way, Morgan, like Darwin, presents opposing interpretations, such as those by Bateson, Castle, and Goldschmidt (see Morgan, 1924, 711–12) and proceeds to show how each has less consistency and/or explanatory power than his own view. Like Darwin, Morgan also freely admits areas where knowledge is circumstantial, contradictory, or lacking altogether. The overall effect leads the reader to conclude that, while the presentation is advocating a particular interpretation (point of view), it is being done in a fair-minded, critical, and rigorous way. The thoroughness with which Morgan covers all the angles of interpretation adds to the ultimate credibility of his argument.[4]

The chapter is divided into fourteen sections, each devoted to a particular set of phenomena that bear on the relationship between genes and chromosomes. In section 1 Morgan explains the basics of Mendelian heredity, including Mendel's two "laws" (Morgan's term), dominance, recessiveness, dihybrid crosses, and the like, using examples from his own lab's work on *Drosophila*. Morgan turns in section 2 to the evidence for linkage, which provides the basis for the exceptions to Mendel's second rule (independent assortment). He illustrates this with a dihybrid cross between vestigial/yellow-body and normal wing/black-body *Drosophila*, which show 83% linkage and 17% recombination between the two parental traits. He then proceeds to point out that the number of linkage groups obtained by breeding analysis equals the number of chromosome pairs observed by cytology: that is, four in each case. For Morgan, these observations explain both Mendel's second rule (when genes are on separate chromosomes), and also the cases of linkage (when they are on the same chromosome): "This interpretation . . . would in itself furnish a strong argument in favor of the view that the phenomenon of linkage is due to genes being carried in the chromosomes" (Morgan 1924, 697–98). Justifying the attempt to relate the Mendelian breeding results to the behavior of the chromosomes, Morgan and his coauthors wrote in the 1915 introduction to *The Mechanism of Mendelian Heredity*, "Why, then, we are often asked, do you drag in the chromosomes. Our answer is that since the chromosomes furnish exactly the kind of mechanism that the Mendelian laws call for, . . . it would be folly to close one's eyes to so patent a relation." More tellingly, they explained that their broader concerns were "as biologists, we are interested in heredity not primarily as a mathematical formulation, but rather as a problem concerning the cell, the egg and the sperm" (Morgan et al. 1915, viii–ix). Morgan always retained this basic biological focus, even as he and his group concentrated more and more in the ensuing decades on their basic methodology of breeding, cytological correlation, and mapping. As philosopher of biology Kenneth Waters (Waters 2004) has noted, the Morgan group always used its breeding and cytological experiments to address wider problems in biology, particularly cell biology (for example, mechanisms of crossing-over, the physiological function of genes, and evolution). As important as the linkage maps were to establishing the material reality of the MCPH, the Morgan group's aim was never merely to construct maps to ever finer degrees of resolution. Broader biological processes were always the underpinning of the genetics itself.

Sections 3 and 4 of Morgan's chapter deal with issues of crossing-over and double crossing-over. Here Morgan reviews the evidence that crossing-over is the most logical and parsimonious interpretation of the recombination ratios seen in the breeding results. He goes on to discuss the hypotheses of double and triple crossovers. Echoing Muller's earlier argument in response to Castle's "rat trap" model, Morgan points out that it would be illogical to limit crossing-over to one event per chromosome pair, especially in cases where two loci are quite far apart. Double crossing-over accords well with the empirical observation that the sums of distances between two genes, *a-b*, and *b-c* is less than the additive distance between *a* and *c*. For example, in one cross in *Drosophila*, the distance between black body (*b*) and vestigial wing (*vg*) is 17%; however, the distances between *b* and an intermediate gene, cinnabar (*cn* for eye color) is 9%, and that between *cn* and *vg* is 9.5%. The total 18.5%, calculated only on the basis of the two individual crosses, gives an erroneous map distance, because it does not consider the frequency of double crossovers.

Thus, Morgan argues, multiple crossovers are not merely an ad hoc hypothesis, but their occurrence provides a rigorous and accurate explanation for the apparent contradiction in map distances when only two genes are considered, versus when a third gene between the first two is included. For genes even further apart, it is necessary to consider the possibility of triple crossovers, which would, in fact cancel out the effects of double crossovers. Thus, postulating multiple crossovers was not only logical but necessary to produce accurate genetic maps. Morgan was obviously concerned to show that the linear arrangement was supported by multiple lines of evidence, since it was the basis on which the whole mapping enterprise was founded.

Sections 5 and 6 deal with interference (in crossing-over), the problem of random or nonrandom chromosome assortment in gametogenesis (that is, do all maternal chromosomes go to one pole of the dividing cell and all paternal chromosomes go to the other pole, or are they randomly assorted so each pole gets a combination of maternal and paternal chromosomes?), and the relationship between number of linkage groups and number of chromosome pairs. Interference is the failure to observe second crossovers in the regions to one side or the other of where a first crossover has taken place, and appears to be due to mechanical constraints that make a second crossover in the area less likely to occur. This does not undermine

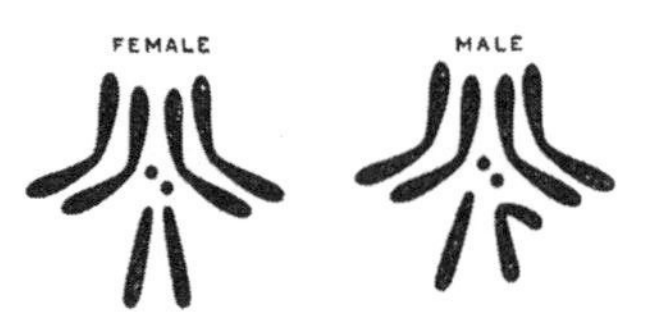

Figure 8.3. Chromosome pairs in *Drosophila melanogaster*. Drawing of actual chromosome complement, showing the four groups in a female (*left*) and male (*right*). The female's two X chromosomes contrast with the male's XY. Chromosome group IV is shown as the two small dots in the center. From Morgan, 1924, fig. 1, 704.

the theory of crossover but simply reflects the way the physical properties of chromosomes pose certain limitations to the process (it might be analogous to the physical problems encountered when you tie a knot in a rope and then try to tie another knot directly adjacent to it: the first structure inhibits, mechanically, the formation of the second). Morgan reviews the evidence originally supplied by the extraordinary cytological work of E. Eleanor Carothers (Carothers 1913), indicating that genes on different chromosomes do in fact assort independently, in accord with Mendel's second law; conversely, those on the same chromosome do not; rather, they show linkage and periodic recombination due to crossing-over. Morgan does not let it go unnoticed that all of this is consistent with the interpretation that genes are indeed physically parts of the chromosomes and arrayed in a linear order.

Bearing on this latter claim, perhaps the most convincing observation for Morgan is the correlation between number of linkage groups, determined by breeding results, and the number of pairs of chromosome revealed by cytology. In *Drosophila melanogaster*, for example, there are four linkage groups and four pairs of chromosomes (fig. 8.3); in *D. virilis* there are six linkage groups and six pairs of chromosomes, while in *D. willistoni* there are three linkage groups and three pairs of chromosomes. Moreover, the number of genes mapped to each chromosome is roughly proportional to the physical size of the chromosome itself. In *D. melanogaster*, the small fourth chromosome pair (shown as dots in the center of the left-hand diagram) had been found to have only three or four genes, compared to the dozens mapped to the three larger sets. There were a few exceptions to the correlation cited at the time—Lathyrus (sweet peas) and Pisum (peas)—in both of which the number of chromosome pairs appeared to be one less than the number of linkage groups; later, however, both were shown to have equal numbers of linkage groups and chromosome pairs, and thus were not exceptions (Stern 1928). It should be pointed out that in these early days of cytogenetics, one of the major problems was simply getting accurate (and consistent) counts of chromosome numbers, especially in organisms with a large diploid number. So far, then, cytology provided substantial evidence for the breeding results found in Mendelian crosses.

In the wider world of biologists in general, the relationship between number of linkage groups and chromosome groups also appears to have been the single most convincing piece of evidence that genes were really associated as physical parts of the chromosomes. Historian of science Stephen Brush carried out a thorough study of the relative importance of the various lines of evidence supporting the MCPH that were most (or least) convincing to British and American biologists (Brush 2002). At the top of the list was the correlation of number of linkage and chromosomal groups. The conclusion to the present chapter includes further discussion on the relative importance of the various lines of evidence in convincing biologists of the reality of the MCPH by the mid-1920s.

It is in section 6 that Morgan confronts the issue of whether the chromosomes maintain their linear arrangement of genes during interphase, when the chromosomes appear to dissolve into a mass of long, stringy fibers. As he notes, "It has been difficult to demonstrate that the chromosomes do remain intact during the resting stages [interphase] of the nucleus; while this assumption would offer the simplest interpretation of the reappearance of the same number of chromosomes, having the same shapes and sizes, at each mitosis, yet this might be due to other relations than that of continuity" (Morgan 1924, 703). This is not a trivial point, since the fundamental hypothesis of the MCPH—namely, the linear arrangements of genes on chromosomes—requires that the chromosomes retain the same overall structure and order of genes from one cell cycle to the next. The issue also bears on determining when in meiosis crossing-over takes place: during this resting or interphase stage or during early prophase.

It is at this point that Morgan bemoans the lack of support for the MCPH that the geneticist would have hoped cytology might provide. Unfortunately, he notes, the cytological evidence is, at best, only circumstantial: "Thus, because cytologists have not been able to prove beyond all question the 'individuality' of the chromosomes,[5] the geneticist is left without the support his evidence calls for." Continuing, he notes that cytologists have not even been able to determine if, or whether, interchange between homologous chromatids (through crossing-over) actually occurs during synapsis, as claimed originally by Janssens in 1909, or at some other period:

"Janssens has attempted to show such an interchange might take place, but his interpretation has been questioned by other cytologists. . . . It is a misfortune for genetics that at present students of the cell are not agreed as to any one period or method by which interchange between members of the pairs of chromosomes might occur" (Morgan 1924, 704). Ever the empiricist, a few pages later Morgan points out that " it would not be profitable to speculate further from . . . genetic evidence to the rather meager facts supplied by cytology" (715).

In the face of this impasse, Morgan takes a unique turn in his argumentation: "For the present, at least, we shall have to reverse the situation, and argue that since there is excellent evidence that the chromosomes carry the genes, the chromosomes must remain intact, except in so far as crossing-over takes place between homologous chromosomes" (Morgan 1924, 704). Since there was no independent cytological evidence bearing on these issues, Morgan argues, we are justified in assuming to be true what the theory demands! While this is not an unusual tactic in scientific argumentation (Darwin uses it constantly), it was the making of these kinds of assumptions that many critics of the MCPH (such as Goldschmidt) saw as major weak points in the whole argument.

Sections 7 and 8 review various theories, from Janssens and others, of the possible mechanisms of crossing-over and how and when it might occur. We need not go into the details of this discussion, except to reiterate that it was precisely in this area that Morgan felt geneticists most needed help from cytology, but that up to this point that had not been forthcoming. Moving on to sections 9 and 10, Morgan again reiterates the point that since the chromosomes are linear or threadlike in appearance, and since genetic maps constructed from breeding ratios are also linear, and gene loci can be mapped consistently, it seems clear that the MCPH is the best explanation available for the relations between genes and chromosomes. This is approximately the twentieth time he has reminded his readers of this point.

Morgan then moves on to discuss the problem of variable chromosome numbers when chromosomes are added to or deleted from the genome—what is known as aneuploidy, including haploidy, diploidy, tetraploidy, and so on. In particular, Calvin Bridges's discovery (1913) of nondisjunction, the failure of homologous chromosome pairs to segregate at anaphase I of meiosis, became another point of interface between genetics and cytology. In this case the relevance spoke to both the claim that genes are located on chromosomes and that the addition or deletion of chromosomes affects physiology and development. Bridges first encountered nondisjunctions in the failure of the two X chromosomes to segregate in a strain of white-eyed females during oögenesis. He used these flies to create a whole array of different sex-chromosome combinations, which, when bred with normal, red-eyed males, produced a variety of offspring whose phenotypic characters reflected their particular chromosomal constitution. For example, an offspring inheriting the two white-eye X chromosomes from the original nondisjunction female, and another X from the red-eyed male, would phenotypically be red-eyed and cytologically XXX. Nondisjunction in males could also be followed and used to create a further array of combinations when bred with normal or nondisjunction females. Strains that have lost one member of a chromosome pair, such as Haplo-IV (loss of one member of the small fourth chromosome pair), were also used in breeding experiments in which the original parents had been heterozygous for one or more genes on the small chromosome. If the Haplo-IV offspring had lost the member of the pair with the dominant gene(s), only the recessive gene(s) would be expressed phenotypically. All of these results, Morgan again reminds the reader, are consistent with, and therefore supportive of, the hypothesis that the genes are physically parts of the chromosomes.

The final two sections of the paper discuss genes in relation to the cytoplasm. Section 13 is concerned with the phenomenon known at the time as "cytoplasmic inheritance," while 14 deals with the mutual interactions between genes and the rest of the cell, especially the cytoplasm. Morgan dispenses quickly with the issue of cytoplasmic inheritance by noting that the only known case of such a phenomenon is that of chloroplasts (the self-replication of mitochondria was not well-recognized at the time, though it was clear that somehow they reappeared in roughly equal numbers in daughter cells after mitosis). At any event, Morgan points out, there is nothing in the self-replication of cytoplasmic components (organelles) that is in opposition to the MCPH. They are to be regarded simply as separate processes.

More important for Morgan to clear up is the issue of the relationship between the genes in the nucleus and the cytoplasm making up the rest of the cell, which is the subject of section 14. Here he employs two rhetorical strategies: (1) He clearly delineates the research questions that are directly related to genetics, as separate from those related to development; and (2) he emphasizes how little we know about the influence of genes on the cytoplasm or vice-versa. With regard to the first issue, Morgan states flatly that "genetics is not directly concerned with the question of the relation of the chromosomes to the cytoplasm—these phenomena belonging to the developmental aspect of biology" (Morgan 1924, 725); and, more explicitly a page later: "The genetic problems in a strict sense are concerned with the shuffling of the genes between generations; nevertheless, the genetic work has thrown some light on certain aspects of the relations of the genes to the cytoplasm, which is one of the problems of embryonic development" (726).

One of these problems is the claim that the cytoplasm determines the most important and general features of the organism: its phyla, class, family, and other higher-level phylogenetic characteristics, while the chromosomes only carry determiners for very specific, largely trivial, individual characters (this point had been argued by Theodor Boveri and Morgan's Woods Hole colleague and friend, Edwin Grant Conklin (1863–1952). Admitting that the most general features, say, of the vertebrate body plan (axial orientation, dorsal-ventral differentiation) are probably determined by the cytoplasm as it is divided up during early cleavage of the embryo, Morgan hastens to add that Mendelian genes impress upon this general *Bauplan* all the other major characters that arise during development. He discounts the claim that Mendelian traits are "trivial" by pointing out that the distinction between "trivial" and "nontrivial" characters is very fuzzy and subjective. Furthermore, since geneticists depend on mutations to follow the inheritance of traits, mutations in major features in early stages of development would likely result in death of the embryo and thus do not form material that can be studied easily from a genetic point of view.

With regard to the question of whether the cytoplasm should be considered as important as the chromosomes in determining hereditary characteristics, Morgan sees this as a meaningless issue: "It is sometimes said that the cytoplasm must be as important as the chromosomes, since no development is known except in the presence of the cytoplasm, and by its activity. Whether the cytoplasm or the chromosome is or is not equally 'important' is a matter that cannot be determined, and is of very little consequence. The statement is an example of obscurantism rather than of profundity" (Morgan 1924, 127). He then goes on to emphasize (once again) that the inheritance patterns of all the traits that have been studied to date can be accounted for by the behavior of the chromosomes. Moreover these traits are inherited in the same manner regardless of the hereditary background of the egg cytoplasm. This is demonstrated, for example, with reciprocal crosses between recessive and dominant traits, where in one case the egg comes from a female dominant for the trait (wild type eyes in *Drosophila*), and the sperm from a male with a recessive trait, such as pink eyes, or vice-versa. The same ratios in the offspring are found in either case. Thus, the determining factors for eye color must be related to the chromosomes, not the cytoplasm. Referring again to the parallelism between chromosome behavior and the outcome of breeding experiments, he points out, "*What genetics has so far discovered . . . is this: . . . All the examples of heredity . . . show that all adult characters . . . are accounted for by the known behavior of the chromosomes. In other words, they 'follow' the chromosomes regardless of the source from which the protoplasm comes*" (Morgan 1924, 727).

Of course, Morgan reiterates, the cytoplasm is important, but it is impossible to say how the chromosomes affect it, except that they *must*, through chemical influences impressed on the chromosomes by the cytoplasm or called forth by the cytoplasm. However, "These questions must be kept entirely free from predilections until we have found out more about the physiological processes in the chromosomes and in the cytoplasm. Whatever the future has in store, . . . the answer does not prejudice the present situation so far as the observed effects of the genes in heredity are concerned" (Morgan 1924, 728). As Morgan hinted, and as his Princeton colleague E. G. Conklin noted explicitly, the relationship between nucleus (chromosomes) and cytoplasm is reciprocal. The hereditary influences of the chromosomes impress themselves on the nucleus by their power to synthesize substances that become a part of the cytoplasm. Conversely, the cytoplasm appeared to affect chromosomal function by eliciting, in different tissues of the developing embryo, the expression of specific genes characteristic of that tissue. The cytoplasm is thus just as active an agent in the expression of phenotype as the genes on the chromosomes. As Conklin put it in 1908: "Neither the nucleus nor the cytoplasm can exist long independently of the other; differentiations are dependent upon the interaction of these two parts of the cell; the entire germ cell, and not merely the nucleus or cytoplasm, is transformed in the embryo or larva and it therefore seems necessary to conclude that both nucleus and cytoplasm are involved in the mechanism of heredity" (Conklin 1908, 93).

Conclusion

Morgan's paper provides a capsule-sized view of the status of the Mendelian-chromosome paradigm of heredity in the mid-1920s. At a time when we might have expected the whole idea of the linear arrangement of genes on chromosomes to be one of the most well-established paradigms of biology, there were still substantial questions and doubts about its overall validity and general applicability. From a survey of British and American biologists, journals, and textbooks, historian Stephen Brush concluded that the basic tenets of the MCPH were well, if not universally, accepted in the United States by 1925 and in Britain by 1930 (Brush 2002). The situation appears to have been different in parts of Europe: France and Germany in particular. In France, the strong, neo-Lamarckian tradition had made it difficult for the MCPH to gain a strong following, to say nothing of its association with chromosomes (Burian, Gayon, and Zallen 1988). In Germany, the lingering tradition of Haeckel, Weismann, and Nägeli, among others, who sought to treat heredity, development, and evolution as a unified theory, the MCPH by itself seemed too narrow (Harwood, 1997). For Morgan and other geneticists, however, uniting Mendelian breeding data with cytology was enough of a synthesis at the moment, though both development and evolution were never far from Morgan's concerns (Maienschein 2016). The great virtue of the MCPH was that it was a materialist conception; that is, it gave Mendel's abstract factors a concrete association with specific cell structures, and it led to predictions that could be tested experimentally. Morgan organized his argument in the Cowdry volume to this end, presenting evidence from every conceivable perspective to support his inference to the best (and simplest) explanation: the Mendelian-chromosome paradigm of heredity.

The argument that Morgan put forth in the Cowdry chapter involved six major lines of evidence. All support the interpretation that genes are arranged in a linear order as parts of chromosomes; Morgan is careful to point out that there is no necessary commitment to the physical reality of this model, but only that it is operationally consistent. (1) Sex-linked inheritance was apparent, where phenotypic traits followed the transmission of the sex-determining chromosome (in *Drosophila*, humans, and most other animals, the X chromosome). (2) The behavior of the chromosomes in anaphase of meiosis provides a concrete mechanism for Mendel's hypothesized process of segregation of paired alleles (the first rule, or law). (3) Researchers were able to construct maps of linked characters from recombination frequencies, whose linearity appeared to match the linear, rod-shaped structure of the chromosomes. (4) The number of linkage groups obtained by breeding data was correlated with the number of chromosome pairs observed cytologically for different species. (5) Nondisjunction and other cytologically visible chromosome alterations (such as deletions) corresponded with observed phenotypic variations. (6) Although it remained a hypothesis, the likelihood that homologous chromatids in chiasma formation, as observed by Janssens, could actually exchange parts provided a plausible mechanism for recombination ratios within a linkage group. Along the way Morgan adduced other lines of evidence, such as H. H. Plough's analysis of when in meiosis chiasmata might occur, but these were usually of less overall significance, though consistent with, the MCPH.

According to Stephen Brush, for general biologists item (3) seems to have been the most effective and easiest to comprehend without a considerable technical background (Brush 2002). Running a close second among nonspecialists, perhaps surprisingly, was the ability to construct genetic maps. As Rheinberger, Gaudlliere, and their colleagues have suggested, a "mapping culture" in general, not limited to biology, may have been so prevalent as to give maps a kind of familiarity and authority that other lines of evidence may have lacked (Rheinberger and Gaudlliere 2004). After all, if you can map something, it must in some sense be "real.". Numbers (4) and (5) appeared to be particularly convincing to geneticists who were familiar with the technicalities of the breeding as well as cytological methodology.

Although in 1924 Morgan was disappointed that the data from cytology was not clear enough to provide the *direct* and unequivocal support for the MCPH he would have liked, he did not have long to wait. In the period 1926–31 several important developments emerged from cytology that provided important corroboration for the genetic data. Between 1927 and 1929 Barbara McClintock, working in R. A. Emerson's group at Cornell, developed several new cytological procedures, including modification of existing aceto-carmine staining techniques and chromatin stains, and began to elucidate the visible structure of the ten chromosome pairs of corn (*Zea mays* or maize). Because the maize chromosomes, like those of most plants, are larger than those of animals, their fine structure became visible cytologically as early as the mid-1920s (Kass and Bonneuil 2004). Particularly through McClintock's work, by 1929 the Emerson group had made considerable progress in identifying the structure and linkage groups in maize. In a manner similar to that of the Morgan group with *Drosophila*, the maize geneticists were developing methods for correlating genetic maps with the cytological structure of the chromosomes.

On the other hand, geneticists studying *Drosophila* and other animals were hampered cytologically by the small size of animal chromosomes and the lack of visible detail of their morphology. All this was dramatically changed through discovery of the giant salivary gland chromosomes. In 1930 Theophilus S. Painter (1889–1969), then at the University of Texas, discovered the large salivary gland chromosomes in *Drosophila* larvae, which provided a wealth of visible structure that allowed geneticists to correlate cytological with genetic maps (fig. 8.4). The detailed banding patterns that were revealed in the giant salivary chromosomes made it possible to observe inversions, translocations, and other chromosomal structural variations that could then be related to alterations in expected phenotypic ratios

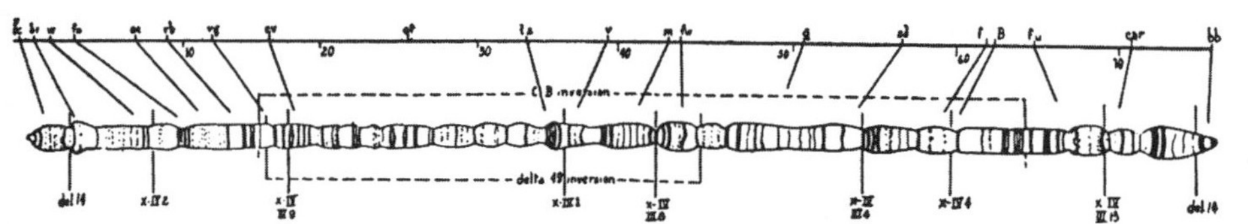

Figure 8.4. Correlation of genetic and cytological maps, made possible by the discovery of the giant salivary gland chromosomes of *Drosophila*, which reveal considerable fine structure of the chromosome. The genetic map at the top was constructed from recombination frequencies in breeding experiments, and thus represents only relative distances of genes from one another. The chromosome map at the bottom was constructed from cytological examination of the large salivary gland chromosomes of *Drosophila* larvae. The correlation of loci on the genetic map to specific bands on the cytological map was determined by using deletions and other visible chromosomal alterations that affected specific phenotypic ratios in offspring. From T. S. Painter, "A New Method for the Study of Chromosome Rearrangements and Plotting of Chromosome Maps," *Science* 78 (1931): 585–86. Reprinted with permission from AAAS.

of offspring. For example, inversions that were now visible in detail were observed to prevent crossing-over in the inverted region, thus altering map distances for genes in that region of the chromosome. The cytological observations for these types of chromosomal alterations provided the physical evidence that was needed to correlate cytology with altered phenotypic ratios observed through breeding experiments.

The other major development between 1930 and 1931 was the cytological proof for crossing-over, provided for maize by Barbara McClintock and Harriet Creighton (Creighton and McClintock 1931) and, just a few weeks later, for *Drosophila* by Curt Stern (Stern 1931). In both cases, identification of visible chromosome markers (knobs in the case of maize) were shown to have changed positions between homologous chromatids during meiosis. This was about as direct proof as one could get for one of the major mechanisms postulated by the MCPH and at the heart of the mapping process. Although many biologists were already convinced of the basic tenets of the MCPH before the appearance of these lines of evidence, the new cytological work put the finishing touches on an already well-supported and elegant paradigm.

Looking back, it seems obvious that Morgan may have had no need to organize his chapter in the Cowdry volume as a persuasive document—he was, as it seems, preaching largely to the converted. Yet, the various criticisms that the MCPH had encountered in the preceding decade, and the authority of some of his critics, may have suggested to him that a wholesale review of all the evidence was necessary, at least in a volume directed primarily to cytologists and general cell biologists. I suspect that his approach may have led to an increased interest among cytologists in providing the more direct, visual evidence to support the genetic hypothesis that had been lacking up to that time.

Acknowledgments

I would especially like to thank Jane Maienschein and Karl Matlin for organizing the two conferences on the history of cell biology, and for some helpful suggestions on the structure and execution of this chapter. Two anonymous reviews also provided useful insights and suggested improvements, for which I am grateful.

Notes

1 I use the term *paradigm* here in Thomas Kuhn's sense of a collection of theories, methodologies, assumptions, and a community of investigators that is more inclusive than the term *theory* connotes. The same inclusiveness is conveyed by Imre Lakatos's idea of a "research program," but I have found Kuhn's terminology more generally applicable.

2 While aneuploidy (duplication of one or several chromosomes among the normal number of pairs), or polyploidy (duplication of the full set of chromosomes) was recognized, especially in plants, it was seen as an exception to the more general rule of constancy of chromosome number.

3 *Chromatid* refers to the replicated chromosomes that are still attached by the centromere as a tetrad (four strands) prior to separation at anaphase of meiosis I. It is the homologous chromatids that intertwine and exchange parts at synapsis that provide the basis for crossover analysis.

4 I am particularly grateful to my colleague Carl Craver for his insights in helping me to understand, from a philosophical perspective, the importance of Whewell's and also John Herschel's views on consilience and other forms of persuasive argument.

5 Morgan's terminology here is a bit confusing, since "individuality of the chromosomes" is usually associated with the work of Theodor Boveri (1862–1915), who carried out a series of experiments, published in 1902, showing that each pair of chromosomes differed in its effects on development from the other sets (that is, each chromosome pair was individually different from the others in the hereditary factors it contained). In the passage quoted above, however, Morgan is referring to the question of whether each chromosome retains its integrity after dissolving into chromatin threads during interphase (see Laubichler and Davidson 2008).

References

Allen, Garland E. 1974. "Opposition to the Mendelian-Chromosome Theory: The Physiological and Developmental Genetics of Richard Goldschmidt." *Journal of History of Biology* 7 (1): 49–92.

———. 1978. *Thomas Hunt Morgan: The Man and His Science*. Princeton, NJ: Princeton University Press.

Amidon, Kevin R. 2008. "The Visible Hand and the New American Biology: Toward an Integrated Historiography of Railroad-Supported Agricultural Research. *Agricultural History* 82, (3): 309–36.

Baltzer, Fritz. 1967. *Theodor Boveri: The Life of a Great Biologist*. Translated by Dorothea Rudnick. Berkeley: University of California Press.
Bateson, William. 1916. "Review of *Mechanism of Mendelian Heredity." Science* 44:536–43.
Browne, Janet. 1995. *Charles Darwin: Voyaging*. New York: Alfred A. Knopf.
Brush, Stephen. 2002. "How Theories Become Knowledge: Morgan's Chromosome Theory of Heredity in America and Britain." *Journal of the History of Biology* 35 (3):471–535.
Burian, Richard M., Jean Gayon, and Doris Zallen. 1988. "The Singular Fate of Genetics in the History of French Biology, 1900–1940," *Journal of the History of Biology* 21 (3): 357–402.
Castle, William E. 1919a. "Is the Arrangement of the Genes in the Chromosome Linear?" *Proceedings of the National Academy of Sciences* 5:25–32.
———. 1919b. "Are Genes Linear or Non-Linear in Arrangement?" *Proceedings of the National Academy of Sciences* 5:500–506.
———. 1920. "Model of the Linkage System of Eleven Second Chromosome Genes of *Drosophila*." *Proceedings of the National Academy of Sciences* 6:73–77.
Carothers, E. Eleanor. 1913." The Mendelian Ratio in Relation to Certain Orthopteran Chromosomes." *Journal of Morphology* 24:487–511.
Churchill, Frederick B. 1974. "Wilhelm Johannsen and the Genotype Concept." *Journal of the History of Biology* 7 (1) :5–30.
———. 2015. *August Weismann: Heredity, Development and Evolution*. Cambridge, MA: Harvard University Press.
Coleman, William. 1970. "Bateson and Chromosomes: Conservative Thought in Science." *Centaurus* 15 (3–4): 228–314.
Conklin, Edwin G. 1908. "The Mechanism of Heredity." *Science* 27:89–99.
Cowdry, Edmund V., ed. 1924. *General Cytology: A Textbook of Cellular Structure and Function for Students of Biology and Medicine*. Chicago: University of Chicago Press.
Creighton, H. B., and McClintock, B. 1931. "A Correlation of Cytological and Genetical Crossing-over in *Zea mays*." *Proceedings of the National Academy of Sciences* 17:492–97.
Darden, Lindley, and Nancy Maull. 1977. "Interfield Theories." *Philosophy of Science* 44:43–64.
Dietrich, Michael. 1995. "Richard Goldschmidt's 'Heresies' and the Evolutionary Synthesis." *Journal of the History of Biology* 28 (3): 431–61.
Falk, Raphael. 2009. *Genetic Analysis: A History of Genetic Thinking*. Cambridge: Cambridge University Press.
Goldschmidt, Richard. 1917. "Crossing-Over ohne Chiasmatypie?" *Genetics* 2:82–95.
Harwood, J. 1997. "The Reception of Genetic Theory among Academic Plant Breeders in Germany, 1900–1930." *Sveriges Utsädesförenings Tidskift* 107:187–95.
Johannsen, Wilhelm. 1911. "The Genotype Conception of Heredity." *American Naturalist* 45:125–45.
Laublichler, Manfred, and Eric Davidson. 2008. "Boveri's Long Experiment: Sea Urchin Merogones and the Establishment of the Role of Nuclear Chromosomes in Development." *Developmental Biology* 314 (1): 1–11.
Maienschein, Jane. 2016. "How T. H. Morgan Misled Garland Allen about Development." *Journal of the History of Biology* 49 (4): 587–601.
Mayr, Ernst. 1991. *One Long Argument: Charles Darwin and the Genesis of Modern Evolutionary Thought*. Cambridge, MA: Harvard University Press.

McClung, C. E. 1902. "The Accessory Chromosome—Sex Determinant?" *Biological Bulletin* 3 (1–2): 43–94.

———.1924. "The Chromosome Theory of Heredity." In Cowdry 1924, 611–87.

Morgan, T. H. 1917. "The Theory of the Gene." *American Naturalist* 51 (609): 513–34.

———. 1924. "Mendelian Heredity in Relation to Cytology." In Cowdry 1924, 691–734.

———, A. H. Sturtevant, H. J. Muller, and C. B. Bridges. 1915. *The Mechanism of Mendelian Heredity*. New York: Henry Holt.

Reznick, David. 2010. *The "Origin" Then and Now: An Interpretative Guide to the "Origin of Species."* Princeton, NJ: Princeton University Press.

Rheinberger, Hans Jörg, and Jean-Paul Gaudlliere. 2004. *Classic Genetic Research and Its Legacy: The Mapping Cultures of Twentieth-Century Genetics*. New York: Routledge.

Richmond, Marsha, and Michael Dietrich. 2002. "Richard Goldschmidt and the Crossing-Over Hypothesis." *Genetics* 161:477–82.

Roll-Hansen, Nils. 1978. "The Genotype Theory of Wilhelm Johannsen and Its Relation to Plant Breeding and the Study of Evolution." *Centaurus* 22 (3): 201–35.

Ruse, Michael. 1975. "Darwin's Debt to Philosophy." *Studies in the History and Philosophy of Science* 6:159–81.

Sutton, Walter S. 1902. "On the Morphology of the Chromosome Group in *Brachystola magna*." *Biological Bulletin* 4 (1): 1247–39.

Stern, Curt. 1928. "Fortschritte der Chromosomtheorie der Vererbung." *Ergebnisse der Biologie* 4:205–359.

———. 1931. "Zytologisch-genetische Untersuchungen als Beweis für die Morgansche Theorie des Faktorenaustausches." *Biologische Zentralblatt* 51:547–87.

Vorms, Marion. 2013. "Models of Data and Theoretical Hypotheses: A Case Study in Classical Genetics." *Synthese* 190 (2): 293–319.

Waters, C. Kenneth. 2004. "What Was Classical Genetics?" *Studies in the History and Philosophy of Science* 35:783–809.

CHAPTER 9

EPIGENETICS AND BEYOND

Jan Sapp

The multiple meanings of *epigenetics* and the difficulty of achieving a unified definition have been discussed for more than a decade. Some have recommended that the term *epigenetics* be strictly defined to denote environmentally induced hereditary changes in gene expression that are transmitted through mitosis and meiosis, such as those based on DNA methylation and histone modification (Holliday 1994; Wu and Morris 2001; Deans and Maggert 2015). Others insist that heritability not be a requirement for the term, but that it include other modes of gene regulation that are comparatively short-lived (Bird 2007).

Discussion of the broader and narrower use of the term *epigenetics* is not new to this century, as is explained below. It occurred during the birth of molecular biology when the term was first used to signify a mode of cellular differentiation based on stable states of gene expression. However, at that time the issue was not about whether epigenetic systems should imply hereditary as well as nonhereditary modes of gene regulation. It was feared that when used to denote states of gene regulation, *epigenetics* might semantically conceal other modes of somatic cell heredity and thereby impede their recognition and investigation. This included hereditary information far removed from genes and gene regulation, such as preformed cell structure and morphogenetic fields, typically excluded from definitions of epigenetics in common use today.

To better understand the historical meanings of these terms and lesser-known phenomena of heredity associated with preformed structure and spatial properties of cells, I situate their discussion first in the context of embryology and genetics in the early twentieth century, then in the context of genetic debates over the significance of cytoplasmic inheritance in the 1950s, and finally within the semantic paradigm of epigenetics today.

Embryology and the Gene

Embryologists had encountered two conceptual problems with the gene theory of classical genetics. One was the paradox of cellular differentiation in the face of nuclear equivalence. Results of experimental embryology indicated that cell differences that arise in development are determined by the action of environmental factors upon individual cells and upon whole groups of cells. Embryologists thus adopted epigenetic theory in direct opposition to preformationist theories, arguing that cells were not necessarily predetermined as different parts but could be primarily alike in constitution. Cells differentiate during development under the influence of their environments, and experiments with tissue cultures indicated that some differences among cells of animals persisted when they were taken out of the body and allowed to grow in a test tube. Yet, there was no evidence of qualitative changes in the chromosomes in the nucleus of the cell, and geneticists knew no way of directing gene mutations.

That nuclear differentiation did not occur during development was supported by the well-known experiments of Hans Spemann, who demonstrated in 1914 the developmental equivalence of nuclei at the eight-cell stage of a newt embryo (Spemann 1938). He had constricted fertilized newt eggs with a ligature, thereby separating the egg into two portions: one with a nucleus and one without. After a series of nuclear divisions, one of the daughter nuclei escaped into the enucleated cytoplasm and there continued its division. If those nuclei had undergone any irreversible differentiation during these early divisions, abnormal development would have been expected; a normal twin, however, developed.

"Each cell inherits the whole germplasm," Thomas Hunt Morgan (1919, 241) said. If so, how could gene theory, which explained so much in terms of sexual transmission of adult traits, account for cellular differentiation? How could identical genes in every cell lead to inherited differences in differentiated cells? To resolve the problem, embryologists logically pointed to the cytoplasm of the egg, where differentiations could actually be seen in the course of development.

There was a second, related difficulty with gene theory or any particulate conception of heredity: If genes were proteins or somehow determined proteins, how could these gene products come together so that they combined at the right time and place to build a cell or organism? This was the problem of morphogenesis. Cellular differentiation was not simply a matter of cells adapting to each other and responding to the environment in various ways. Epigenetic theory had limitations: although the effects of the environment mediated through the metabolic activities of the cell could account for modifications in individual development, such processes alone could not be responsible for the specific form of an organism. Some sort of spatial principle was required to organize the developmental process in time and space.

Many embryologists had rejected the idea that an organism was simply a community of individual cells bound together by interactions and mutual dependence based on a physiological division of labor. Arguing against the idea of the "organism as a beehive" of separate cellular individualities, they often referred to "the organism as a whole." They maintained that organisms made cells; cells did not make organisms, but were instruments not agents of differentiation and morphogenesis. Charles Otis Whitman (1893, 649) put it this way: "It is not division of labor and mutual dependence that control the union of the blastomeres. It is neither functional *economy* nor social instincts that binds the two halves of an egg together, but the constitutional bond of *individual organization*. It is not simple adhesion of independent cells, but integral structural cohesion." There was a "pre-organization," "a grade of organization as a result of heredity," a "structural foundation" that preceded cell formation and regulated it. The fate of cells was roughly prescribed in a submicroscopic organization, a "principle of unity," manifested in the organization of the egg cytoplasm, which Whitman's former student Frank Lillie (1906, 251) said was "a part of the original inheritance."

For Jacques Loeb, the whole materialistic outlook on life rested on the assumption that there was a preformed organization in the egg that persisted from one generation to the next. It provided a guide for gene products and determined the early differentiation and pattern of the organism—where and when cell differences appeared in the course of development. "Without structure in the egg to begin with," he said, in his book *The Organism as a Whole*, "no formation of a complicated organism is conceivable" (Loeb 1916, 39). Indeed there was a structure observable in the egg cytoplasm of various organisms before fertilization that foreshadowed the pattern of the embryo.

That preformed "ground plan" was sometimes revealed by differences in pigmentation that could be followed visually in the early stages of development. When the egg divided, each daughter cell obtained different amounts and types of cytoplasmic materials. Three different germ layers (endoderm, ectoderm, and mesoderm) that give rise to the basic body plan and differentiate into the many tissues and organs of the adult body were traceable back to visible substances in specific locations in the egg. Sometimes characteristic organizational patterns of cytoplasmic materials could be recognized in different phyla.

The spatial structure, revealed by the polarity and organization of the egg cytoplasm, provided the basis for the orderly pattern of morphogenesis.

Substances in the vicinity of the animal pole typically gave rise to ectoderm; substances near the vegetal pole became the endoderm; and the axis of the egg became the chief axis of the adult animal. The bilateral symmetry of many animals was also foreshadowed in the egg cytoplasm. All the early features of development up to gastrulation: polarity, symmetry, type of cleavage, and pattern or relative positions of future organs seemed to be cytoplasmically determined. "The facts of experimental embryology strongly indicate the possibility that the cytoplasm of the egg is the future embryo (in the rough)," Loeb wrote, "and that the Mendelian factors only impress the individual (and variety) of characters upon this rough block."

No one elucidated the structure of the egg cytoplasm more than Edwin Conklin. In a chapter entitled "Cellular Differentiation" in Edmund V. Cowdry's *General Cytology* and elsewhere, he explained how the polarity of the egg was the earliest recognizable and most fundamental differentiation of morphogenesis, and it existed independently of the substratum upon which it acted (Conklin 1924, 581). When the location of the visible substance in the egg was changed experimentally by pressure or centrifugal force, the displaced parts would soon return to normal: cell polarity persisted in the cytoplasm after the positions of the nuclei, mitotic figures, and cleavage planes were altered (548).

That observation was critical to understanding the basis of cell polarity, for which there was a plurality of viewpoints. Some embryologists saw polarity in terms of a magnetic field; it existed independently of the cellular substratum upon which it acted. Others thought of it in terms of a microstructure similar to a liquid crystal located in the "ground substance" of the cytoplasm (Haraway 1976; Sapp, 1987). Conklin (1924, 581), like many others, pointed to the cell cortex as "the ground substance": "The cause of polarity applies also to egg pattern," he said. "It must be found in some substance of the egg which does not change when the yoke, pigment, and other substances are dislocated, and the only material substances of the egg which fulfill these conditions are the ectoplasmic layer and the spongioplasmic framework."

The implication from studies of the organization of the egg was clear to embryologists: the chromosomal genes of geneticists were concerned with characteristics that "topped off" the fundamental organismal features, those that define higher taxonomic groups. Theodor Boveri also distinguished between preformed characters and epigenetic characters. The former were blocked out or prelocalized in the organization of the egg, independent of the nucleus, and involved the general character of the embryo. "Epigenetic" characters resulted from the interactions between the nucleus and cytoplasm and reciprocal interactions among parts of the embryo, which were superimposed on the cytoplasmic organization (Wilson 1925, 1102–08). Conklin put it this way:

> We are vertebrates because our mothers were vertebrates and produced eggs of the vertebrate pattern; but the colour of our skin and hair and eyes, our sex, stature and mental peculiarities were determined by the sperm as well as by the egg from which we came. There is evidence that the chromosomes of the egg and sperm are the seat of the differential factors or determiners for Mendelian characters, while the general polarity [animal–vegetal], symmetry, and pattern [localization of ectoderm, endoderm, mesoderm] of the embryo are determined by the cytoplasm of the egg. (Conklin 1915, 176)

Although the earliest fundamental differences were indeed maternally inherited, it was not certain if these egg characteristics were actually "determined" by the cytoplasm of the egg or if they instead developed under the influence of the chromosomal genes during oogenesis (Sapp 1987). Geneticists were not prepared to make any compromises. "The whole case of the supporters of any theory which views the cytoplasm as determinative," L. C. Dunn (1917, 299) wrote, "rests on either their refusal to go back and inquire [into] the source of this cytoplasm, or on their refusal to give due emphasis to the source, even though they recognize it." The mechanism of heredity is known, he said, "if heredity be only properly concerned with the way in which hereditary factors are distributed in the germ cells. For development, the mechanism is but grossly known, but we have learned enough . . . to foster a suspicion that one day the governance of the chromosomes over development will be explained in physico-chemical terms." Morgan (1926, 491) offered this final and terse comment on the matter: "In a word, the cytoplasm may be ignored genetically."

From an embryological perspective, such claims for genetics were little more than hubris. Frank Lillie (1927, 367) responded to geneticists: "Those who desire to make genetics the basis of [the] physiology of development will have to explain how an unchanging complex can direct the course of an ordered developmental stream." Ross Harrison (1937, 372) had a similar view: "The prestige of success enjoyed by the gene theory might easily become a hindrance to the understanding of development by directing our attention solely to the genome, whereas cell movements, differentiation, and in fact all developmental processes are actually effected by the cytoplasm."

The rift between embryology and genetics would remain for decades. For geneticists, the meaning of heredity was restricted to the study of the sexual transmission of traits. As genetics, based on cross-breeding analysis, grew rapidly, the concept of heredity drifted so as to become synonymous with genetic practice as the embryological conception of heredity as a developmental process was overshadowed (Sapp 1987).

Plasmagene Theory

It was difficult to discuss cellular differentiation without considering the relative importance of the nucleus and cytoplasm in heredity and evolution. "The usual and most probable view is that cellular differentiation is cytoplasmic and must therefore persist and be transmitted to daughter cells by cytoplasmic heredity," as Sewall Wright (1941, 501) commented. "The chief objection is that it ascribes enormous importance in cell lineages to a process which is only rarely responsible for differences between germ cells, *at least within a species* (my italics)."

A new generation of geneticists addressed the problem of cellular differentiation in the face of genomic equivalence after World War II, when genetic evidence of cytoplasmic inheritance in microorganisms emerged. Various cytoplasmic entities had long been suggested to be self-reproducing: chloroplasts, mitochondria, centrioles, and kinetosomes, or basal bodies (a version of the same organelle as centrioles, at the base of cilia). A small band of biologists, including Tracy Sonneborn, Boris Ephrussi, André Lwoff, and Jean Brachet brought together evidence for cytoplasmic inheritance under the rubric of "plasmagenes" and emphasized their importance for cellular differentiation (Sapp 1987). Their problematic was identical to that of embryologists: "*Unless development involves a rather unlikely process of orderly and directed gene mutation, the differential must have its seat in the cytoplasm*" (Ephrussi 1953, 4; italics in orig.).

Responding in various ways to the influence of the environment and interacting among themselves and nuclear genes, plasmagenes would account for the inheritance of differences among somatic cells. The behavior of Kappa particles in *Paramecium*, as studied by Sonneborn at Indiana University, provided the exemplar for understanding the relations between the plasmagenes and environment. When the cells were grown in a medium in which cell reproduction was rapid, they tended to multiply more rapidly than the Kappa, and cells had a different phenotype. In other words, the concentration of Kappa and therefore the character of the cell was controlled by environmental conditions.

At the Institut Pasteur, Lwoff championed the importance of kinetosomes (basal bodies) as plasmagenes in his book *Problems of Morphogenesis in Ciliates*. "The morphogenesis of a ciliate," he said "is essentially the multiplication, distribution, and organization of populations of kinetosomes and the organelles which are the result of their activity" (Lwoff 1950, 241). Ciliates could be considered as cells, or as whole organisms possessing complex morphological differentiation without cellularization. In the later view, Lwoff compared the behavior of kinetosomes within a ciliate to that of cells in a multicellular organism. Kinetosomes, he said, were instruments, not agents of differentiation; they did not control their own destinies. They possessed what embryologists had called "prospective potencies," and their movement and their fate were determined by "some mysterious and powerful field of forces." They produced various structures according to their position within their "hosts": cilia, trichocysts (cylindrical rods beneath the cell surface that elongate inward), and other fibers and organelles.

The research on cytoplasmic inheritance became politicized in the context of Lysenkoism during the 1940s and 1950s, when Soviet biologists cited it in a crude attempt to dismiss all of "Western genetics" (Sapp 1987). Sonneborn addressed the issue head on in an article titled "Heredity, Environment and Politics" published in *Science*: "The work on Kappa in *Paramecium* and other plasmagenes shows that acquired characteristics can be inherited if the characters fall in a certain sub-division of the non-Mendelian category" (Sonneborn 1950, 535).

There were still other cases of non-Mendelian inheritance, such as mating type and serotype specificity in *Paramecium* not associated with any cytologically visible bodies. Sonneborn attributed them to submicroscopic plasmagenes; others suggested they were due to self-perpetuating metabolic states that affected the expression of nuclear genes (Beale 1954). Ephrussi, who studied respiratory-deficient (*petite*) mutations in yeast associated with mitochondria brought the cytoplasmic genetic data together in a small book, *Nucleo-Cytoplasmic Relations in Microorganisms: Their Bearing on Cell Heredity and Differentiation*. Appearing on the eve of the double helix and the molecular biology of the gene, it offered a synthesis of the possibilities in regard to cell heredity and differentiation: The non-living environment can induce changes of the concentration of Kappa particles and of antigenic type in Paramecia, and loss of cytoplasmic particles in yeast. Lastly we find that nucleus and cytoplasm affect each other's activity. The cytoplasmic particles of yeast are activated by a nuclear gene. In turn, in Paramecia, definite cytoplasmic states permit the expression of definite nuclear genes. Here is a set of facts that ought to help explain development" (Ephrussi 1953, 100).

The evidence that cytoplasmic genetic entities were involved in "fundamental" organismal functions such as photosynthesis, respiration, and morphogenesis raised the question again of whether nuclear gene mutations affected only "superficial" characteristics of the organism. "I think that the question is today in need of serious reconsideration," Ephrussi (1953, 119) remarked, "and that it should not be answered by metaphors or by the usual counter-questions."

Epigenetics against "the Master Molecule"

The demise of the plasmagene theory of cellular differentiation began at a symposium entitled "The Chemical Basis of Heredity" at Johns Hopkins University in June 1956, when one of Sonneborn's former students, David Nanney, argued against a hereditary classification based on cell location—nucleus versus cytoplasm—and argued instead for the importance of steady states of gene regulation (whether nuclear or cytoplasmic) in cellular differentiation. He noted that the concept of self-perpetuating metabolic patterns, or the concept of hereditary steady states, had been formulated in general terms and published earlier in the century by Sewall Wright (1941) to account for some cases of cytoplasmic inheritance involving environmentally directed changes: "Persistence may be based on interactions among constituents which make the cell in each of its states of differentiation a self-regulatory system as a whole, in a sense, a single gene, at a higher level of integration than the chromosomal genes" (Wright 1945, 198).

Nanney adopted this steady-state model to protest against the concept of the gene as a dictatorial "master molecule" directing the activities of the cell, whether nuclear or cytoplasmic, as an adequate explanation of cellular differentiation. In contrast to the master molecule concept, which he likened to a "totalitarian government," he lent his support to a "democratic organization" in the cell, "composed of cellular fractions operating in self-perpetuating patterns (Nanney 1957, 136).

Nanney also articulated a new state of affairs in genetics by pointing to the reports of gene regulation in bacteria by Joshua Lederberg, and research indicating that nuclear differentiation occurred in metazoa. Lederberg and colleagues had developed the concept of stable states of gene expression and feedback to explain hereditary variations in *Salmonella* serotypes whose rates of change were too high to be mutations. In their model, phase variation in the flagella antigens were due to a single pair of alternative genes that would exist in two states, active and inactive. There was "no mutation in antigenic specificity, only a choice of which of the two alternatives will be expressed" (Lederberg and Lederberg 1956, 114).

Addressing the relevance of this model to cellular differentiation, Lederberg and Iino said, "The concept of local states may provide a more acceptable hypothetical basis for nuclear differentiation" than "orderly and directed gene mutation" (Lederberg and Iino 1956, 755). They also pointed to the experimental results of Thomas King and Robert Briggs (1955), who transplanted nuclei from blastula cells into enucleated frog eggs and reported that irreversible nuclear differentiation had occurred, as well as to the reports of Barbara McClintock (1956) indicating that transposable controllers could regulate activity at a number of loci in maize.

Nanney followed suit. "It might appear that the dichotomy between germinal and somatic inheritance, between cytoplasmic and nuclear bases was after all a mistake, and that investigations may now converge with a unified perspective" (Nanney 1957, 143). In so doing, the concept of heredity itself had to be freed from its restricted reference to particulate genes, based on cross-breeding analysis. An older, broader definition was called for: the term *heredity*, Nanney said, "may be used to describe the more general capacity of living material to maintain its individuality (specificity) during proliferation. . . . 'Heredity' in this sense is a type of homeostasis, similar to physiological homeostasis but implying more, since it includes regulation during protoplasmic increase" (1957, 134).

Nanney used the expression "epigenetic systems" for the first time, in the fall of 1957 at a conference on "extra-chromosomal heredity" in Gif-sur-Yvette, organized by Ephrussi and moderated by Jacques Monod. The term *epigenetics* had been used by Conrad Waddington (1942) for the study of development, to replace "*Entwicklungsmechanik*" and "experimental embryology," and to link genetics and development. *Epigenetics* was essentially a synonym for developmental biology: "Perhaps the most satisfactory expression would be 'epigenetics.' This is derived from the Greek word epigenesis, which Aristotle used for the theory that development is brought about through a series of causal interactions between the various parts; it also reminds one that genetic factors are among the most important determinants of development" (Waddington 1956, 10).

Nanney (1958, 712) referred to Waddington's use of the term and chose it for cases of self-perpetuating regulatory states "to emphasize the reliance of these systems on the genetic systems and to underscore their significance in developmental processes." It was important to distinguish "epigenetic systems that regulate the expression of the genetically determined potentialities" from "genetic systems" that regulate the maintenance of structural information based on DNA replication by a template, whether nuclear or cytoplasmic (713). Some epigenetic systems, based on self-perpetuating metabolic states, were located in the cytoplasm, as indicated by mating type and serotype specificities in ciliates. Others were in the chromosomes themselves, Nanney said (714), as suggested by the research of Lederberg and colleagues on *Salmonella*. The concept of epigenetic systems did not mean that cytoplasmic entities played no important role in cellular differentiation; only that it would not be by virtue of their location in the cell.

Ephrussi adopted Nanney's terminology the following year, when the molecular basis of somatic cell variation was the subject of crowded sessions of the Eleventh Annual Biology Research Conference sponsored by the Oak Ridge National Laboratory in Gatlinburg, Tennessee, in April 1958. It was important, he said, to distinguish "truly genetic mechanisms," based on "the transmission of particles carrying their own structural information" from "epigenetic mechanisms involving functional states of the nucleus. This has a been a major source of confusion in the past and it is not going to be easy to avoid in the future because we have all been trained to regard the problem of differentiation as a nucleus/cytoplasmic dilemma" (Ephrussi 1958, 49). Although this was admittedly a concession on his part, he emphasized that it also entailed a new conception of heredity: "Many of my geneticist friends will, I am sure, enjoy the shift of my stand. Unfortunately, I must remind them that, as a corollary, we must admit that not everything that is inherited is genetic" (49).

Language Constrains Thought

The genetic/epigenetic dichotomy was thus proposed to replace the nucleus/cytoplasm dichotomy in resolving the paradox of cellular differentiation in the face of genomic equivalence. The term *epigenetic* in reference to gene regulation would not be widely adopted for decades, after new molecular mechanisms in eukaryotes were proposed and discussions about the confused and often conflicting meanings of the term followed. But such discussions were not new.

Lederberg had objected to the term *epigenetic* to refer to the inheritance of cell variations during ontogeny in his concluding lecture at the meeting in Gatlinburg in which he pointed up the hazards of conflating Waddington's term with states of gene expression. The problem, in his view, was that using the term *epigenetics* in the way Nanney and Ephrussi did would semantically conceal other forms of somatic cell variation: not every form of cell heredity that is "epigenetic" in Waddington's sense of the term was included in the more restricted usage for heritable states of gene regulation.

Epigenetic changes (in Waddington's sense) could involve, for example, somatic cell mutations due to changes in nucleotide sequences. The generation of cell diversity by gene mutations, Lederberg (1958, 398–400) argued, may be involved in the ability of the immune systems of animals to produce specific antibodies in response to new antigens. He explained this concept further the following year, arguing that antibody variability could be based on a high frequency of somatic mutations during lymphocyte proliferation (Lederberg 1959). Accordingly, there would be a continuous evolution of antibodies resulting from a repetitive alternation of gene diversification and antigen-mediated selection. Frank McFarlane Burnet adopted this important concept in his later formulation of the clonal selection theory of acquired immunity (Burnet 1964; Neuberger 2008). Lederberg also pointed to the possibility of hereditary information far removed from genes and nucleic acid regulation. Thus, he offered a different vocabulary based on three kinds of information.

The first kind was "nucleic" information, which is dependent "on the *sequence of nucleotides* in a nucleic acid" (Lederberg 1958, 385; italics in orig.). This would apply to genetic systems in the nucleus and to organelles, plasmids, and symbionts in the cytoplasm. Then there was "*epinucleic information*"—"an aspect of nucleic acid configuration other than nucleotide sequence [e.g.,] polypeptide or polyamine adjuncts to the polynucleotide" (385). It "regulates the manifestation of nucleic potentialities in the dynamic, temporally responsive functioning of actual development." It would include cytoplasmic steady states and also "dynamic equilibria at chromosome loci, and involving genes and their products" (386).

Epinucleic chromosome variation in multicellular organisms, he said, was admittedly an "entirely speculative hypothesis designed to leave leeway for differentiation in the chromosome without . . . determinate changes in nucleic acid sequences." Nothing could be said about the precise mechanisms, but he recommended that geneticists "should perhaps look for variations in nucleic acid structure that do not alter the fundamental sequence" (Lederberg 1958, 386).

In the 1960s, studies of gene regulation culminated with the lac operon model in *E. coli* proposed by François Jacob and Jacques Monod (1961). Genes could be switched on and off. Accordingly, Jacob and Monod asserted in 1961, just as Nanney and Ephrussi had several years earlier in regard to epigenetic systems, that "the biochemical differentiation (reversible or not) of cells carrying an identical genome does not constitute a paradox as it appeared to do for many years to both embryologists and geneticists" (397). They recognized that variable gene action was critical to the new molecular paradigm of cellular differentiation. Still, one could not simply assume that mechanisms of gene regulation in bacteria were the same in eukaryotes.

Lederberg's third category was *extranucleic information*—"that residing in molecules or reaction cycles not directly connected with nucleic acid" (1958, 385). He applied it to self-perpetuating metabolic patterns, which had been invoked to explain mating type and antigen specificity in ciliates. Though Lederberg's trichotomous terminology was not widely adopted, the category "extranucleic information" was applied to hereditary phenomena based on new research programs that emerged in the 1960s and were focused on the inheritance of cell structure (Nanney 1966; Landman 1993; Sapp 2003).

Spatial Principles

Studies of what had been called cytoplasmic heredity moved in two main directions. One was the genetics of mitochondrial and chloroplast genomes. The symbiotic origin of those organelles would be at the center of vibrant research programs with the rise of molecular phylogenetics in the 1970s and 80s (Sapp 1994; Sapp 2009). The other focused on the hereditary and morphogenetic role of cell structure, led by Sonneborn and his school of ciliate protistologists.

The spatial properties of the cell became lost to view in the new concepts of messages, codes, feedback regulation, and cell circuitry. Plasmagene theorists of the 1950s had long emphasized that the cell was not a "bag of enzymes," or of "self-reproducing entities," whether genes or organelles. Plasmagenes were considered to be instruments, not agents, of cellular differentiation. Ephrussi had addressed the problem from the point of view of an embryologist who had long pointed to an organizing principle in the cytoplasm of eggs expressed by its polarity, which was responsible for the pattern of cellular change in time and space during ontogenesis. The primary cause of differentiation, he said, resided in the initial anisotropy of the egg: "*Development is an orderly process: it follows a 'plan'* which dictates when and where the instruments of differentiation come into action. Experimental embryology has taught us that a more or less rough outline of this 'plan' is engraved already in the cytoplasm of the undivided egg. Sometimes it is indicated by the visible distribution of cytoplasmic materials; sometimes it can be revealed only by experiment" (Ephrussi 1953, 101; italics in orig.).

Sonneborn had phrased the problem for cellular morphogenesis: If genes were in exclusive "control" of heredity, then it would have to be concluded that a genome isolated under conditions that permit its multiplication would be capable of reconstituting cells of the kind from which it is taken. Otherwise it would have to be concluded "that the cell, including the cytoplasm, somehow serves as a necessary model for the formation of new cellular material in essentially the same sense as the genes are necessary models for the formation of new genes" (Sonneborn 1951, 310).

Put in molecular terms, the question of morphogenesis was this: How can linear nucleotide sequences determine living forms in three-dimensional space and time? Molecular biologists sometimes spoke of some sort of "program," asserting that the "blueprints" for organisms were encoded in their DNA. François Jacob expressed it in his book, *The Logic of Life* (1973, 254): "The whole plan of growth, the whole series of operations to be carried out, the order and the site of synthesis and their co-ordination are all written down in the nucleic acid message." For ciliate protistologists, such statements were little more than molecular biology hubris, just as classical geneticists' claims for the "governance of the chromosomes" appeared to be for embryologists of the 1920s. Cell organization had to be preserved; it was a hereditary property.

Ciliates were the right organisms for the job of investigating the morphogenetic role of preformed cell structure. The complex patterns on their cell surface, or cortex, are composed of linear arrays of a large number of ciliary units arranged in a repeating pattern that is reproduced through a sequence of events during growth and cell fissions (fig. 9.1). As Vance Tartar (1961, 1–2) remarked, "A cytoarchitecture which has been repeatedly postulated as necessary to explain the orderly development of eggs is visibly displayed in Stentors and does in fact play a cardinal role in their morphogenesis."

During the 1960s, when Sonneborn turned to studying the role of preformed cell structure in morphogenesis and heredity, the concept of random self-assembly of gene products was gaining ground among molecular biologists, according to which cellular order was generated from the properties of proteins, their random collision, and the ionic and molecular constitution of the cell "soup" (Sonneborn 1964, 924). That concept was strengthened by the fact that a linear genetic code could be translated

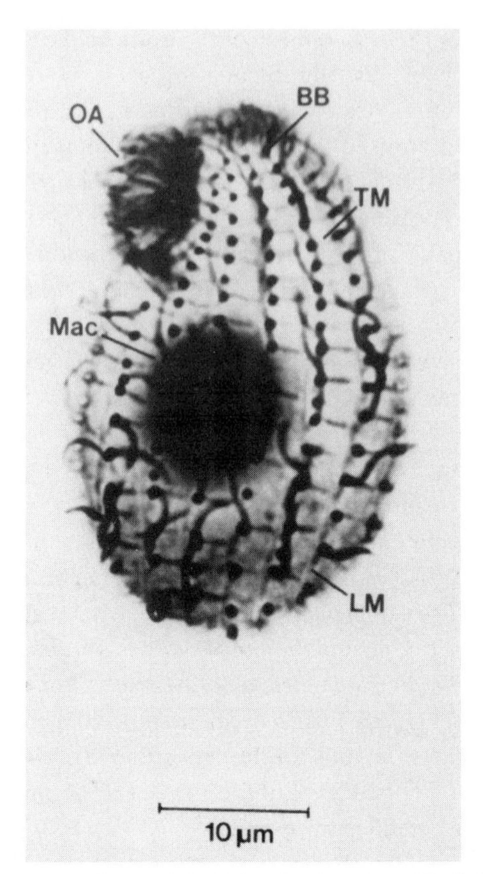

Figure 9.1. *Tetrahymena thermophilia* stained by the protargol technique. *BB*, basal bodies; *TM*, transverse microtube bands; *OA*, oral apparatus; *LM*, longitudinal bands; *Mac*, macronucleus. Reproduced with permission from figure 3.1 in Joseph Frankel, *Pattern Formation: Ciliate Studies and Models* (New York: Oxford University Press, 1989), 45. By permission of Oxford University Press, USA.

into the three-dimensional structure of viruses. But viruses are not cells, and they do not grow and divide like cells. Their nucleic acid is replicated and the other structures are separately formed, and come together in the final organization. Unlike the structure of viruses, cell structure persists throughout growth and division. Cells are not constructed *de novo*. Cells arise from preexisting cells. This was one of the greatest generalizations of nineteenth-century biology.

Sonneborn began to study the cortex experimentally, based on cross-breeding analysis and grafting experiments. Cortical differences bred true to type through sexual and asexual reproduction, free from genic influence. The first case he studied was the inheritance of "double monsters." When paramecia and other ciliates conjugate, they do not always separate, and doublets are formed, basically two cells fused back to back with twice the number of surface structures. Sonneborn found that he could generate doublets at will by treating cells with antiserum. Doublets give rise at fission to new doublets through sexual and asexual reproduction. The hereditary basis of the doublets lay in the cortical structure itself. "Preformed cell structure," he concluded, acts as a scaffolding or template for the assembly of new cell structure.

Preexisting cortical structures would play a role in determining where some gene products go in the cell, how these combine and orient, and what they do. The inheritance of preformed cell structures showed that they could be decisive in cellular differentiation. He coined the term "cytotaxis" for "the ordering and arranging of new cell structure under the influence of preexisting cell structure." He considered this a "second principle of cellular differentiation, one that is quite distinct from variable genic activity; . . . there is more than 'self' to mechanisms of assembly; it includes pre-existing and independently modifiable assembly" (Sonneborn 1964, 925–26).

In grafting experiments, Janine Beisson and Sonneborn reported that even a portion of the cortex, when rearranged, acted as an element of inheritance. They inverted a small patch of ciliary units, each unit comprised a kinetosome, cilium, and a variety of fibers and specialized membranes, and that patch grew during cell division until it extended the full length of the body surface (fig. 9.2). Since the variation occurred by grafting, no change would have occurred in DNA, but progeny inherited the inverted row for hundreds of generations. The important conclusion was that structural information could be maintained in and transmitted by complex supramolecular assemblages: the cortex carried information for its own gross organization and transmitted it to progeny. "Our observations on the role of existing structural patterns in the determination of new ones in the cortex of *P. Aurelia*," they said, "should at least focus attention on the information potential of existing structures and stimulate explorations, at every level, of the developmental and genetic roles of cytoplasmic organization" (Beisson and Sonneborn 1965, 281).

The evidence that DNA was contained in centrioles/kinetosomes was on again and off again throughout the 1960s and 1970s (Sapp 1998). But centriolar/kinetosomal DNA, even if it had existed, would not explain the orderly

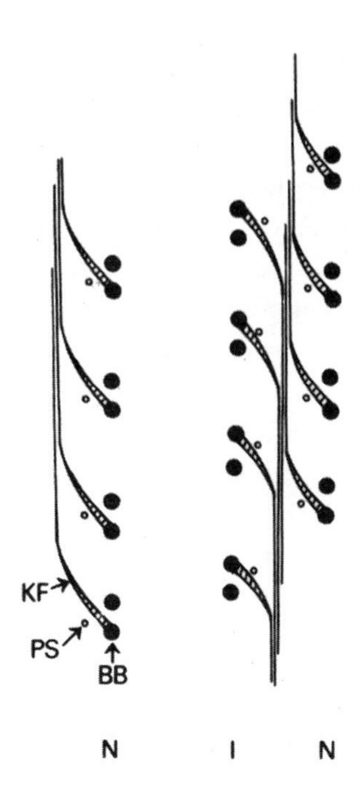

Figure 9.2. Normal (*N*) and inverted (*I*) ciliary rows of *Paramecium tetraurelia*. *BB*, Basal bodies; *PS*, parasomal sacs; and *KF*, kinetodesmal fibers. Reproduced with permission from figure 4.3 in Joseph Frankel, *Pattern Formation: Ciliate Studies and Models* (New York: Oxford University Press, 1989), 75. By permission of Oxford University Press, USA.

pattern of cortical inheritance. The inherited changes were not in the organelle itself, but in the spatial orientation of the organelles. The reproduction of centrioles/ kinetosomes, microtubule based-organelles, is still not fully understood. They had long been imagined to reproduce by division, like mitochondria and chloroplasts, but electron microscopy revealed that was not so. In the cells of some species, including *Paramecium*, they arise only in relation to, and orthogonal to, preexisting centrioles which act as scaffolding for the formation of new ones. In other taxa, centrioles appear to arise *de novo* in the absence of preexisting centrioles, and centrioles are absent in flowering plants and most fungi. Jeremy Picket-Heaps (1969) proposed the concept of "microtubule organizing centers" (MTOCs) as a unifying principle to explain this apparent paradox. According to this, in the cells of some organisms, MTOCs are visible as centrosomes containing two centrioles, and in others the MTOCs are less concentrated.

Organismic Crystals and Morphogenetic Fields

The causes underlying large-scale cell structural inheritance in ciliates were similarly not all fully understood. During the 1970s and 1980s the study of cortical inheritance was extended to other ciliates by several of Sonneborn's former students, especially by Nanney, who investigated several variations, or "corticotypes," in *Tetrahymena*. Each corticotype possessed characteristic patterns and variations in a number of cortical properties, including number and positions of vacuole pores, numbers of oral apparatuses, and patterns of stomatogenesis (Nanney 1966; Nanney 1985).

The success of molecular biology, Nanney argued, had led to a prejudice in favor of linear information sources, omitting "a multidimensional information storage and transmission system whereby the pattern, in a sense, maintains itself." Pondering the shifts in concepts in the cultural revolution of the 1960s, he remarked, "Now that we are encountering a cultural revolution in which the medium is becoming the message, our academic progeny may be more susceptible to a broader view of information structure. Certainly information can be stored and transmitted by supramolecular mechanisms" (Nanney 1968, 502).

Supramolecular information was as fundamental to evolution as it was to morphogenesis. Sonneborn favored the idea of "a parallel, independent, and selectively correlated evolution of genome and cortex" (1963, 213). Nanney (1968, 497) posited that nucleic acids might specify only the appropriate protein building blocks, but that the preexisting cell structures were the cellular architects that determine the nature of the edifice that was to be made. "Because of this morphogenetic role of preexisting structure," he later commented, "the cell has some of the properties of an organismic crystal" (Nanney 1980, 173).

Still, it remained uncertain whether all forms of pattern inheritance and cellular information in ciliates could be explained in terms of templating by supramolecular structures. Others have considered the idea that some of the multidimensional supramolecular patterns inherited in ciliates may lie beyond the reach of molecular biology's principles of templating. Phage geneticist Alfred Hershey invoked the venerable idea of a magnetic field underlying cell polarity, just as some embryologists had decades earlier:

> If cells draw on an extragenic source of information, a second abstraction must be invoked, another vital principle superimposed on the genotype. A likely candidate already exists in what is usually called cell

polarity, which tradition places in a rigid ectoplasm for good reason—it's a spatial principle and as such requires mystical language. Seemingly independent of the visible structures that respond to it, polarity pervades the cell much as a magnetic field pervades space without the iron filings that bring it to light. Biological fields are species-specific, as seen in the various patterns and symmetries of growing things. (Hershey 1970, 700)

The iron filings in this case were the cortical structures that make longitudinal stripes on the surface of *Stentor*, investigated by Tartar, which, when grafted in reverse orientation, may rotate to restore normal polarity or develop into their own longitudinal mosaic stripe pattern. This experiment and others showed that "polarity resides in all parts of the cell cortex." Perhaps, Hershey said, "as many people think, polarity represents something that was invented only once and evolved since on its own" (1970, 700). There was experimental evidence that there was indeed something else beside the cortical structures themselves that was involved in pattern inheritance: structural changes in the cortex could be inherited when microtubules and cortical organelles such as kinetosomes were completely absent. This evidence came from experiments on the inheritance of doublets in another ciliate, *Oxytricha fallax*, by another of Sonneborn's students, electron microscopist Gary Grimes.

Oxytricha doublets are inherited through sexual and asexual reproduction as a cortical trait, just as are doublets in *Paramecium* and other ciliates. But something other than preexisting cell structures was involved. The evidence came from taking complete and incomplete doublets through a cyst stage (fig. 9.3). *Oxytricha* form cysts when they are starved, and no discernible microtubules or visible cortical organelles could be seen using electron microscopic techniques. Yet, doublets emerged from doublet cysts. Grimes (1973, 66) concluded, "The doublet phenotype is inherited through sexual and asexual reproduction as a cortically determined trait. The trait is also inherited through cystment, independent of cyst size. Prior work shows an absence of all visible cortical organelles." Thus, the visible structures are not themselves determinative. Based on these and other observations, he concluded that "at least two levels of cytotactical control of cell patterning are operative on the ciliate cortex; one is dependent upon visible ciliature, whereas the other is dependent upon an as yet ultra-structurally unidentifiable molecular architecture" (Grimes and Hammersmith, 1980, 19).

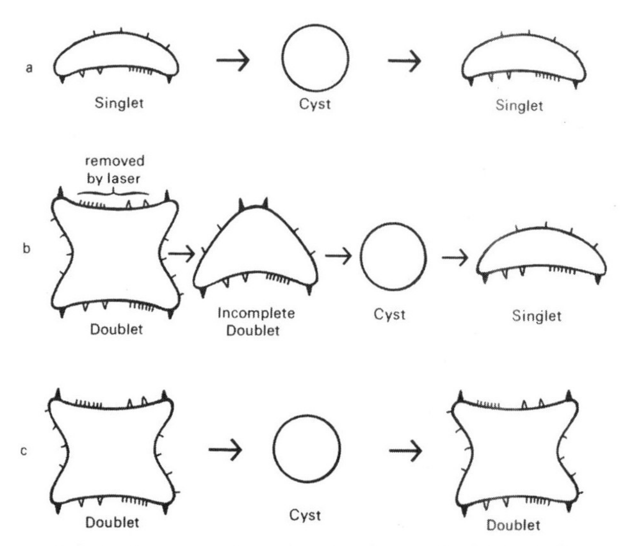

Figure 9.3. Schematic cross sections showing the results obtained after passing (*a*) singlets, (*b*) incomplete doublets, and (*c*) complete doublets through cysts. Reproduced with permission from figure 4.11 in Joseph Frankel, *Pattern Formation: Ciliate Studies and Models* (New York: Oxford University Press, 1989), 89. By permission of Oxford University Press, USA.

That ultra-structurally unidentified system can retain information concerning the nature, number, and large-scale asymmetry of the organelle sets that are to be formed following excystment.

What "the unidentifiable molecular architecture" might be remains a matter of guesswork. Some ciliate biologists suspect that Grimes's results might be attributed to "a filamentous scaffold of still unknown composition," which has remained undetected by microscopic observation (Beisson and Jerka-Dziadosz 1999, 374). Others argued that Grimes's observations deny that possibility and instead show "the self-organizing and regulating characteristic of developmental fields—a morphogenetic field of some kind somehow persists in cysts" (Frankel 1989, 90). As Joseph Frankel saw it, cell structure resulted from at least two processes: a local constraint involving microscopically visible structure acting as a scaffold, and a more global level of pattern development based on morphogenetic field. The nature of the field remains obscure. Frankel called the inheritance of large-scale regulative fields "structural inheritance" to imply "as few mechanistic connotations as possible" (92).

Disturbing the Conceptual Consensus

Sonneborn (1974) and his followers had emphasized the importance of their findings for cell theory, evolution, and for the development of metazoans. During the 1980s, discussions of structural inheritance in ciliates remerged in the context of debates over morphogenetic fields and developmental constraints on evolutionary change with the rise of evolutionary developmental biology. A main focus was on the cause of "pattern formation": that is, the process by which cells acquire different characteristics during development as a function of their relative position in the embryo such that tissues and organs develop in the right place and orientation in the body.

Frankel entered debates between Lewis Wolpert and Brian Goodwin. Whereas Wolpert (1982) adopted a gene-centered view, maintaining that genes specify pattern, Goodwin (1984, 226) argued to the contrary that "generative principles" governing morphogenetic fields provide developmental constraints. He pointed to the phenomena of cortical inheritance in ciliates to argue against the notion that "all aspects of organismic form are determined by hereditary particulars encoded in DNA." Frankel (1983, 312) agreed with Goodwin, and he posited that "a cytactically propagated pattern transition might occasionally be the first step in evolutionary change."

Nevertheless, leading evolutionists, including John Maynard Smith (1983, 45), considered cases of cortical inheritance in ciliates as "exceptions" and to be "the only significant threat" to neo-Darwinian views. "Heredity without genes" continued to "disturb the experimental and conceptual foundations of the modern consensus," as Nanney (1984, 365) observed. It was uncertain what biology was to make of these "curious exceptions." Perhaps, he said, "keeping theoreticians from unbearable arrogance is use enough. On the other hand, this rejected stone just might have a place in some edifice not yet imagined" (Nanney 1985, 287).

Three decades later it still is not clear what to make of their place in biological theory. Although ciliates are well suited to study cytotactic phenomena, they are not unique (Frankel 1989, 92). Structural inheritance has been observed in studies of the form of the cytoskeleton in mammalian cells (Albrecht-Buehler 1977), the asymmetry of *Chlamydomonas* (Holmes and Dutcher 1989), and orienting polarity and cell division in yeast cells (Chen et al. 2000).

The general importance of preexisting structures functioning as templates in the formation of new structures has also been shown for the reproduction of mitochondria (Luck 1963), chloroplasts (Mullet 1988), centrioles (Meunier and Spassky 2016), the morphogenesis of the flagellum of trypanosomes (Flávia Moriera-Leite et al. 2001), and cell membrane heredity (Cavalier-Smith 2000). It also applies to prions (Alper et al. 1967; Griffith 1967), which propagate by transmitting their misfolded protein (Aguzzi 2008). Some philosophers of biology have also viewed all cases of cortical inheritance in ciliates within the templating paradigm of molecular biology; they mention the inheritance of morphogenetic fields in which no cytological visible structure is involved (Jablonka and Lamb 1995; Moss 2003). Whether all cases of structural inheritance can be accounted for by templating or not, all were excluded from a new definition of epigenetics as it emerged into a new specialty in the twenty-first century.

Semantic Drift and Conceptual Constraints

Sonneborn sometimes referred to cytotaxis as a form of epigenetic inheritance, knowing well of Lederberg's caution of 1958 that referring to everything under that neologism might conceal more than it revealed. "Calling these cytotactic events epigenetic or epigenic," Sonneborn (1964, 926) wrote, "should not obscure their degree of independence or their decisiveness for the end result of cellular differentiation." Nanney (1966) referred to cytotaxis both as "epigenetic" and "extranucleic." Despite their use of the term *epigenetic* in reference to heredity, it had remained more or less dormant for three decades until it arose anew in the late 1980s.

But when the term *epigenetic* had a new beginning in the late 1980s, its meaning narrowed in reference to newly proposed molecular mechanisms that switch genes on and off in eukaryotes by such processes as DNA methylation and histone modification. There was no place for the inheritance of cortical patterns and no mention of the inheritance of morphogenetic fields in the new conceptual framework of epigenetics when it emerged as a new molecular genetic specialty in the twenty-first century. The central problem of morphogenesis was not included.

Widespread use of *epigenetics* in reference to somatic cell heredity followed its reintroduction in a paper by British molecular biologist Robin Holliday in 1987, "The Inheritance of Epigenetic Defects." He argued therein that heritable changes in gene expression could be responsible for cancer. Holliday and those who followed him were unaware of Nanney and Ephrussi's use of the term for heritable stable states of gene expression several decades earlier and its subsequent use by Sonneborn and Nanney for self-perpetuating metabolic states and cortical inheritance in ciliates.

Holliday (1987, 163) referred to Waddington's definition of epigenetics as the processes by which "genotype gives rise to phenotype." At first he took that to mean gene regulation in somatic cells, whether transient or permanent. But he restricted the meaning of epigenetics further a few years later, and suggested an even narrower definition of epigenetics as "nuclear inheritance, which is not based on differences in DNA sequence" (Holliday, 1994, 454).

There followed much discussion over the meanings of the term. Should it apply only to hereditary states of gene regulation, and if so, should the same term for somatic cell heredity be used for transgenerational heredity not involving changes in DNA? Some writers grouped together all forms of transgenerational inheritance by various mechanisms of gene regulation in metazoans as representing epigenetic inheritance, and as a "Lamarckian" component in evolution, in contradistinction to neo-Darwinian tenets (Jablonka and Lamb 1995; Jablonka and Lamb 2002; Chong and Whitelaw 2004).

Others have warned of the confusions that arise from using the same term for hereditary states in somatic cells and those that occur in germinal inheritance (Müller and Olsson 2003). Still others insist that heritability, whether somatic or germinal, not be a requirement. Instead they recommend that the term include other modes of gene regulation, such as chromatin marks that are comparatively short-lived (Bird 2007, 398).

Such discussion notwithstanding, epigenetics began to take the form of a new specialty based on the study of nuclear gene regulation, and it would accordingly take on that more restricted meaning as the "study of phenomena and mechanisms that cause chromosome-bound, heritable changes to gene expression that are not dependent on changes to DNA sequence" (Deans and Maggert 2015, 809; see also Wu and Morris 2001). Lederberg (2001) entered the discussions over the meaning of the term to warn again of the conceptual traps resulting from the semantic drift from Waddington's definition of the process by which genotype gives rise to phenotype to Holliday's "nuclear inheritance which is not based in DNA sequence." Methylation corresponded perfectly with his unused category of "epinucleic information."

Yet, other important epigenetic (in Waddington's sense), heritable processes do depend on changes in nucleic acid sequences ("nucleic information"): cellular senescence following telomere-shortening or immunocyte diversification based on recombination and mutagenesis. That developmental changes were never to be traced back to changes in nucleic acid sequence, Lederberg (2001, 6) said, was a "dogma," which "greatly delayed the formulation and acceptance of the clonal selection theory of immunity; and who knows what also may be impeded by it?"

Nucleo-centric definitions of epigenetics also exclude the involvement of mitochondria and chloroplasts as well as symbionts, which play roles in phenotypic development. Lederberg proposed the term *the microbiome* earlier in 2001 "to signify the ecological community of commensal, symbiotic, and pathogenic microorganisms that literally share our body space and have been all but ignored as determinants of health and disease" (Lederberg and McCray 2001). All of it would be included in a general concept of epigenetics as a developmental process.

Concepts of epigenetics and *epigenome* based on heritable states of nuclear gene regulation say nothing of how parts become integrated in space and time, the information contained in the spatial properties of cells, complex supramolecular patterns, and the enigmatic mechanism underlying the inheritance of morphogenetic fields. The perceptions of chromosomes as "governing" bodies and the gene as "master molecule" have long faded, but a restricted definition of epigenetics nevertheless reinforces a gene-centered conception as the fundamental problem of morphogenesis becomes semantically hidden. Only a cell can make a cell.

References

Aguzzi, A. 2008. "Unraveling Prion Strains with Cell Biology and Organic Chemistry." *Proceedings of the National Academy of Sciences* 105:11–12.

Albrecht-Buehler, Guenter. 1977. "Daughter 3T3 Cells: Are They Mirror Images of Each Other?" *Journal of Cell Biology* 72:595–603.

Alper, T., W. A. Craper, D. A. Haig, and M. C. Clark. 1967. "Does the Agent of Scrapie Replicate without Nucleic Acid?" *Nature* 214: 764–66.

Beale, Geoffrey. 1954. *The Genetics of Paramecium Aurelia*. New York: Cambridge University Press.

Beisson, Janine, and Maria Jerka-Dziadosz. 1999. "Polarities of the Centriolar Structure: Morphogenetic Consequences." *Biology of the Cell* 94:367–78.

———, and T. M. Sonneborn. 1965. "Cytoplasmic Inheritance of Organization of the Cell Cortex in *Paramecium Aurelia*." *Proceedings of the National Academy of Sciences* 53:275–82.

Bird, A. 2007. "Perceptions of Epigenetics." *Nature* 447:396–98.

Burnet, F. M. 1964. "The Clonal Selection Theory of Acquired Immunity: A Darwinian Modification." *Australian Journal of Science* 27:6–7.

Cavalier-Smith, T. 2000. "Membrane Heredity and Early Chloroplast Evolution." *Trends in Plant Science* 5 (4): 174–82.

Chen, T., T. Horiko, A. Chaudhuri, F. Inose, M. Lord, S. Tanaka, J. Chant, et al. 2000. "Multigenerational Cortical Inheritance of the Rax2 Protein in Orienting Polarity and Division in Yeast." *Science* 290:1975–78.

Chong, Suyinn, and Emma Whitelaw. 2004. "Epigenetic Germline Inheritance." *Current Opinion in Genetics and Development* 14:692–96.
Conklin, E. G. 1915. *Heredity and Environment in the Development of Men*. Princeton, NJ: Princeton University Press.
———. 1924. "Cellular Differentiation." In *General Cytology: A Textbook of Cellular Structure and Function for Students of Biology and Medicine*, edited by Edmund V. Cowdry, 537–608. Chicago: University of Chicago Press.
Deans, Carrie, and Keith Maggert. 2015. "What Do You Mean 'Epigenetic'?" *Genetics* 199:887–96.
Dunn L. C. 1917. "Nucleus and Cytoplasm as Vehicles of Heredity." *American Naturalist* 51:286–300.
Ephrussi, Boris. 1953. *Nucleo-Cytoplasmic Relations in Microorganisms*. Oxford: Clarendon Press.
———. 1958. "The Cytoplasm and Somatic Cell variation." *Journal of Cellular and Comparative Physiology* 52:35–54.
———. 1972. *Hybridization of Somatic Cells*. Princeton, NJ: Princeton University Press.
Frankel, Joseph. 1983. "What Are the Developmental Underpinnings of Evolutionary Change in Protozoan Morphology." In *Development and Evolution*, edited by B. C. Goodwin, N. Holder, and C. C. Wylie, 279–314. Cambridge: Cambridge University Press.
———. 1989. *Pattern Formation: Ciliate Studies and Models*. New York: Oxford University Press.
Goodwin, B. C. 1984. "A Relational or Field Theory of Reproduction and Its Evolutionary Implication." In *Beyond Neo-Darwinism*, edited by M. W. Ho and P. Saunders, 219–41. London: Academic Press.
Griffith J. S. 1967. "Self-Replication and Scrapie." *Nature* 215:1043–44.
Grimes, Gary. 1973. "An Analysis of the Determinative Differences between Singlet and Doublets of *Oxytricha fallax*." *Genetic Research* 21 (1): 57–66.
———, and R. L. Hammersmith. 1980. "Analysis of the Effects of Encystment and Excystment on Incomplete Doublets of *Oxytricha fallax*." *Journal of Embryology and Experimental Morphology* 59 (1): 19–26.
Haraway, Donna. 1976. *Crystals, Fabrics and Fields: Metaphors of Organicism in Twentieth-Century Biology*. New Haven, CT: Yale University Press.
Harrison, Ross. 1937. "Embryology and Its Relations." *Science* 85:369–74.
Hershey, A. D. 1970. "Genes and Hereditary Characteristics." *Nature* 226:697–700.
Holliday, Robin. 1987. "The Inheritance of Epigenetic Defects." *Science* 238:163–70.
———. 1994. "Epigenetics: An Overview." *Developmental Genetics* 15:453–57.
Holmes, Jeffrey, and Susan Dutcher. 1989. "Cellular Asymmetry in *Chlamydomonas Reinhardtii*." *Journal of Cell Science* 94 (2): 273–85.
Jablonka, E., and M. J. Lamb. 1995. *Epigenetic Inheritance and Evolution: The Lamarckian Dimension*. Oxford: Oxford University Press.
———, and M. J. Lamb. 2002. "The Changing Concept of Epigenetics." *Annals of the New York Academy of Sciences* 981:82–96.
———, and Ehud Lamm. 2012. "Commentary: The Epigenotype—Dynamic Network of Development." *International Journal of Epidemiology* 41 (1): 16–20.
Jacob, François. 1973. *The Logic of Life: A History of Heredity*. Translated by Betty Spillman. New York: Random House.

———, and Jacques Monod. 1961. "Teleonomic Mechanisms in Cellular Metabolism, Growth and Differentiation." *Cold Spring Harbor Symposium for Quantitative Biology* 21: 389–401.
King, T. J., and R. Briggs. 1955. "Changes in Nuclei of Differentiating Gastrula Cells as Demonstrated by Nuclear Transplantation." *Proceedings of the National Academy of Sciences* 41:321–25.
Landman, Otto. 1993. "Inheritance of Acquired Characteristics Revisited." *BioScience* 43:696–705.
Lederberg, J. 1958. "Genetic Approaches to Somatic Cell Variation: Summary Comment." *Journal of Cellular and Comparative Physiology* 52:383–401.
———. 1959. "Genes and Antibodies." *Science* 129:1649–53.
———. 2001. "The Meaning of Epigenetics." *Scientist* 17 (18): 6.
———, and Tetsuo Iino. 1956. "Phase Variation in Salmonella" *Genetics* 41:743–57.
———, and Esther Lederberg. 1956. "Infection and Heredity." In *Cellular Mechanisms in Differentiation and Growth*, edited by D. Rudwick, 101–24. Princeton, NJ: Princeton University Press.
———, and Alexa McCray. 2001. "Ome Sweet Omics—A Genealogical Treasury of Words." *Scientist* 15 (7): 8.
Lillie, Frank R. 1906. "Observations and Experiments Concerning the Elementary Phenomena of Embryonic Development in *Chaetopterus*." *Journal of Experimental Zoology* 3 (2): 153–267.
———. 1927. "The Gene and the Ontogenetic Process." *Science* 64:361–68.
Loeb, Jacques. 1916. *The Organism as a Whole*. New York: Putnam Sons.
Luck, David. 1963. "Formation of Mitochondria in *Neurospora Crassa*: A Quantitative Radioautographic Study." *Journal of Cell Biology* 16:483–99.
Lwoff, André. 1950. *Problems of Morphogenesis in Ciliates: The Kinetosomes in Development, Reproduction and Evolution*. New York: John Wiley and Sons.
Maynard Smith, John. 1983. "Evolution and Development." In *Development and Evolution*, edited by B. C. Goodwin, N. Holder, and C. C. Wylie, 33–47. Cambridge: Cambridge University Press.
McClintock, B. 1956. "Intranuclear Systems Controlling Gene Action and Mutation." *Brookhaven Symposia in Biology* 8: 58–71.
Meunier, Alice, and Nathalie Spassky. 2016. "Centriole Continuity: Out with the New, in with the Old." *Current Opinion in Cell Biology* 38: 60–67.
Morgan T. H. 1919. *The Physical Basis of Heredity*. Philadelphia: Lippincott.
———. 1926. *The Theory of the Gene*. New Haven, CT: Yale University Press.
Moriera-Leite, F. F., T. Sherwin, L. Kohl, and K. Gull. 2001. "A Trypanosome Structure Involved in Transmitting Cytoplasmic Information During Cell Division." *Science* 294:610–12.
Moss, Lenny. 2003. *What Genes Can't Do*. Boston: MIT Press.
Mullet, J. E. 1988. "Chloroplasts Development and Gene Expression." *Annual Revue of Plant Physiology and Plant Molecular Biology* 39:475–502.
Müller, Gerd, and Lennart Olsson. 2003. "Epigenesis and Epigenetics." In *Keywords and Concepts in Evolutionary Developmental Biology*, edited by Brian Hall and Wendy Olson, 114–23. Cambridge, MA: Harvard University Press.
Nanney, D. L. 1957. "The Role of the Cytoplasm in Heredity." In *The Chemical Basis of Heredity*, edited by W. D. McElroy and B. Glass, 134–64. Baltimore: Johns Hopkins University Press.

———. 1958. "Epigenetic Control Systems." *Proceedings of the National Academy of Sciences* 44:712–17.
———. 1966. "Corticotypes in *Tetrahymena pyriformis.*" *American Naturalist* 100:303–18.
———. 1968. "Cortical Patterns in Cellular Morphogenesis." *Science* 160:496–502.
———. 1980. *Experimental Ciliatology*. New York: Wiley and Sons.
———. 1984. "Review of *Development and Evolution.*" *Journal of Protozoology* 31:365.
———. 1985. "Heredity without Genes: Ciliate Explorations of Clonal Heredity." *Trends in Genetics* 1: 295–98.
Neuberger, Michael S. 2008. "Antibody Diversification by Somatic Mutation: From Burnet Onwards." *Immunology and Cell Biology* 86 (2): 124–32.
Pickett-Heaps, Jeremy D. 1969. "The Evolution of the Mitotic Apparatus: An Attempt at Comparative Ultrastructural Cytology in Dividing Plant Cells." *Cytobios* 3: 257–80.
Sapp, Jan. 1987. *Beyond the Gene: The Struggle for Authority in the Field of Heredity*. New York: Oxford University Press.
———. 1994. *Evolution by Association. A History of Symbiosis*. New York: Oxford University Press.
———. 1998. "Freewheeling Centrioles." *History and Philosophy of the Life Sciences* 20 (3): 255–90.
———. 2003. *Genesis: The Evolution of Biology*. New York: Oxford University Press.
———. 2009. *The New Foundations of Evolution: On the Tree of Life*. New York: Oxford University Press.
Sonneborn, T. M. 1950. "Heredity, Environment and Politics." *Science* 111: 529–39.
Sonneborn, T. M. 1951. "The Role of the Genes in Cytoplasmic Inheritance." In *Genetics in the 20th Century*, edited by L. C. Dunn, 291–314. New York: Macmillan.
———. 1963. "Does Preformed Cell Structure Play an Essential Role in Cell Heredity?" In *The Nature of Biological Diversity*, edited by J. M. Allen, 165–221. New York: McGraw Hill.
———. 1964. "The Differentiation of Cells." *Proceedings of the National Academy of Sciences* 51:915–29.
———. 1974. "Ciliate Morphogenesis and Its Bearing on General Cellular Morphogenesis." *Actual Protozoology* 1:327–55.
Spemann, Hans. 1938. *Embryonic Development and Induction*. New Haven, CT: Yale University Press.
Tartar, Vance. 1961. *The Biology of Stentor*. Oxford: Paramount Press.
Waddington, C. H. 1942. "The Epigenotype." *Endeavour* 1:18–20. Reprinted in 2012. *International Journal of Epidemiology* 41 (1): 10–13.
———. 1956. *Principles of Embryology*. New York: Macmillan.
Whitman, Charles O. 1893. "The Inadequacy of the Cell Theory of Development." *Journal of Morphology* 8: 639–58.
Wilson, E. B. 1925. *The Cell in Development and Heredity*. New York: Macmillan. Orig. pub. as *The Cell in Development and Inheritance* in 1896; 2nd ed., 1900.
Wolpert, Lewis. 1982. "Pattern Formation and Change." In *Evolution and Development*, edited by J. T. Bonner, 43–55. Berlin: Springer-Verlag.
Wright, Sewall. 1941. "The Physiology of the Gene." *Physiological Reviews* 21:487–527.
———. 1945. "Genes as Physiological Agents." *American Naturalist* 79:289–303.
Wu, C., and J. R. Morris. 2001. "Genes, Genetics, and Epigenetics: A Correspondence." *Science* 293:1103–5.

CHAPTER 10

HEADS AND TAILS

MOLECULAR IMAGINATION AND THE LIPID BILAYER, 1917–1941

Daniel Liu

That the cell has a membrane is perhaps one of its most obvious features: more than any other part, the membrane defines the cell, sets its outer boundary, and determines how the cell as an individualized unit interacts with its environment. A schematic picture of the cell membrane is a staple of any introductory biology textbook, in part because it does more than any other illustration to show that cells can be pictured as being composed of molecules large and small, with all manner of shapes and functions, a complex sandwich of lipids, studded with potato-like protein globules and wispy carbohydrate chains. The membrane binds the cell into a single entity, and today it is almost impossible to imagine that anyone could have doubted its existence.

Yet until the late 1910s the existence of the cell membrane was a matter of considerable debate and controversy, and even Edmund Cowdry's *General Cytology* in 1924 had a few hints of ambiguity and doubt regarding the cell membrane's existence and composition. An early chapter by Albert Mathews cheerfully suggested that "limiting membranes wherever they occur" might be made of oriented graphite rods, a suggestion made largely through his idiosyncratic analogy to electric battery construction (Cowdry 1924, 43, 68–71).[1] Merle Jacobs's chapter on cell permeability spent several pages defending the existence of a cell membrane that could allow for differential diffusion, yet he also noted a great deal of disagreement about the membrane's composition, writing that "the whole subject is of too speculative a nature to make further discussion profitable; . . . what is most needed in the field of cell permeability at the present day is facts" (156).[2] Robert Chambers even briefly noted the possibility that some cells might not possess a membrane, but have instead a thick, "cement-like substance" holding cells together in some tissues (241). If Cowdry and his collaborators were largely convinced of the reality of the cell membrane, in 1924 it would still have been a relatively novel and fraught position to take, and any theory of the membrane's structure would have remained entirely a matter of speculation.

By the 1930s, however, the membrane was not only a positive fact of science, but the idea that it primarily consisted of a lipid bilayer and associated proteins was quickly accepted as a likely molecular structure for the cell membrane. The so-called Danielli-Davson model of the cell membrane from 1935 is now often cited as the first time a lipid bilayer was proposed as the basic structural element of the cell membrane, though James Danielli (1911–1984) found this attribution irritating. The bilayer concept has also often been attributed to Evert Gorter and François Grendel's 1925 paper, "On Bimolecular Layers of Lipoids on the Chromocytes of the Blood." It appears that Gorter and Grendel's membrane hypothesis was not well known until the late 1930s, by which time Danielli's theory had achieved broad recognition, and credit was retroactively given (Lombard 2014, 10–11). At least later in life, Danielli stressed that the lipid bilayer was *not* his idea, and he argued, without a hint of doubt, that the lipid bilayer "would have been obvious to any competent physical chemist," and that such an idea "flowed almost automatically" from the basic physical chemistry of the 1930s (Danielli 1973, 64). Indeed he and his colleague Hugh Davson never explained why they thought the cell membrane had a lipid bilayer at all; since the lipid bilayer was so obvious, their attention was on the permeability of the protein layer they thought was adsorbed on either side of the lipid (fig. 10.1; Danielli and Davson 1935). Danielli's later irritation might have come from the fact that he and Davson were trying to articulate a functional or physiological theory of cell permeability, but were misread as having "discovered" a biological-structural principle that they claimed no credit for.

How was it decided that the cell's membrane and interior lamellar structures were composed of phospholipids, arranged with their heads facing outwards and tails facing inwards? And how did such structure go from an unprofitable speculation in 1924 to an obvious matter of fact in 1935? In this chapter I argue that many biologists arrived at the lipid bilayer structure largely through a schematic, graphical iconography, one that was originally developed as a strictly heuristic analogy or conceptual aid for the abstract physical concept of *molecular orientation*. The ball-and-stick image that was eventually used to represent lipid lamellar structures in living cells was not just a schematized representation of a chemical formula: it allowed biologists to imagine that living matter was composed of molecules of definite size, shape, and orientation, and that those molecules could construct a complex, living cell strictly by sorting, aggregating, and segregating themselves through physical forces. In other words, biologists in the mid-1930s were developing an essential part of a biological microworld

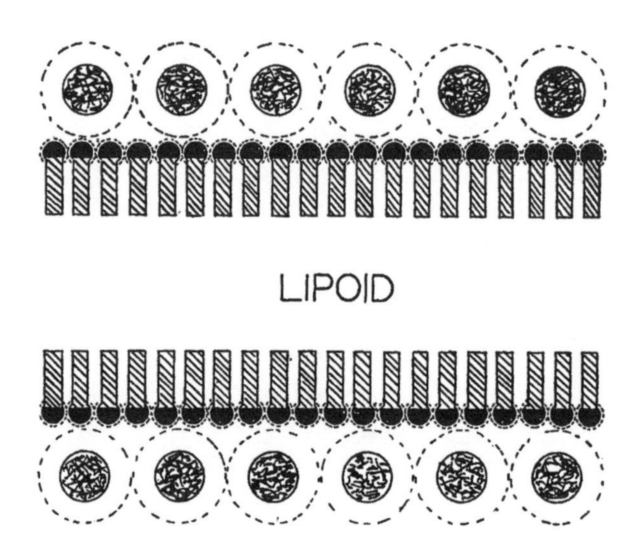

Fig. 1 Schema of molecular conditions at the cell surface.

Figure 10.1. Danielli-Davson model of the cell membrane, "of between unimolecular and trimolecular thickness," with spherical protein molecules adsorbed to both surfaces. Reprinted from Danielli and Davson 1935, 498. With permission of John Wiley and Sons.

not necessarily through mathematical physics or a deep understanding of structural chemistry, but by understanding a diagrammatic convention as a realistic representation of molecular reality.

Recent work in the history of physics and the history of chemistry has stressed the roles of imagination and visual culture in constructing theories of the microworld of submicroscopic atoms, molecules, and otherwise invisible particles and forces. Ursula Klein (2003) and David Kaiser (2005) have each argued that "paper tools," mathematical symbols, diagrams, and even doodles can play a part in directing and keeping account of unruly and abstract scientific thought. And building upon the work of Klein and Kaiser, Alan Rocke (2010) has recently written about the role of imagination in the sciences of atoms, molecules, and forces that are fundamentally beyond the reach of human senses. Rocke argues that mental images were essential in turning work with flasks and analytical balances into an entire metaphysics of molecular structures. The psychic and mental lives of scientists work in large part through symbols and images, and Rocke, Klein, and Kaiser alike argue that paper tools and diagrams can be thought of as pale shadows of scientists' dreams and flights of fancy about the microworld—dreams and images that are often not condoned in "proper" scientific settings like scholarly journals or monographs. In this chapter I take a more limited approach to imaginings and images of the microworld, if only because a full exploration into the inner psychic lives of long-dead and ill-recorded scientists is frighteningly difficult, as Rocke himself has admitted.

Here I look most carefully at the more didactic genres of physical-chemical writing and image-making, because diagrams and invocations of imagination or visual analogy are often used to communicate difficult theories to audiences of varying degrees of impressionability. This is somewhat in contrast to the also-growing literature on models and modeling, the enthusiasm for which has been met by historians with increasing suspicion, as many have noticed slippage between actors' and analysts' use of the words *model* and *modeling* (Creager, Lunbeck, and Wise 2007; de Chadarevian and Hopwood 2004). Didactic genres of scientific writing carry the weight of intentional transmission and translation, and I would like to entertain the idea that images and analogies are among the more potent and portable parts of the genre. Even Aristotle in *De Anima* identified the human imagination's capacity for creative image-making beyond common perception, as a place for invention and free association, and as a heuristic guide to both the senses and to reason. Situated between different kinds and degrees of mastery of abstract physical theories, the imagination is a place where heuristic guides and assumptions about reality can slip—and this slippage between nominalist and realist representations of the microworld became easier in the tricky transmission and translation of difficult theories across disciplines. I argue that in the 1920s and 1930s, the cell membrane and especially the "molecule" were precisely such underdetermined concepts, for which these kinds of translations between disciplines could happen without any clearer pattern of citation or other historically traceable intellectual descent.

More specifically in this chapter I seek to show how the concept of molecular orientation emerged out of physical chemistry in the 1910s and transformed from a relatively difficult synthesis of mathematical models, empirical facts, and abstract physical theories to an easily manipulated image or icon on paper and in the imagination. By looking for the graphical and imaginary origins of the lipid bilayer in biology, I show how the lipid bilayer became "obvious" to a small number of biologists in the mid-1930s. One of these biologists, the Giessen zoologist Wilhelm J. Schmidt (1884–1974), went further than most, imagining and then mobilizing the image of self-orienting lipids to render a dazzling world of submicroscopic atoms and molecules, all delicately assembled through no forces foreign to physics and chemistry.

The "Molecule" up to 1924

The word *molecule* itself was an underdetermined concept in the nineteenth century, despite its common use, and it was only in the years after World War I that the molecule was clearly conceived as an assemblage of atoms with definite shape.[3] The word *molecule* has its origins in Pierre Gassendi's *Syntagma Philosophicum* (published posthumously in 1658), a speculative work on Epicurean mechanical philosophy, and is thus allied with René Descartes' corpuscular metaphysics; for Gassendi, the Latinate neologism *molecule* would simply have meant "little mass" (OED Online, s.v. "molecule"). Even a century and a half later, Cartesian vortex theory could still be deployed in biology and natural history, with little change to the neo-Epicurean meaning of the word *molecule*: for example, "Life, then, is a vortex, more or less rapid, more or less complicated, the direction of which is constant, and which always carries along molecules of the same kind, but into which individual molecules [*les molécules individuelles*] are continually entering, and from which they are constantly departing" (Cuvier 1817, 13).

If corpuscular and discontinuous theories of matter had little bearing on biology in the nineteenth century, it was perhaps in part due to chemists' and physicists' continuing disagreement over the nature of the molecule as well: the physicists' "atom" and "molecule" were nearly incommensurable with those of the chemist, well into the twentieth century (Schütt 2002; Gavroglou and Simões 2012). Even if chemists were essentially united in a practical or pragmatic understanding of molecular *identity* by the 1860s—that is, a minimal unit of a distinct chemical species that could be identified by specific molecular weight—then exactly how this could be reconciled with physicists' views of molecular *forces* remained an open question.

Thus, on the one hand, chemists could disagree over whether atoms and molecules were real, indivisible particles or merely formulaic conventions on paper alone (Nye 1972; Nye 1993; Nye 1996; Ramberg 2003; Rocke 2010). On the other hand, physicists' formal mathematical equations left a great deal open to interpretation, and, on paper at least, the physicists' mathematics had little to do with the chemists' increasingly elaborate written formulas for molecules, reactions, and products. James Clerk Maxwell's physical definition of molecules in thermodynamics and gas law, for example, hypothesized that molecules might alternately be "portions of [a gas] which move about as a single body," or "pure centers of force endowed with inertia, or the capacity of performing work while losing velocity" (Maxwell 1868, 136). By the end of the nineteenth century, even as physicists and chemists were knitting together kinetic theory and the behavior of specific chemical substances, physicists found themselves again embroiled in tough metaphysical debates about the continuity or atomicity of matter, tussling over whether thermodynamic equations ontologically privileged either energy, on the one hand, or a statistical understanding of atomic or molecular behavior, on the other (Porter 1994; Staley 2008). Especially for physiologists with a clear physicalist bent, the absolute primacy of the Second Law of Thermodynamics could suggest that "molecules" were necessarily indeterminate, statistical, wandering beings, rather than clearly defined structural members of a living machine (Gray 1931, 14).

Despite the centrality of thermodynamics in physicists' and physicalist physiologists' understanding of the molecule, the physical chemistry of fats played a very different and genuinely outsized role in changing how molecules were conceived. Partly by historical accident, the physical investigation into fats began physicists' attempts to quantify both surface tension and molecular dimensions. Quite famously, in the early 1880s, while caring for convalescent parents, Agnes Pockels (1862–1935) noticed that the surface tension of her dishwater changed dramatically when it became slicked with oil. Using tin from a can of Liebig's meat extract and her father's pharmaceutical balance, Pockels built the first instrument to quantitatively measure the surface tension of thin liquid films: a broad rectangular trough, the scale measuring how much weight was required to separate a 6mm tin disk from the surface of water contaminated with oil, and the degree of contamination adjustable by a long tin or paper strip that scraped the water's surface, stretching or compressing the oil slick (Al-Shamery 2011; Beisswanger 1991; Rayleigh 1899; Ostwald 1932).

Meanwhile, in 1889 Lord Rayleigh (John William Strutt, 1842–1919) had begun to investigate the well-known phenomenon of camphor dancing upon water, and the interruption of that dancing by even a minute amount of oil. Using a "sponge bath of extra-size," Rayleigh, likely working at home with his wife, Evelyn Balfour (Opitz 2012), drew a bath thirty-three inches in diameter and placed camphor flakes on the surface; then, using a loop of platinum wire, he deposited tiny amounts of olive oil, which he claimed to be able to measure down to a twentieth of a milligram (Rayleigh 1890). By measuring the amount of olive oil required to stop the camphor from moving, and dividing that volume by the diameter of the tub, Rayleigh estimated that the maximum thickness of the oil film was 1.63nm ($\mu\mu$ in late nineteenth-century notation)—and, by extension, that this measurement might estimate the diameter of a single molecule of olive oil. By January 1891, Pockels had read of his interest in thin oil films in the *Naturwissenschaftliche Rundschau*, and wrote a twelve-page letter to Rayleigh, describing her tin trough apparatus and the variability of the surface tension of contaminated water. Rayleigh immediately forwarded the letter to *Nature* for publication, securing Pockels's high standing among physicists (Pockels 1891; Al-Shamery 2011).

Remarkably, for a bathtub experiment, Rayleigh's measurement for the diameter of an oil molecule was only slightly refined in the next two decades. This measurement, and this confluence of experiments on surface tension and molecular dimensions, happened in a relatively lowly domain of physics, far from the rarified realms of abstruse thermodynamic equations or metaphysical debates: in France, for example, Henri Devaux performed research on the camphor point, surface tension, and molecular diameters with a tiny toy boat (Devaux 1888; Devaux 1913). What they had in common, however, was a continuing operative assumption that molecules were perfect spheres—after all, this is the only way one could assume a molecule has a diameter, rather than a length, width, and height (Garber 1978). The physical assumption of spherical molecules in turn affected the way Rayleigh interpreted Pockels's discovery of the effects of oil on the surface tension of water. Pockels had found in the 1880s that the surface tension of water dropped when contaminated with oil, but surprisingly there was no clear linear or geometrical relationship between the amount of oil and the decrease in surface tension. As Pockels slowly added oil to the water's surface, surface tension remained unchanged until a certain amount of oil was on the surface; then it plummeted sharply in relation to the amount of oil, but before long the drop in surface tension leveled off, decreasing only slowly. Rayleigh suggested that the sudden drop in surface tension was due to the effects of packing the spherical molecules in an increasingly tight space, as well as the different forces at work between the oil molecules and the water's surface. The sharp decrease in surface tension "must depend upon the forces supposed to be operative between the molecules of oil. If they behave like the smooth rigid spheres of gaseous theory, no forces will be called into play until they are closely packed." Rayleigh's well-hedged conclusion was that the sharp drop in surface tension occurred as the oil film on the water's surface transitioned from being one molecule thick to two molecules thick. Any heterogeneity in the olive oil might then explain the differences across measurements, "whereby some molecules would mount more easily than others" in the chaotic, jumbled transition state (Rayleigh 1899, 337).

Interpreting Surface Tension: Molecular Orientation

It was this confluence of the clear facts of surface tension measurements and the tentativeness of molecular hypotheses that would lead two different American physical chemists to independently and simultaneously develop the theory of molecular orientation in 1917.[4] Irving Langmuir (1881–1957) and William Draper Harkins (1873–1951) knew each other professionally, and the timing of their announcements in the *Journal of the American Chemical Society* (*JACS*) a mere five months apart led to a bitter priority dispute and accusations against Harkins of intellectual theft (Coffey 2008, 128–34).[5] Even though Harkins and Langmuir eventually agreed on the principle and theory of molecular orientation, their approaches to molecular orientation were quite different and addressed to slightly different scientific communities. Harkins, a relatively traditional university chemist, wrote and spoke in part to colloid chemists, a new and rapidly growing discipline that counted many biologists in its ranks. Langmuir, on the other hand, cemented his reputation as an iconoclastic and revolutionary chemist who endeavored to unify and clarify differences between physical and chemical approaches to atoms and molecules.

Langmuir had trained in Walther Nernst's eclectic physical laboratory in Göttingen, but in 1909 he joined General Electric's new research laboratory in Schenectady, New York, eschewing a traditional academic career (Süsskind 2008; Kohler 1974). At GE, Langmuir was free to pursue whatever interested him (unusual for a corporate scientist), and this would eventually include research on thin films and atomic structure in light bulb design (Wise 1980; Wise 1983; Reich 1983). His most important agenda in the 1910s and 1920s was bridging what he saw as a yawning chasm between chemical and physical theories of molecular behavior, asserting that chemists' structural formulae—formulae that did not suggest perfect, spherical symmetry—ought to have a greater bearing on theories of physical structure and behavior. The experiments of Rayleigh and Pockels with oil films provided the opportunity to build that bridge.

Langmuir (1917) used what was essentially a more elaborate version of Pockels's tin trough, repeating many of Pockels's and Rayleigh's experiments on surface tension. The key difference was that Langmuir used very specific and chemically pure oils, rather than whatever olive oil happened to be in the kitchen, as Lord Rayleigh had in 1889. Langmuir observed that most oils decreased the surface tension of water by the same amount when they were laterally compressed, but that this ability to lower surface tension with uncompressed films depended on exactly what kind of oil was being used. He believed that the specific composition of a fatty acid's hydrocarbon chain and the number of double bonds in that chain corresponded with the ability to stretch a monomolecular film without breaking it—and indeed Langmuir found that the saturated stearic acid covered a maximum area that was less than half of a film covered by the monounsaturated oleic acid.[6] Langmuir concluded by arguing that a single molecule of oil resting on a water surface had its carboxyl group and any unsaturated carbon double bonds chemically bonded to the water, while the CH_3 hydrocarbon tails flopped around freely on the surface.[7] Thus, when the oil was compressed, only the carboxyl groups remained stuck to the surface of the water, while the hydrocarbon tails stood vertically upright. In other words, Langmuir found an experimental system that could show that fats with different chemical formulas could be found to have different lengths, and that there were two different kinds of relationships between surface tension and length: there was the relationship between the length of the fatty acid and the changes in surface tension, but there was also a less direct relationship between the level of chemical saturation in the fatty acid and the changes in surface tension. The specific chemistry of fats, Langmuir argued, seemed to override the more general assumptions made in physics.

Langmuir's series in the *JACS* was brilliant in synthetic scope, but also difficult to understand in all of its details unless one had as wide-ranging a command of chemical and physical theory as Langmuir had. In contrast, Harkins's work on surface tension relied less on synthesizing a wide range of theories and more on tackling a specific problem: the relationship of surface tension to solubility. For example, the theory suggested that urea and water enter into solution very easily because they have extremely low surface tension, while oil and water are so insoluble you can see the surface tension working with your naked eye. However, Harkins and his laboratory team at the University of Chicago discovered that surface tension alone was a poor predictor of solubility, especially of fats and other organic acids. Using a much more precise set of instruments than the Pockels-Langmuir trough (Harkins and Brown 1916; Harkins and Humphery 1916), Harkins's team surveyed surface tension data for 336 different substances in both air and water and noticed that, for many substances, the surface tension of one substance in the air was drastically different than if it had an interface with water (Harkins, Brown, and Davies 1917; Harkins, Davies, and Clark 1917). Furthermore, the differences seemed to be roughly related to the presence of carboxyl (COOH) groups and the relative saturation of any hydrocarbon chains. However, rather than make any general argument about the length of a molecule or hydrocarbon chains flopping around on water, Harkins proposed a very physicalist thought experiment, asking, How much work would it take to separate two substances, say, benzene and water, at their interface?

> If it is imagined that a single liquid is divided into two parts by a horizontal plane, and that when this imaginary plane is lifted the upper layer rises with it, then, where before there was no surface, two surfaces now appear . . . since the surface tension of water at 20° is, according to our measurements, 72.8 dynes per cm., the free energy per square cm. is 72.8 ergs. The total energy of the two surfaces, each of which may now be supposed to have an area of 1 square cm., will be 145.6 ergs. If the two surfaces now approach and meet one another, this free energy disappears, since there is now no surface energy at the imaginary interface. (Harkins, Brown, and Davies 1917, 335)

Or, stated in more formal terms: If the independent surfaces of two separate substances are maintained by a certain amount of energy, then what is the decrease in energy if two substances approach and touch one another? This gave the following mathematical expression:

$$(\gamma_a + \gamma_b - \gamma_{ab} = -\Delta\gamma),$$

where γ_a and γ_b are the surface tension measurements of substances *a* and *b* independently in air or water, and γ_{ab} is the surface tension of *a* and *b* when they are in contact with each other. If there is difference remaining, $-\Delta\gamma$, it would suggest that there is something about the interface of the two liquids that is very different from the behavior of the two liquids acting independently of one other. Harkins argued that if there was any non-zero value for $-\Delta\gamma$, then in order to make the transition from *a* to *b* less abrupt, the molecules could be imagined to orient themselves in a way that lowered the tension at the interface. As he put it in a more general way: the boundary of any homogenous liquid with another must have some structure to make the boundary less energetic, if possible.

Harkins concluded that "at the interface between another liquid and water, the molecules in the surface of the liquid set themselves in such a way as to turn their more active or polar groups toward the surface of the water. At such surfaces liquids therefore show a structure" (363). Harkins's explanation for the energetic difference at the interface was thus the same as Langmuir's explanation of the relationship between surface tension and the maximum area of a monomolecular oil film: there must be some shape, or structure, or other kind of polarity in molecules that causes them to orient at the interface, and this orientation works to reduce surface tension.

Colloid Chemistry and the Iconography of Molecular Orientation

Conceivably, Langmuir's position at General Electric insulated him from other scientists who needed to understand how his theory might be generally applicable: he was a lone genius given free rein in a corporate laboratory, and the truly eclectic nature of his writings seems to reflect the wide range of interests he held in a somewhat undisciplined fashion. Harkins's writings and lectures were only slightly less difficult, but he was to prove more capable than Langmuir in speaking and writing to audiences who did not have much use for either mathematical physics or the details of organic chemical theory.

Not only was Harkins less dogmatic in his views, but he was more closely engaged with the interests and concerns of colloid chemistry. In the 1920s colloid chemistry was a discipline ascendant, propelled by effective evangelists, promises of wide industrial application, and catholic epistemological standards (Ede 2007, 78–101). Colloids were defined from the mid-nineteenth century as heterogeneous aggregates that defied the usual methods of chemical analysis by sublimation or crystallization; colloid chemistry was thus a science of unruly and mixed materials like soaps, blood, rubber, soil, mucus, sewage, and, crucially, cells and protoplasm.

Colloid chemistry was typically instrumentalist or nominalist in its methods, in large part because of the wide range of materials classified as "colloids"; typical experimental topics included viscosity, flow, opacity, behavior in changing temperatures, response of a colloid to mechanical forces, and response to electrical fields and charges. This focus on techniques of measurement and description of materials at hand allowed colloid chemists to communicate across vastly different specialties, despite working with a diverse range of colloidal materials. John Heilbron (1982), Ted Porter (1994), and others have called this general tendency in fin de siècle physics "descriptionism," and Porter in particular has argued that this epistemological remove from specific objects of inquiry allowed physics to broaden its scope and influence—that "descriptionism aimed to make physics almost impregnable, to confer on it something like the degree of certainty normally associated with mathematics. . . . The release of physics from all particular objects helped to dissolve the boundaries that confined physics to one aspect of the natural world" (Porter 1994, 130). The diversity of topics in colloid chemistry journals, symposia, and international meetings meant that publishing in *Kolloid-Zeitschrift* or *Protoplasma*, or attending a meeting of the Faraday Society, gave an individual scientist potentially broad reach.

It was perhaps this kind of wider engagement that led Harkins to give those less mathematically or theoretically inclined colloid chemists a series of verbal and graphical analogies for molecular orientation, starting in his June 1924 lecture to the National Colloid Symposium hosted by Northwestern University.[8] Harkins's lecture, "The Orientation of Molecules in the Surfaces of Liquids," has the first graphical representation of molecules as a ball and stick, to illustrate his surface structure principle from 1917. The sheer novelty of the concept of molecular orientation, however, gave cause for Harkins to elaborate two analogies in the lecture. One was verbal: "The ordinary observation of large scale objects, such as logs or ships, as they lie on the surface of a body of water, indicates that these objects exhibit a characteristic orientation with respect to the surface. Thus logs, when not too closely crowded together lie flat upon the water, that is the longitudinal axis is parallel to the surface. However, if one end of each log is loaded with a mass of iron or brass of the proper weight, it floats upon the surface and the longitudinal axis becomes vertical" (Harkins 1924b, 141).

This exercise in imagination was then accompanied by a visual and material analogy, physically dragged out onto the stage in front of the audience at Northwestern. As the published text in the *Colloid Symposium Monograph* described the scene parenthetically,

> (These phenomena were illustrated by the use of a large number of cylindrical sticks of wood 3 mm. in diameter and 14 cm. long, weighted by a small cylinder of brass placed at one end. These were thrown upon the surface of the water in a large glass cylinder. This is represented in a diagrammatic way in Fig. 1. One of the vertical sticks was taken from

the water, the brass weight removed, and the stick dropped upon a vacant space upon a water surface. At once this assumed a horizontal position, thus exhibiting another type of orientation.) (Harkins 1924b, 141–42)

The first figure in the lecture is static (fig. 10.2*A*), and claims to represent the analogy of weighted logs floating on water.

By equal measure, Harkins also emphasized that his diagrams were "highly conventionalized" (Harkins 1924b, 149), and in some of them it is not clear whether the diagrams were supposed to illustrate the molecules themselves or to illustrate dissymmetrical fields of molecular and surface forces. Yet the potential for slippage into realism was clear, and some of Harkins's other figures (fig. 10.2*B*) seem to show how a jumbled mass of butyric acid molecules really could behave—individual molecules plunging into the water and tumbling back out, some molecules curved and other straight, most of the surface molecules neatly oriented, and a few molecules left out of the orientation party. Such a figure was supposed to illustrate Harkins's argument that "disorder has been overemphasized" in thermodynamic conceptions of molecules in liquids. Yet in attempting to illustrate a semi-ordered system, structured at the surface but unstructured in the greater body, Harkins managed to produce schematic diagrams that were realistically suggestive precisely because of their liveliness.

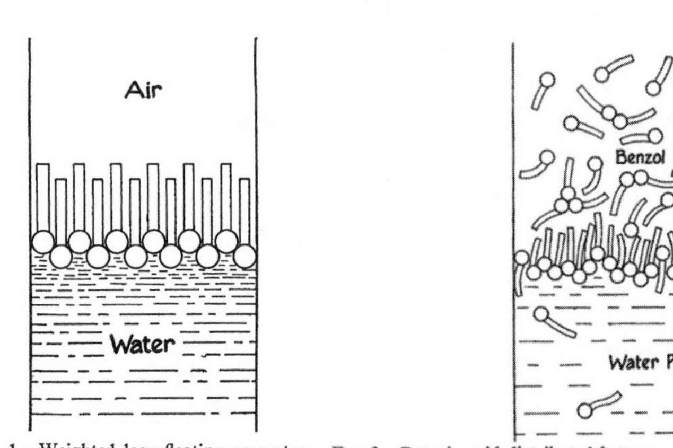

Figure 10.2. William Harkins's diagrammatic representation of sticks with brass weights on one end, thrown in a container of water. *A*, a strictly schematic analogy. *B*, introducing an element of realism in a purportedly schematic diagram. From Harkins 1924b, 142, 151.

It is not clear exactly how or when Harkins's diagrams began to make their way through other parts of colloid chemistry. By now he was well known as a leading authority on surface forces, and versions of the Colloid Symposium lecture found their way into two colloid chemistry textbooks. In the first textbook Harkins even mentions that polar molecules "have been represented in this laboratory for many years" by the ball-and-stick symbol (Harkins 1924a, 154), though this is the only place where he makes this claim. (This is also the only place where Harkins credits his student Ernest B. Keith with the illustration.) In the second textbook, part of the very influential multivolume series edited by the colloid chemist Jerome Alexander, Harkins not only reproduces all of the diagrams from 1924, but ceases to refer to them as "conventions" (Alexander 1926, 192–264). Langmuir as well wrote a chapter for Jerome Alexander's textbook (525–46), and this chapter seems to have been the first time Langmuir resorted to using a diagrammatic representation for molecular shape and dissymmetry, at least in print. Rather than use a version of Harkins's diagram, Langmuir here used a small black dot connected to a fat, elongated tube, like a caper stuck to one end of a sausage, with the tubes varying in length to represent the real length of the molecule in question (fig. 10.3). Few if any later diagrams look like Langmuir's 1926 diagram, which would have been more useful in illustrating molecular dimensions than the larger-scale, aggregate effects of molecular orientation.

Harkins was more than just an authority on surfaces, however. By the mid-1920s, surfaces became a central organizing theory in colloid chemistry, with "colloids" themselves being redefined as systems that were composed of a vast number and amount of surfaces. Earlier in the twentieth century, colloids had been redefined from an operational state (e.g., inability to crystalize, inability to pass through parchment paper) to being a "disperse, polyphase system," a mixture of multiple substances with different chemical identities (e.g., mud is a mixture of a watery "continuous" phase and a "disperse," mineral particulate phase).[9] The physicist Herbert Freundlich (1880–1941) quickly recognized that this definition of colloids as disperse, polyphase systems meant that a colloid was generalizable as a gigantic surface: each particle of the disperse phase would thus have an exterior surface that remained in contact with the continuous phase, and with the total surface between the two phases measurable in the range of tens to hundreds of square meters within a single cubic centimeter of a colloid substance (Freundlich 1907). By June of 1926 the soap and colloid chemist James W. McBain (1882–1953) stood as the keynote speaker of another Colloid Symposium, now hosted at MIT, and argued that surface tension was

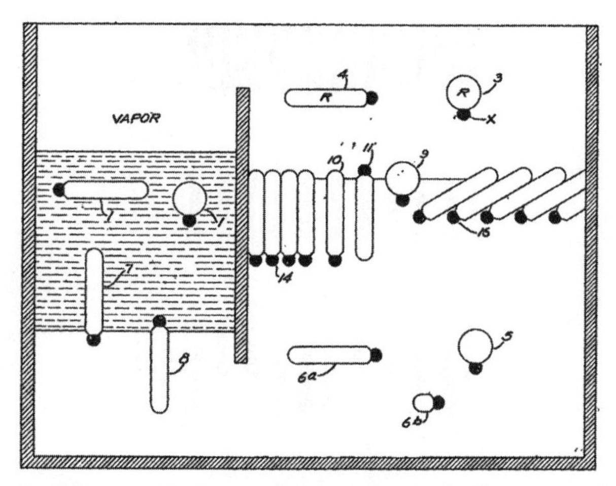

Figure 10.3. Irving Langmuir's diagram explaining "molecular dissymmetry." The small black dots represent a radical active group, and the cylinders represent hydrocarbon chains. From Irving Langmuir, "The Effects of Molecular Dissymmetry on Properties of Matter," in *Colloid Chemistry: Theoretical and Applied*, ed. Jerome Alexander (New York: Chemical Catalog Co., 1926), 525–46, on 538.

the ultimate determinant of colloidal stability: "It is not the nature of the interior," he declared, "but the composition of the exterior of the particle that determines [the colloid's] chief properties and degree of stability. . . . The motto of the colloid is, 'Save the surface, and you save all' " (McBain 1926, 9).

The very first case I have found where the ball-and-stick image was used by someone other than Harkins dates from just one month prior to McBain's Colloid Symposium address in 1926. This was also by McBain, in May of 1926, in a very technical physical lecture entitled "An Experimental Test of the Gibbs Adsorption Theorem" (fig. 10.4; McBain and Davies 1927). McBain used a single, four-part "diagrammatic representation" (2231) of a monomolecular film, copying Harkins's diagrams, not Langmuir's. McBain and his student George Davies created this diagram to compare some of the discrepancies between Langmuir's 1917 basic theory (*a* in the diagram), other explanations coming from thermodynamic theory (*b* and *c*), and attempted measurements of how many molecules actually seemed to be adsorbed to the surface, as well as how *deep* the surface layer could be (*d*). Harkins is not cited as a source for the image, and McBain and his student George Davies only note that Harkins and several others, had offered "a clear picture of the structure of films of insoluble materials resting upon a solvent such as water" (2230).

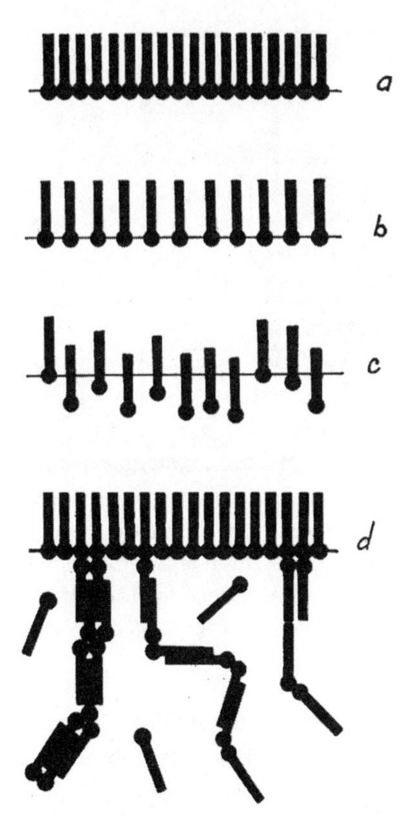

Figure 10.4. Ball-and-stick models of monomolecular films. This may have been the first time the ball-and-stick model was used after Harkins. Reprinted with permission from McBain and Davies 1927, 2231. Copyright 1927 American Chemical Society.

McBain probably meant "picture" figuratively as a "conception" rather than literally as an "image" or "visual representation"; he cites Rayleigh, Adam, Devaux, Langmuir, and Harkins, and of these scientists by May 1926, only Harkins had published an image of surface molecules. McBain's use of the ball and stick to represent a molecular film is quite casual and unattributed, so it is impossible to specify exactly the source from which he might have borrowed the image, or whether he invented the image himself. However, it seems very likely that the images have the same provenance, given the importance of Harkins's and Langmuir's writing, and given that McBain was a contributor to both of the textbooks for which Harkins had also written. McBain's own work in soap chemistry offers another route of transmission with Harkins: many of Harkins's 1924–25 articles engaged with soap chemistry, and in this context he briefly suggested that the ball-and-stick model actually represented molecular "wedges" capable of orientation. In 1925, in a popular lecture at the Royal Institution, McBain had described colloidal soap particles as, "like a pair of military hair brushes, in which the bristles represent the hydrocarbon chains of the molecules arranged parallel to each other in sheets, two such layers being put together hydrocarbon to hydrocarbon. The two backs of the brushes on the outside represent the hydrate layer and the un-ionised electric double layer" (McBain 1925, 581).

This picture of an opposing pair of brushes was accompanied by an overly detailed chemical diagram that suggested a precise location for every atom and valence bond, a mesmerizing arrangement of capital *H*s and *C*s in neat, parallel zig-zags and rows—an image useful for showing detailed structure but less so for illustrating orientation (fig. 10.5). This connection between Harkins and soap chemistry was also probably not an accident:

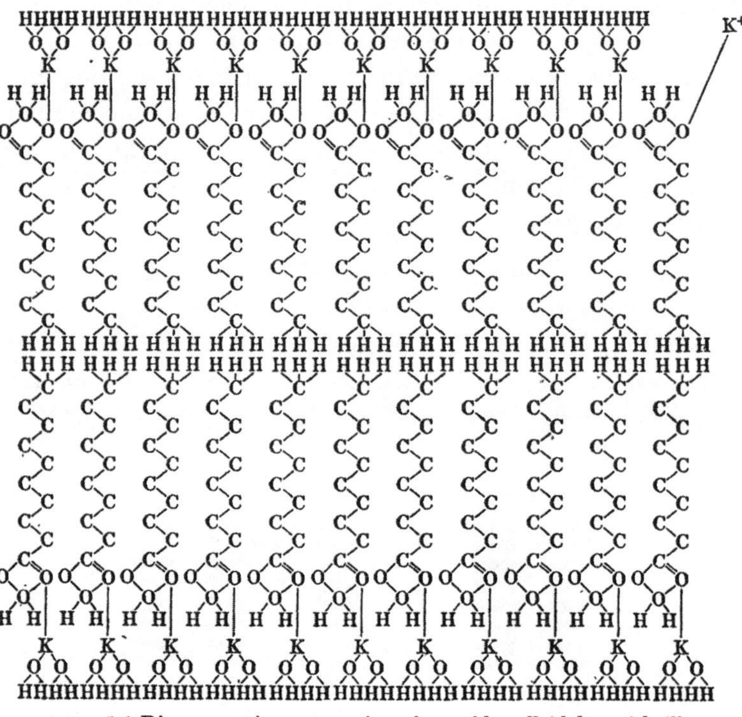

FIGURE 1·1 Diagrammatic cross section of a stable colloidal particle illustrating the principle of "like to like."

Figure 10.5. An overly complicated attempt at chemical realism, the "pair of military hair brushes." This image originally accompanied James McBain's 1925 lecture at the Royal Institution, but was only published later in McBain 1950.

McBain saw the study of soap and soap production as an especially rich area for colloid chemistry, since soaps were chemically simple substances that were but poorly understood in their manifold physical behaviors (Stadler 2009).

Even more evocatively, McBain also cheerfully suggested that the colloidal particles of soap "resemble a group of, say, less than a dozen eels tied together by the tails, and pointing outwards in all directions from the common centre" (McBain 1925, 582). Although there was some precedent to describing fat molecules as having hydrocarbon "tails" before 1925 (Langmuir 1917, 1864), the verbal convention of referring to lipids as having "heads" as well had become common enough in the 1920s that an older soap chemist thought it merited some disparaging comments:

> The individuality of soap molecules is so peculiar that they may be described as eccentric. By various workers they have been credited with heads and tails, although they prefer to stand upon the former. Indeed, they appear to try to emulate the ostrich and bury their heads in the most unlikely surfaces while the rest of their body, which only consists of a tail, sticks up in the air. This type of anthropomorphic familiarity, however picturesque, should only be indulged in with caution, . . . [and] the implied endopsychic endowment of the molecules is quite unjustifiable. (Lawrence 1929, 132)

This particularly ill-tempered soap chemist was probably the first to publish an illustration of a "sandwich" of fat molecules (fig. 10.6), with tails oriented toward each other, and using the ball-and-stick convention (Stadler 2009, 73–74). This image by A. S. C. Lawrence came from "certain

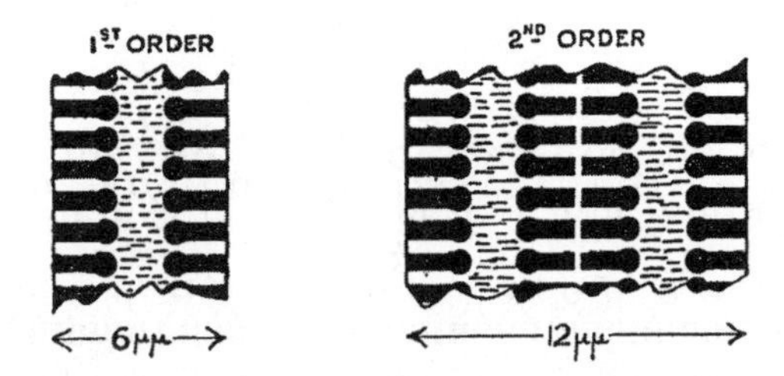

Fig. 59.—Sectional view of black film, diagrammatic.

Figure 10.6. A. S. C. Lawrence's static diagram of very thin soap films, perhaps the first to be shown as a lipid sandwich. This image originally appeared in Lawrence 1929, 128, but it was reproduced and likely more commonly seen in Adam 1930, 137.

a priori possibilities": that is that since a monolayer of fats seems to only exist on a surface of water, two surfaces of water (the shaded regions in fig. 10.6) could support the existence of a bilayer (Lawrence 1929, 11–12). This image was copied and cited in 1930, without mention of the "sandwich" metaphor, by Neil Kensington Adam (1891–1973), the physical chemist who was the mentor and advisor to the "inventor" of the lipid bilayer cell membrane in 1935.

Lipids and the Biological Microworld

When James Danielli proposed his cell membrane model in 1935—a layer of protein adsorbed onto the lipid bilayer that "would have been obvious to any competent physical chemist"—he had already spent seven years under Adam's tutelage at University College, London, having gone to Adam for chemistry lessons since 1928, at the precocious age of seventeen (Stein 1986). So it should be no surprise that Danielli thought a bilayer of lipid molecules was an "obvious" structure that needed no citation. The closest citation for a lipid bilayer in Danielli and Davson's short and quite speculative 1935 paper is to Adam's 1930 textbook, *The Physics and Chemistry of Surfaces*, where the only molecular diagram was the one borrowed from Lawrence (Adam 1930, 136–37).

Danielli may have been aware that, in the early 1930s, schematic diagrams of lipids and the structure of fats were slowly spreading across to France and Germany, where the study of fats had become associated with the biology of nerve cells in addition to the physics of soaps. Since 1924, the Giessen biologist Wilhelm J. Schmidt had made a reputation for himself by arguing that animal tissues and cells were made of "building blocks" (*Bausteine*) of submicroscopic, crystalline particles (Schmidt 1924). This was an unusual position for a biologist to take in the 1920s, when most biologists had just recently embraced colloid chemistry as the future for cell research—and in so doing, they had made the decision to avoid microphysical or submicroscopic speculations. The structure of the cell and protoplasm had widely been acknowledged as being a colloid since the late 1890s, giving the view that the cell was a dynamic, heterogeneous aggregate of living slime. For example, through the 1920s, the plant physiologist D. T. MacDougal (1865–1958) was engaged in building artificial plant cells out of gelatin, fats, and filter paper, attempting to create the colloidal structures that mimicked the way plants absorb water (MacDougal 1924). MacDougal's diagram of what he thought was the gradual transition from the colloidal cell wall to the colloidal protoplasm offers an exceptionally

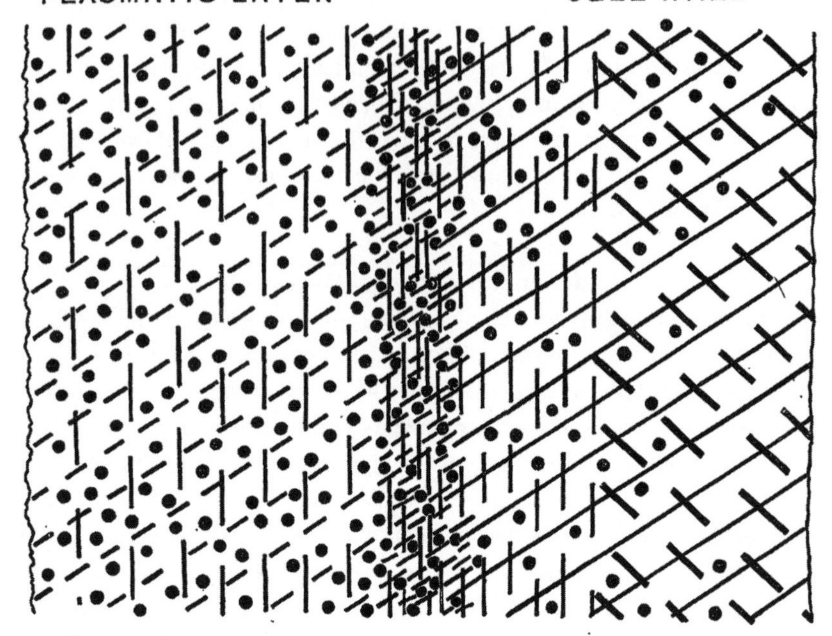

Fig. 1. Diagram of the arrangement of material in the plasmatic layer and cell-wall. / Cellulose. \ Pectin. | Mucilages. / Proteins. • Fatty substances.

Figure 10.7. D. T. MacDougal's schematic of the colloidal arrangement of the cell wall, lipoidal membrane, and protoplasmic body, showing the three systems as one continuous system, rather than three discrete layers. This is one of the only schematic, didactic images of the colloidal structure of the protoplasm, and its rarity is perhaps accentuated by the emphasis on instrumental measurement in colloid chemistry. From MacDougal 1924, 77.

clear (and exceptionally rare) visual insight into how biologists in the 1920s envisioned the cell as a colloidal aggregate: not as a series of clearly delimited anatomical parts, like walls, membranes, or chromosomes, but all as part of a dynamic, polyphase colloidal system, each part blended into the others (fig. 10.7).

Up to 1937, Schmidt was apparently still unaware of Langmuir, Harkins, Adam, and certainly not aware of Danielli and Davson, and none of those names appeared in any of his citations until 1938. Schmidt had been trained as a zoologist in the relatively old-fashioned zoological institute at Bonn in comparative anatomy and natural history. Of those who published in the *Kolloid-Zeitschrift*, Schmidt was perhaps the most naive about physical chemistry, and in none of his writings does Schmidt show more than a passing acquaintance with topics like surface chemistry or colloidal theory. As a university student, he had taken classes in physics and chemistry, but his interests were in philosophy, art, and classics; and as a graduate student in zoology, he had passions for reptiles, mollusks, and a sunshine-filled life at the various marine research stations along the Mediterranean (Schmidt 1964).

It was around 1910 and probably at the Naples Zoological Station where Schmidt learned polarized light microscopy in order to study oyster shells and mother of pearl, and it was through studying the technique that Schmidt became committed to seeing living cells as being composed of crystalline building blocks rather than unstructured colloidal slime. Polarized light microscopy was a well-known technique to detect anisotropy—directionality or orientation—in crystals and minerals, and it had long been used in mineralogy and geology to identify rocks. Initially, Schmidt began to use polarized light microscopy to study teeth, shells, scales, hair, hard excrescences, and bones. His 1924 comparative anatomy project, *The Building Blocks of Animal Bodies in Polarized Light*, was essentially five hundred pages of detailed examinations of the hard, solid parts of many animals, in the tradition of nineteenth-century comparative anatomy and zoology.

Through the 1920s and into the 1930s, however, Schmidt began to immerse himself in the technical and theoretical approaches to polarized light microscopy that were being promoted by the botanists Hermann Ambronn (1856–1927) and Albert Frey (1900–1988).[10] Ambronn and Frey's ideas promised to give biologists the ability to make reasonable guesses about the submicroscopic structure of the soft, colloidal parts of living cells, such as the unlignified cell wall or the chromatin in chromosomes and nuclei. Their technique relied on a set of optical theories developed by the physicist Otto Wiener (1862–1927), and known as the "Wiener Mischkörper" or "Wiener mixed bodies" (fig. 10.8; Wiener 1904; Wiener 1909). This theory suggested that two idealized colloidal structures would show very specific kinds of birefringence and colorful interference patterns when viewed under cross-polarized light: rodlets arranged in parallel columns within a fluid system would show positive birefringence, while platelets stacked in alternating layers of the same fluid medium would show negative birefringence. Ambronn and Frey (1926) proposed immersing the colloid (or cell or tissue) in a fluid whose refractive index could cancel out the optical properties of the medium surrounding these rods or platelets, thus making any intrinsic optical properties of the rodlets or platelets directly

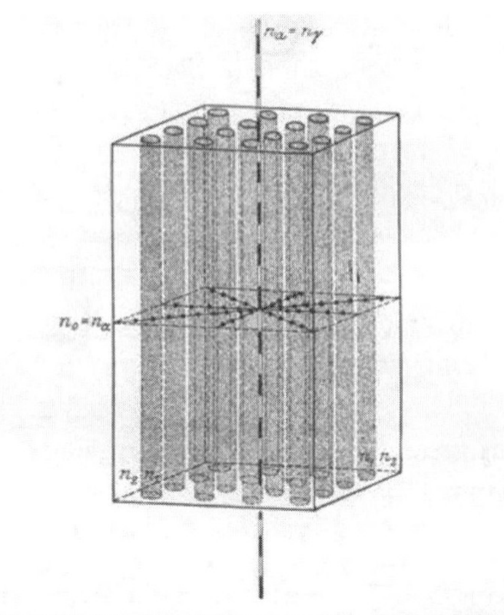

Fig. 25. Zylindermischkörper. $n_a > n_o$, optisch positiv.
$\text{▬▬▬} = n_\gamma$, $\text{•-•-•} = n_\alpha$.

A

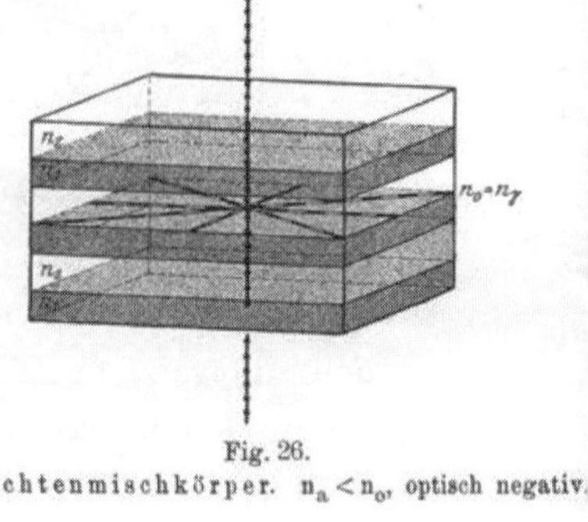

B

Figure 10.8. Wiener mixed bodies: *A*, rodlet mixed body; *B*, platelet mixed body. If the rodlets and platelets in the two systems were of identical materials, and the media were identical to each other as well, then the rodlet mixed body would always show positive form birefringence parallel to the direction of the rodlets, while the platelet mixed body would always show negative form birefringence perpendicular to the direction of the platelets. From Ambronn and Frey 1926, 114, 119.

accessible. This "imbibition method" would allow the biologist to determine exactly how the submicroscopic rodlets or platelets were arranged within the living tissue, unobscured by the continuous colloidal phase. Ambronn and Frey argued that a good polarization microscopist could separate the "form birefringence" (*Formdoppelbrechung*) of the whole system from the "intrinsic birefringence" (*Eigendoppelbrechung*) of the underlying submicroscopic parts (e.g., parallel rodlets or stacked platelets).

The technique required a great deal of patience, but it had a crucial advantage over traditional cytological fixation and staining: it did not require preserving and killing the cells, which would alter their delicate, submicroscopic, colloidal structure. The first soft tissue Schmidt tackled using Ambronn and Frey's technique was frog eye retina (Schmidt 1935). Schmidt knew that the rod cells especially were delicate complexes of fatty and proteinaceous layers, and at a minimum, he wanted to see if polarization microscopy could allow him to see how they were intertwined. In this 1935 paper, the schematic illustrations are all aimed at working out not only where the fats and proteins are located but whether the proteins and fats showed orientation with respect to one another (fig. 10.9). At this point in 1935, Schmidt believed that living matter was ultimately composed of atoms and molecules, but with the polarized light microscope he could only hint at the directionality of any molecular or supramolecular structures with long dashes.

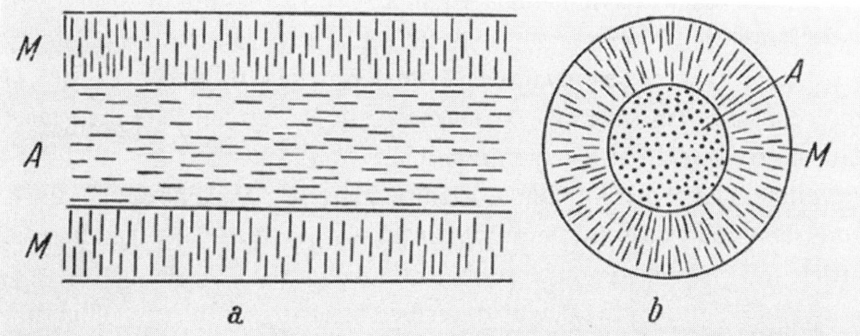

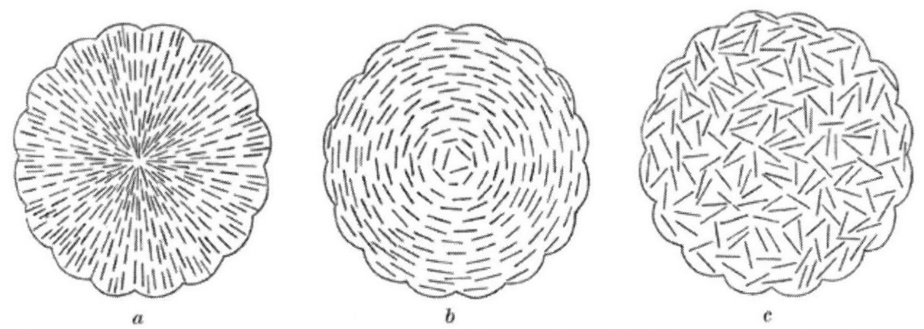

Figure 10.9. Wilhelm J. Schmidt's early attempts to decipher the fine structure of frog eye rod cells. Note that "molecule" and "colloidal particle" are construed as synonyms in the captions, while the images only schematically show oriented particles. The pattern of arrangement of the linear particles in all of these diagrams is meant to guide and predict what kinds of form and intrinsic birefringence might be seen under polarized light. From Schmidt 1935, 513. With permission of Springer.

Between 1935 and 1938, Schmidt began to read more widely on the optical properties and molecular structures of fats and lipids, the same area of physical investigation where surface tension and molecular orientation were built. According to his citation patterns in 1938, Schmidt relied especially heavily on two works by two French scientists. The first was the thin-film chemist Henri Devaux (1931), who had written a comprehensive review article on thin oil films and molecular orientation. Second was the neuroanatomist Jean Nageotte (1936), who had written a monograph on the morphology and polarization optics of lipid gels. Devaux pointed to Langmuir and Harkins's theoretical work, while Nageotte had also incorporated recent French and German X-ray crystallographic research on soap structures. Nageotte was especially attentive to bimolecular lamellar structures, and the only molecular diagram he reproduced was of a bimolecular soap micelle by P. A. Thiessen and R. Spychalski (1931)—each molecule rendered as a very thin ball and stick, arranged in a crystalline rectangle, a fairly distant relative to the icon used by Harkins or McBain. In addition to Nageotte and Devaux, Schmidt was aware of the very influential article in *Protoplasma* by the Dutch colloid chemist H. G. Bungenberg de Jong and J. Bonner (1935), who described birefringent bilayers of lecithin that could self-organize under the right electrostatic conditions. By 1938 Schmidt was beginning to use an iconography of the lipid bilayer structure that would have been very familiar to the surface tension theorists of the 1920s. Schmidt likely came to the ball-and-stick representation through following Nageotte and Devaux's citations—though perhaps he had seen the iconography of molecular orientation at a conference or when chatting with a colleague.

What would have been foreign to workers in the 1920s, however, was Schmidt's complete reliance on a visually inspired language and drawings of shapes of molecules, and his nearly complete abstinence from the complicated physics behind the lipids' shapes and configurations. The phrase "surface tension" ("*Oberflächungspannung*" and variations thereof) appears only three times and only very briefly in Schmidt's first article (1938b) featuring lipid molecules in *Die Naturwissenschaften* (a general audience journal similar to *Science* and *Nature*). When he reprised the article for *Kolloid-Zeitschrift* (1938a) a few months later, he took out any mention of surface tension entirely—an odd move, given the journal. Instead of explanations of fluid or molecular forces, Schmidt provided ten pages of informed guesswork about what kinds of arrangements and materials give rise to specific birefringence patterns, paying close attention to signs of

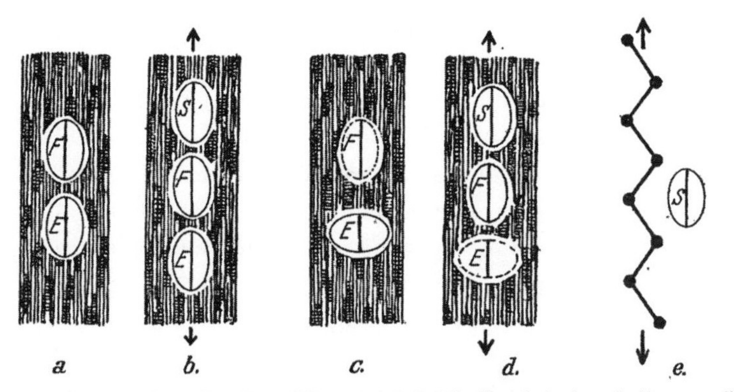

Abb. 48. Spannungsdoppelbrechung (*S*), erzeugt durch Zug bei einer kollagenen Faser (*a*, *b*) mit positiver Form- (*F*) und Eigendoppelbrechung (*E*) und bei einer sumachgegerbten kollagenen Faser (*c*, *d*) mit positiver Form- und negativer Eigendoppelbrechung; *e* = Beanspruchung der Valenzbindungen einer Fadenmolekel durch Zug.

Figure 10.10. One of Wilhelm Schmidt's exercises for learning how to interpret structure from polarization optical observation: observing changes in the sign of birefringence as a collagen fiber is pulled. On the left, *a* and *b* are ordinary collagen, showing positive form birefringence and intrinsic birefringence, as well as positive uniaxial strain birefringence. On the right, *c* and *d* are collagen tanned with sumac, showing positive form birefringence and negative intrinsic birefringence, the latter due to the shrinkage caused by the sumac; *e* is a molecular-atomic schematic of the fiber expanding by pulling, resulting in the positive strain birefringence. From Schmidt 1939, 371. With permission of Springer.

physical polarity and directionality he had seen in various kinds of tissues and cells.

This might have been more biology than readers of *Kolloid-Zeitschrift* were used to, but within the discipline Schmidt was fast becoming known as a leading expert on the optical properties of complex colloids. Further, in a long article explaining his techniques to the German Zoological Society in 1939, Schmidt essentially offered any reader a manual to work out how different molecular structures appeared under polarized light. Rather than immediately asking the reader interested in his methods to look at living tissues, Schmidt (1939) offered a few hypothetical diagrams for protein-lipid structures before taking the reader on a series of exercises with chitin, collagen, and lecithin smears. The exercises using exemplary materials were aimed at training the novice polarization microscopist to notice what kinds of materials and under what conditions certain birefringence patterns could appear. The diagrams in the article were then meant to illustrate the fine structural details that were causing the birefringence patterns (fig. 10.10). For Schmidt in 1939, molecular structures could be "seen" by inference and even manipulated on a large scale, regardless of whether the individual molecules were visible or yet rendered on the page.

The last time Schmidt was to write about his methods in depth came in 1941. Soon afterward, World War II left the Giessen zoological institute devoid of all but a few graduate students; the American firebombing campaign on December 6, 1944, leveled most of the city, including Schmidt's library, laboratory, and much of the rest of the university as well (Frey-Wyssling Briefe, 8 May 1946, HS 0443:1059). In 1941 Schmidt now had the experience and confidence to freely draw and diagram what he thought were the behavior and structural inclinations of proteins and lipids. The realism of these images also represented what he imagined was the fine structure of lipid membranes (fig. 10.11). "*Strong hydrophilic lipoids* such as *lecithin* order themselves automatically in the presence of water into bimolecular layered systems, so-called *myelin figures*: attracted by the hydrophilic groups, water penetrates into the material and gives the molecules

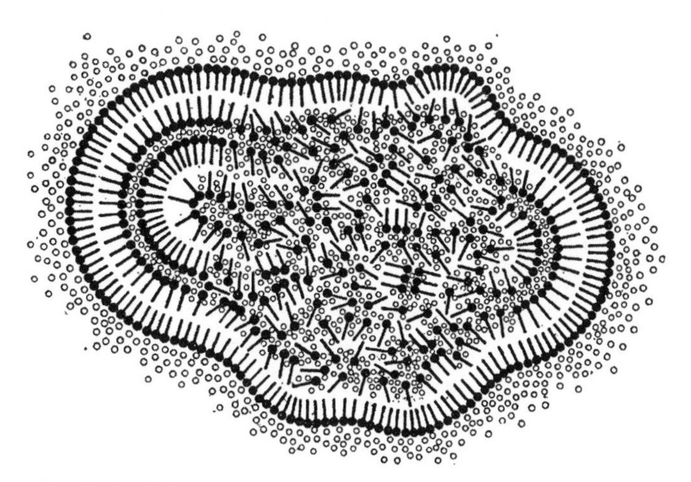

Abb. 6. Lecithinmasse in Wasser (Kreise): Das Wasser ist in das Lecithin eingedrungen; die oberflächlichen Molekeln haben sich mit ihren hydrophilen Polen gegen das umgebende Wasser gekehrt; die so entstandene unimolekulare Lamelle ordnet die anliegenden Molekeln usw.

Figure 10.11. Wilhelm Schmidt's realistic, diagrammatic image of a cross section of a lecithin droplet in water. "The water has invaded the lecithin; the outer surface molecules have turned against the surrounding water with their hydrophilic poles; the developing unimolecular lamella arranges the adjacent molecules, and so on." From Schmidt 1941, 44. With permission of Springer.

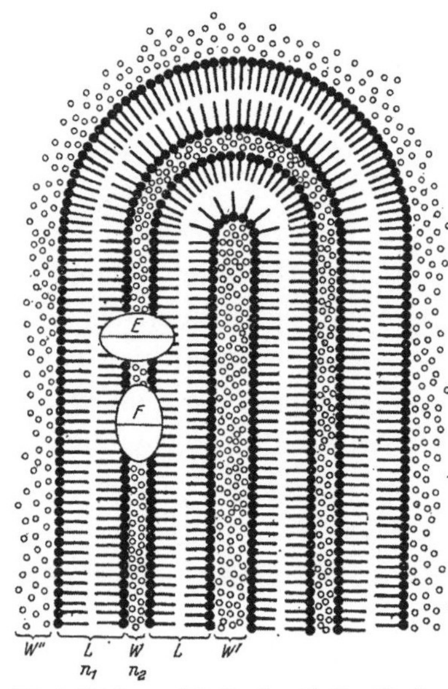

Abb. 7. Feinbau und Optik eines *Myelinschlauches* (oben das freie Ende) am Längsschnitt: *L* bimolekulare Lamelle des Lipoids mit der Brechzahl n_1; *W* Wasserschicht mit der Brechzahl n_2; *W'* axialer Wasserfaden; *W''* umgebendes Wasser. *E* Eigen-, *F* Formdoppelbrechung.

A

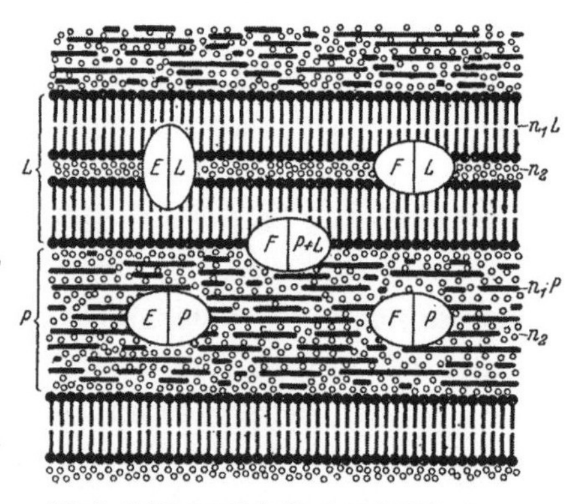

Abb. 8. Feinbau und Optik eines *Schichtsystems* aus *Proteinfolien* (*P*) und *bimolekularen Lipoidlamellen* (*L*) am Durchschnitt. *EP* Eigen-, *FP* Formdoppelbrechung des Proteins; *EL* Eigen-, *FL* Formdoppelbrechung des Lipoids; *F*: *P* + *L* Formdoppelbrechung des Protein-Lipoid-Schichtkörpers; n_1P Brechzahl des Proteins; n_1L Brechzahl des Lipoids; n_2 Brechzahl der Zwischenmasse (Wasser).

B

Figure 10.12. Wilhelm Schmidt's diagrams of the myelin figure (*A*) and a layered lipid-protein system in cross-section (*B*). Note the optical index-ellipses on both diagrams, which indicate the kind of form (*F*) and intrinsic (*E*) birefringence characteristics one should expect from such a system. Such expectations would only be possible if these diagrams were conceived as realistic representations of the biological microworld. From Schmidt 1941, 45–46. With permission of Springer.

freedom of movement. . . . The ones at the surface turn their hydrophilic poles against the water and parallelize themselves; the resulting unimolecular lamellae produce the structure of a second one with a reversed orientation of its molecular poles (see left and right sides of the illustration), and in this way the process continues" (Schmidt 1941, 44).

Not only is the diagram of a mass of lecithin in water especially evocative in its dynamics: the language Schmidt used to animate the lipid molecules was built on reflexive verb constructions to give the molecules agency and individuality. The ball-and-stick lipids are actively sorting themselves out, "*parallelisieren sich*," from a chaotic jumble in the middle of the mass and into orderly bi- and tri-layers at the outer edge—a droplet of lecithin rendered in fine molecular detail.

But it was Schmidt's molecular image of the myelin figure and the protein-lipid system that shows how far the iconography of lipids had come as a scientific tool (fig. 10.12). The ovals laid on top of each figure were meant to indicate form (F) and intrinsic (E) birefringence of the system. With up to four bilayers in the system, Schmidt indicated that, at first glance, the myelin tube would show form birefringence that indicated anisotropy and orientation along the axis of the myelin tube. In fact, Schmidt argued through the image that the intrinsic birefringence of the system—the real arrangement of the individual molecules—is actually perpendicular to the axis of the myelin tube, because of the way lipid molecules orient and arrange themselves into bilayers. And in the case of the protein-lipid system, Schmidt explained that not only did the lipid system (L) have its own form and intrinsic birefringence patterns (hence the labels E|L and F|L), but so too did the protein layer (E|P and F|P), and the entire lipid-protein system as well (F|P+L).

Conclusion

The most crucial feature of the two images of lipid systems in figure 10.12 is that they expect an exact correspondence to nature, at a scale where forces and entities are fundamentally inaccessible to direct observation. Polarized light microscopy could only shows signs of directional orientation and distinguish between material systems with patterns of darkness or through flashes of color; it was at best an indirect method of seeing fine structure, a theory-laden vision that relied heavily on the microscopist's intuition and experience. In the background were the hard-won, measurable, empirical facts: the birefringence of cells, surface tension measurements, and the lipids' chemical formulas. Transforming these facts into an argument about the cell's molecular structure needed these clear facts, but these alone were not sufficient. Schmidt's images, perhaps even more than his observations, were arguments that the biological microworld really *was* structured in the ways he described and illustrated on paper. Having accepted Schmidt's images as true reflections of nature, any other observer could see the patterns of birefringence under the polarized light microscope as affirming the molecular reality shown on the page. Schmidt had to be sure that this biological microworld was both real and (inferentially) visible before he could make any scientific claim to the usefulness or veracity of these images. He could confidently rely on the image of the self-orienting lipid molecule to show his grasp of the laws of physics and chemistry, while also feeling no need to actually address the complex physical forces and dynamics that governed that molecule's individual behavior. The microworld would be close enough to what he drew on paper, because Schmidt not only imagined that cellular structures looked like this: he assumed cellular structures really *were* this. Arguably, the colorful flashes of light seen under the polarized light microscope could never be interpreted without accepting the reality of the images on paper as an expression of the scientist's imagination of the biological microworld.

So lipid molecules and their orientation were, in a way, obvious to Schmidt, at least by 1939 or 1940—and lipids and molecular orientation were obvious to Schmidt in a rather different way than they were obvious to James Danielli in 1935, the latter guided by his deep education and work in physics and chemistry. Whether or not Wilhelm Schmidt "received" the exact ball-and-stick image of lipid structures from Harkins, McBain, Lawrence, or Danielli, I would argue, does not matter as much as the various meanings and possibilities of molecular orientation and colloidal structure that were bound up in the ball and stick.

Schmidt's use of this iconography was a clear departure from the epistemological standards of the communities that originally generated it: the physical chemists insisted first on the factual and mathematical rigor of their theories, with images and molecular diagrams useful only in pedagogy or as a heuristic. Schmidt and many biologists and biochemists who followed his example could safely assume that the physics and mathematics were given, embracing the images and other illustrations first and foremost as representations of reality. This departure transformed the idea of the molecule into an entity with both a clear physical identity and, crucially, also stripped of much of the complex physics. This metaphysical distance between the physicalist abstraction of colloid chemistry and the realism of molecular biology can be seen easily by comparing figure 10.7 and figures 10.11–12—the former an exceedingly rare illustration seen as having dubious scientific value, and the latter two quite common and seen as essential in a scientific method. In physical chemistry and colloid chemistry, not only had there been strong injunctions against structural determinacy at the molecular level, but any images used were necessarily second-class citizens: in physical and colloid chemistry, instrumental measurement and mathematical modeling were supposed to provide the primary validation of a theory. In the biology of the cell and the search for the fine molecular structure of the protoplasm, it was important to know the physics and chemistry, but it was just as important to be able to imagine and draw on paper the living molecular world.

Coda: Molecular Imagination and Postwar Technological Progress

Wilhelm Schmidt's evangelism of polarized light microscopy helped inaugurate molecular-scale research of the whole cell, and his writings—and illustrations—from 1935 to 1941 became minor classics in cell research. While he was able, Schmidt established and led a structuralist turn in cell biology, pushing biologists to explore the cell's molecular structure as an alternative to the colloidists' orientation toward physiological function. What Schmidt referred to as the "building blocks" of animal bodies, others variously called this "ultrastructure" research or "submicroscopic morphology," depending on who was asked. By the centenary celebrations of cell theory that began in 1938, submicroscopic morphology was a well-defined specialty in the life sciences, with a significant presence in cytology and technical microscopy journals, two textbooks, and a great deal of energy and excitement (Schmidt 1937; Aschoff, Küster, and Schmidt 1938; Frey-Wyssling 1938). Before the refinement of the electron microscope for biological use, X-ray diffraction was the only method available to examine isolated molecular structures, and polarized light microscopy was the only method available to place molecular structures in the context of the whole cell (Schmitt 1939). Late in life, the American ultrastructural biologist Francis O. Schmitt (1903–1995) could be heard complaining that Watson and Crick had falsely claimed the mantle of "molecular biology" in the 1950s, when, in fact, Schmitt, Schmidt, and others using only polarized light microscopes had been working at the molecular level decades before. "They call all this 'molecular biology,' " Schmitt grumbled. "Well now that's a very, *brooaaad* feeling, and it's in a sense preemptive terminology to those of us who started the field more than a half century ago [in the 1930s and 40s]. We were molecular biologists *then*" (1990).

Electron microscopy began to replace polarized light microscopy in ultrastructure research soon after the war, but the transition took well over a decade and varied depending on the kind and style of research (Rasmussen 1997; Strasser 2006). In August 1938, Helmut Ruska (1906–1988) presented one of the first electron micrographs of a cytological object in public, at the Fifth International Congress for Cell Research in Zürich (Ruska 1939; Frey-Wyssling 1964). The existence of this early electron micrograph was more remarkable than the image itself: excitement for the possibility to directly observe the molecular structure of the cell was tempered by concerns about the preparation methods needed to make the technique work. The electron microscope itself developed faster than biologists' ability to section cells thinly enough to achieve molecular resolutions, and the fixation and metallic staining regimes required to gain contrast were far harsher than the accepted preservation methods in ordinary light microscopy.

Through the 1950s, polarization microscopes were still used to set up expectations for what the electron microscope could see (Schmitt 1960). Even in the best electron micrographs at the time, molecular structures had to be interpreted from the image: for example, the lipid bilayer was visible only by looking for the parallel contrast lines created by the molecular stain of the phosphate group, separated by the measurable length of the two sets of hydrocarbon chains between them. As Rudolf Oldenbourg shows in chapter 12 of this volume, polarized light microscopy itself has become an instrument for precise measurement of molecular dimensions; polarized light microscopy has always had the benefit of not requiring lethal or injurious preparation techniques.

Today the creation and manipulation of images is not only a paper activity for other scientific purposes: model-making is essential, perhaps *the* essential part of structural-molecular theorizing. However, the use of molecular models in biochemistry and molecular biology rarely scales up to the level of whole cells. The epistemological gap between the observation and the molecular-structural theory remained, mediated by the theory and expectations created by the schematic image of molecules in the biologist's mind. Illustrations were widely available across several physical-chemical and biophysical specialties: in the early years of electron microscopy, the theory and illustrations served to confirm the results of the instrument, and not the other way around. The kind of technological progress that made postwar cell biology possible was thus arguably enabled by biologists' expectations and enthusiasms, themselves aided by a healthy molecular imagination.

Acknowledgments

I owe many thanks to Nick Jacobson and Stephen Neal at the University of Wisconsin–Madison's History of Science department, for their edits, ideas, and suggestions in the finishing stages of this essay. Many thanks as well as to Lijing Jiang, Karl Matlin, and Jane Maienschein for inviting me to participate in this collaborative project.

Notes

1 Mathews uses the battery analogy quite indiscriminately to describe any contained, directional sequence of redox reactions, going so far as to argue that "the living cell is in fact a battery" (Cowdry 1924, 68). The "graphite rods" in Mathews' analogy would be carbon

chains whose ends oxidize, providing energy to the cell. This is obviously a biochemical theory, rather than a structural theory based either on observational or physical evidence.

2 Among the significant targets of Jacobs's skepticism was Ernst Overton's famous "lipoid" theory of membrane permeability, which suggested that a lipid-impregnated boundary layer could serve to explain many problems of the protoplast's selective permeability. While Overton's lipoid theory has been repeatedly cited as an origin for modern cell membrane theories, historically Overton was but one of many scientists working on the broader problem of permeability in living cells, artificial membranes, and colloidal precipitates (Lombard 2014, 10–11).

3 Again, by way of example, the seventh section of Mathews's chapter in *General Cytology* is titled "What Is a Molecule?" suggesting that the meaning of "molecule" may have been far from obvious to a novice reader (Cowdry 1924, 38–43). However, Mathews also argues that molecules are held together by gravitation and magnetic moment, and this is just one example of the very strange physics Mathews seems to have embraced—another being a long digression about the four-dimensional luminiferous ether, which "for practical purposes . . . we have called space and time, [and] may be referred to as Infinity and Eternity" (20–25).

4 By 1917, surface tension was understood mathematically as a proxy for the free energy of a physical system, but surface tension remained as the focus of measurements and experiments.

5 The bitterness of the priority dispute between Harkins and Langmuir lasted for quite some time, and signs of the dispute can be seen in many of their publications and citations. At one point Harkins (1924a, 153) was so intent on bolstering his priority claim that in a textbook chapter he reproduced a page from one of his student's lecture notes from 1914, which is far from convincing. By 1918 Langmuir (1364) was writing in the *JACS* that he had developed the idea in 1916, but that Harkins "elaborated" the theory of molecular orientation in March 1917, at least suggesting he thought Harkins's work was neither insubstantial nor unoriginal. Harkins preferred to point out that the British colloid chemist William Bate Hardy had glancingly suggested the idea of molecular orientation in print five years previous (1912, 634).

Patrick Coffey (2008) has shown that several of Harkins's contemporaries thought that Harkins showed a pattern of intellectual theft, although Coffey is intent on highlighting the discord between American scientists in this period. In my judgment, Coffey's claim for Harkins's dishonesty rings true, but many outsiders happily cited Langmuir and Harkins together (and occasionally Hardy as well) as developing and elaborating the theory of molecular orientation; these included James W. McBain (McBain and Davies 1927) and Henri Devaux (1931). This may have been either out of ignorance or out of support for Harkins; a few physical chemists, including Neil Kensington Adam (1930) conspicuously avoided citing Harkins and his team, while showering Langmuir with praise.

6 For example Langmuir (1917, 1865) reported that a molecule of oleic acid ($C_{17}H_{33}COOH$) occupied an area of $46\times10^{-16}cm^2$, while a molecule of the stearic acid ($C_{17}H_{35}COOH$) covered a surface area of only $22\times10^{-16}cm^2$.

7 Today we would consider such contact due to "physical" van der Waals forces, but in 1917 Langmuir firmly believed that these forces were due to chemical valence, because they were related to the specific chemical formulae of the oil.

8 This was the second such symposium organized by the National Research Council, and topics for the eight symposia held between 1923 and 1930 varied widely from theoretical considerations to instruments and applications of colloid theory in engineering and biology. The 1924 Colloid Symposium, for instance, had papers on the rubber industry, new instruments, soil science, theories of emulsification, iodine, bacteriology, physiology, and an extensive rebuttal of Jacques Loeb's recent work on the Donnan equilibrium in protein solutions.

9 This redefinition of colloids was promoted by Wolfgang Ostwald (1907), along with eight classifications for two-phase colloidal systems: gas-liquid (mist), gas-solid (smoke), liquid-gas (foam), liquid-liquid (emulsion, sol, or gel), liquid-solid (suspension, sol, or gel), solid-gas (solid foam), solid-liquid (sol), solid-solid (gel).

10 Albert Frey (later Albert Frey-Wyssling, after he married Margrit Wyssling in 1928) was in many ways Schmidt's counterpart in botany, one of the leaders among biologists pushing for a "submicroscopic morphology" of cells and protoplasm (Schmidt's preferred term was *Feinbau*). Frey, as Schmidt somewhat wistfully remarked, "had the luck of being in the presence of the great masters" of Swiss and German physical chemistry in his education at Zürich and Jena (Schmidt 1964, 224; Häusermann et al. 1960, 7–12).

References

Adam, Neil Kensington. 1930. *The Physics and Chemistry of Surfaces*. Oxford: Clarendon Press.

Al-Shamery, Katharina. 2011. "Agnes Pockels (1862–1935)." In *European Women in Chemistry*, edited by Jan Apotheker and Livia Simon Sarkadi, 35–38. Weinheim: Wiley-VCH Verlag.

Alexander, Jerome, ed. 1926. *Colloid Chemistry: Theoretical and Applied*. Vol. 1. New York: Chemical Catalog Co.

Ambronn, Hermann, and Albert Frey. 1926. *Das Polarisationsmikroskop: Seine Anwendung in der Kolloidforschung und in der Färberei*. Kolliodforschung in Einzeldarstellungen, Bd. 5. Leipzig: Akademische Verlagsgesellschaft.

Aschoff, Ludwig, Ernst Küster, and Wilhelm J. Schmidt. 1938. *Hundert Jahre Zellforschung*. Protoplasma-Monographien, Bd. 17. Berlin: Gebrüder Borntraeger.

Beisswanger, Gabriele. 1991. "Das Portrait: Agnes Pockels (1862–1935) und die Oberflächenchemie." *Chemie in unserer Zeit* 25 (2): 97–101. doi:10.1002/ciuz.19910250206.

Bungenberg de Jong, H. G., and J. Bonner. 1935. "Phosphatide Auto-Complex Coacervates as Ionic Systems and Their Relation to the Protoplasmic Membrane." *Protoplasma* 24: 198–218. doi: 10.1007/BF01605663.

Chadarevian, Soraya de, and Nick Hopwood, eds. 2004. *Models: The Third Dimension of Science*. Stanford, CA: Stanford University Press.

Coffey, Patrick. 2008. *Cathedrals of Science: The Personalities and Rivalries That Made Modern Chemistry*. London: Oxford University Press.

Cowdry, Edmund V., ed. 1924. *General Cytology: A Textbook of Cellular Structure and Function for Students of Biology and Medicine*. Chicago: University of Chicago Press.

Creager, Angela N. H., Elizabeth Lunbeck, and M. Norton Wise, eds. 2007. *Science without Laws: Model Systems, Cases, Exemplary Narratives*. Durham, NC: Duke University Press.

Cuvier, Georges. 1817. *Le règne animal distribué d'après son organisation, tome I, contenant l'introduction, les mammifères et les oiseaux*. Paris: Chez Déterville.

Danielli, James F. 1973. "The Bilayer Hypothesis of Membrane Structure." *Hospital Practice* 8 (1): 63–71.

———, and Hugh Davson. 1935. "A Contribution to the Theory of Permeability of Thin Films." *Journal of Cellular and Comparative Physiology* 5 (4): 495–508. doi:10.1002/jcp.1030050409.

Devaux, Henri. 1888. "Mouvements spontanés des certains corps a la surface de quelques liquides." *La nature: Revue des sciences et de leurs applications aux arts et à l'industrie* 16 (777): 331–34.

———. 1913. "Oil Films on Water and on Mercury." *Annual Report of the Smithsonian Institution, 1913*, 261–73.

———. 1931. "Les lames très minces et leurs propriétés physiques." *Journal de physique et le radium* 2 (8): 237–72. doi:10.1051/jphysrad:0193100208023700.

Ede, Andrew. 2007. *The Rise and Decline of Colloid Science in North America, 1900–1935: The Neglected Dimension*. Burlington, VT: Ashgate.

Freundlich, Herbert. 1907. *Kapillarchemie und Physiologie: Habilitations-vorlesung gehalten am 29. Oktober 1906*. Dresden: Steinkopff & Springer.

Frey-Wyssling, Albert. 1938. *Submikroskopische Morphologie des Protoplasmas und seiner Derivate*. Protoplasma-Monographien, Bd. 15. Berlin: Gebrüder Borntraeger.

———. 1964. "Frühgeschichte und Ergebnisse der submikroskopischen Morphologie." *Mikroskopie* 19: 2–12.

———. Briefe. W.J. Schmidt to Frey-Wyssling, 8 May, 1946. HS 0443. Hochschularchiv der ETH Zürich, ETH-Bibliothek.

Häusermann, Elsa, Kurt Mühlethaler, Fritz Ruch, and Hans Heinrich Bosshard, eds. 1960. *Festschrift Albert Frey-Wyssling: Professor für Allgemeine Botanik an der Eidgenössischen Technischen Hochschule, Zürich : Zur Vollendung seines sechzigsten Lebensjahres 8. November 1960*. Zürich: Bühler.

Garber, Elizabeth. 1978. "Molecular Science in Late-Nineteenth-Century Britain." *Historical Studies in the Physical Sciences* 9: 265–97. doi:10.2307/27757380.

Gavroglou, Kostas, and Ana Simões. 2012. *Neither Physics nor Chemistry: A History of Quantum Chemistry*. Cambridge: MIT Press.

Gorter, Evert, and Francois Grendel. 1925. "On Bimolecular Layers of Lipoids on the Chromocytes of the Blood." *Journal of Experimental Medicine* 41 (4): 439–43.

Gray, James. 1931. *A Text-Book of Experimental Cytology*. Cambridge: Cambridge University Press.

Hardy, W. B. 1912. "The Tension of Composite Fluid Surfaces and the Mechanical Stability of Films of Fluid." *Proceedings of the Royal Society of London*, Series A, 86 (591): 610–35.

Harkins, William D. 1924a. "Surface Energy in Colloid Systems." In *The Theory and Application of Colloidal Behavior*, edited by Robert Herman Bogue. New York: McGraw-Hill.

———. 1924b. "The Orientation of Molecules in the Surfaces of Liquids." *Colloid Symposium Monograph* 2: 141–73.

———, and F. E. Brown. 1916. "A Simple Apparatus for the Accurate and Easy Determination of Surface Tension, with a Metal Thermoregulator for the Quick Adjustment of Temperature." *Journal of the American Chemical Society* 38 (2): 246–52. doi:10.1021/ja02259a007.

———, F. E. Brown, and E. C. H. Davies. 1917. "The Structure of the Surfaces of Liquids, and Solubility as Related to the Work Done by the Attraction of Two Liquid Surfaces

as They Approach Each Other (Surface Tension V)." *Journal of the American Chemical Society* 39 (3): 354–64. doi:10.1021/ja02248a003.
———, Earl C. H. Davies, and George L. Clark. 1917. "The Orientation of Molecules in the Surfaces of Liquids, the Energy Relations at Surfaces, Solubility, Adsorption, Emulsification, Molecular Association, and the Effect of Acids and Bases on Interfacial Tension (Surface Energy VI)." *Journal of the American Chemical Society* 39 (4): 541–96. doi:10.1021/ja02249a002.
———, and E. C. Humphery. 1916. "Apparatus for the Determination of the Surface Tension at the Interface between Two Liquids (Surface Tension II)." *Journal of the American Chemical Society* 38 (2): 236–41. doi:10.1021/ja02259a005.
Heilbron, John L. 1982. "*Fin-De-Siècle* Physics." In *Science, Technology, and Society in the Time of Alfred Nobel*, edited by Carl Gustaf Bernhard, Elisabeth Crawford, and Per Sörbom, 51–73. Oxford: Nobel Foundation.
Kaiser, David. 2005. *Drawing Theories Apart: The Dispersion of Feynman Diagrams in Postwar Physics*. Chicago: University of Chicago Press.
Klein, Ursula. 2003. *Experiments, Models, Paper Tools: Cultures of Organic Chemistry in the Nineteenth Century*. Stanford, CA: Stanford University Press.
Kohler, Robert E. 1974. "Irving Langmuir and the 'Octet' Theory of Valence." *Historical Studies in the Physical Sciences* 4: 39–87. doi:10.2307/27757327.
Langmuir, Irving. 1917. "The Constitution and Fundamental Properties of Solids and Liquids. II. Liquids." *Journal of the American Chemical Society* 39 (9): 1848–1906. doi:10.1021/ja02254a006.
Lawrence, A. S. C. 1929. *Soap Films: A Study of Molecular Individuality*. London: G. Bell and Sons.
Lombard, Jonathan. 2014. "Once upon a Time the Cell Membranes: 175 Years of Cell Boundary Research." *Biology Direct* 9 (32): 1–35. doi:10.1186/s13062-014-0032-7.
MacDougal, D. T. 1924. "The Arrangement and Action of Material in the Plasmatic Layers and Cell-Walls of Plants." *Proceedings of the American Philosophical Society* 63 (1): 76–93. doi:10.2307/984443.
Maxwell, James Clerk. 1868. "On the Dynamical Theory of Gases." *Philosophical Magazine*, ser.4, 35 (235/256): 129–45, 185–217.
McBain, James W. 1925. "Soaps and the Theory of Colloids." *Notices of the Proceedings at the Meetings of the Members of the Royal Institution* 24:579–84.
———. 1926. "A Survey of the Main Principles of Colloid Science." *Colloid Symposium Monograph* 4: 7–18.
———. 1950. *Colloid Science*. Boston: Heath.
———, and George P. Davies. 1927. "An Experimental Test of the Gibbs Adsorption Theorem: A Study of the Structure of the Surface of Ordinary Solutions." *Journal of the American Chemical Society* 49 (9): 2230–54. doi:10.1021/ja01408a016.
Nageotte, Jean. 1936. *Morphologie des gels lipoïdes; myéline, cristaux liquides, vacuoles*. Actualités scientifiques et industrielles, 431–434. Paris: Hermann.
Nye, Mary Jo. 1972. *Molecular Reality: A Perspective on the Scientific Work of Jean Perrin*. London: Macdonald.
———. 1993. *From Chemical Philosophy to Theoretical Chemistry: Dynamics of Matter and Dynamics of Disciplines, 1800–1950*. Berkeley: University of California Press.
———. 1996. *Before Big Science: The Pursuit of Modern Chemistry and Physics, 1800–1940*. New York: Twayne.

Opitz, Donald L. 2012. " 'Not Merely Wifely Devotion': Collaborating in the Construction of Science at Terling Place." In *For Better or For Worse? Collaborative Couples in the Sciences*, edited by Annette Lykknes, Brigitte van Tiggelen, and Donald L. Opitz, 33–56. Basel: Springer Basel. doi:10.1007/978-3-0348-0286-4_3.

Ostwald, Wolfgang. 1907. "Zur Systematik der Kolloide." *Zeitschrift für Chemie und Industrie der Kolloide* 1 (10): 291–300.

———. 1932. "Die Arbeiten von Agnes Pockels über Grenzschichten und Filme." *Kolloid-Zeitschrift* 58 (1).

Pockels, Agnes. 1891. "Surface Tension." Translated by Lord Rayleigh. *Nature* 43 (1115): 437–39.

Porter, Theodore M. 1994. "The Death of the Object: Fin de Siècle Philosophy of Physics." In *Modernist Impulses in the Human Sciences, 1870–1930*, edited by Dorothy Ross, 128–51. Baltimore: Johns Hopkins University Press.

Rayleigh, Lord 1890. "Measurements of the Amount of Oil Necessary in Order to Check the Motions of Camphor upon Water." *Proceedings of the Royal Society of London* 47 (March): 364–67.

———. 1899. "Investigations in Capillarity." *Philosophical Magazine*, ser. 5, 48 (293): 321–37. doi:10.1080/14786449908621342.

Ramberg, Peter J. 2003. *Chemical Structure, Spatial Arrangement: The Early History of Stereochemistry, 1874–1914*. Aldershot: Ashgate.

Rasmussen, Nicolas. 1997. *Picture Control: The Electron Microscope and the Transformation of Biology in America, 1940–1960*. Stanford, CA: Stanford University Press.

Reich, Leonard S. 1983. "Irving Langmuir and the Pursuit of Science and Technology in the Corporate Environment." *Technology and Culture* 24 (2): 199–221. doi:10.2307/3104037.

Rocke, Alan J. 2010. *Image and Reality: Kekulé, Kopp, and the Scientific Imagination*. Chicago: University of Chicago Press.

Ruska, Helmut. 1939. "Uebermikroskopische Darstellung organischer Struktur (vom Größenbereich der Zelle bis zum Ultravirus)." *Archiv für experimentelle Zellforschung* 22:673–80.

Schmidt, Wilhelm J. 1924. *Die Bausteine des Tierkörpers in polarisiertem Lichte*. Bonn: Friedrich Cohen.

———. 1935. "Doppelbrechung, Dichroismus, und Feinbau des Aussengliedes der Sehzellen vom Frosch." *Zeitschrift für Zellforschung und mikroskopische Anatomie* 22 (4): 485–522.

———. 1937. *Die Doppelbrechung von Karyoplasma, Zytoplasma, und Metaplasma*. Protoplasma-Monographien, Bd. 11. Berlin: Gebrüder Borntraeger.

———. 1938a. "Polarisationsoptische Analyse eines Eiweiß-Lipoid-Systems, erläutert am Außenglied der Sehzellen." *Kolloid-Zeitschrift* 85 (2–3): 137–48.

———. 1938b. "Molekulare Bauweisen tierischer Zellen und Gewebe und ihre polarisationsoptische Erforschung (I)." *Naturwissenschaften* 26 (30): 481–90.

———. 1939. "Polarisationsoptische Erforschung des submikroskopischen Baues tierischer Zellen und Gewebe: Der experimentelle Weg und einige Beispiele." *Verhandlungen der deutsche zoologische Gesellschaft* 41: 303–89.

———. 1941. "Die Doppelbrechung des Protoplasmas und ihre Bedeutung für die Erforschung seines submikroskopischen Baues." *Ergebnisse der Physiologie,*

biologischen Chemie und experimentellen Pharmakologie 44 (1): 27–95. doi:10.1007/BF02118046.
———. 1964. "Aus meiner Werkstatt." *Bericht der oberhessischen Gesellschaft für Natur- und Heilkunde zu Gießen, neue Folge, naturwissenschaftliche Abteilung* 33 (4): 217–37.
Schmitt, Francis O. 1939. "The Ultrastructure of Protoplasmic Constituents." *Physiological Reviews* 19 (2): 270–302.
———. 1960. "Electron Microscopy in Morphology and Molecular Biology." In *Vierter internationaler Kongress für Elektronenmikroskopie*. Berlin 10.–17. September 1958. Edited by W. Bargmann, G. Möllenstedt, H. Niehrs, D. Peters, Ernst Ruska, and C. Wolpers, 2:1–16. Berlin: Springer-Verlag.
———. 1990. Francis O. Schmitt: Microscopy Society of America Oral History Project, Interview by Sterling Newberry. VHS. Online at http://www.microscopy.org/about/interviews.cfm.
Schütt, Hans-Werner. 2002. "Chemical Atomism and Chemical Classification." In *The Modern Physical and Mathematical Sciences*, edited by Mary Jo Nye, 5:237–54. The Cambridge History of Science, 5. Cambridge: Cambridge University Press.
Stadler, Max. 2009. "Assembling Life: Models, the Cell, and the Reformations of Biological Science, 1920–1960." PhD diss. London: Imperial College, University of London.
Staley, Richard. 2008. "The Fin de Siècle Thesis." *Berichte zur Wissenschaftsgeschichte* 31 (4): 311–30. doi:10.1002/bewi.200801340.
Stein, W. D. 1986. "James Frederic Danielli: 13 November 1911–22 April 1984." *Biographical Memoirs of Fellows of the Royal Society* 32:117–35.
Strasser, Bruno J. 2006. *La fabrique d'une nouvelle science la biologie moléculaire à l'âge atomique (1945–1964)*. Firenze: Leo S. Olschki.
Süsskind, Charles. 2008. "Langmuir, Irving." In *Complete Dictionary of Scientific Biography*, 8:22–25. Detroit: Charles Scribner's Sons.
Thiessen, Peter Adolf, and R. Spychalski. 1931. "Anordnung der Moleküle in Seifenmicelen." *Zeitschrift für Physikalische Chemie* 156:435–56.
Wiener, Otto. 1904. "Lamellare Doppelbrechung." *Physikalische Zeitschrift* 5 (12): 332–39.
———. 1909. "Zur Theorie der Stäbchendoppelbrechung." *Berichte über die Verhandlungen der königlich sächsischen Gesellschaft der Wissenschaften zu Leipzig, mathematisch-physische Klasse* 61: 113–16.
Wise, George. 1980. "A New Role for Professional Scientists in Industry: Industrial Research at General Electric, 1900–1916." *Technology and Culture* 21 (3): 408–29. doi:10.2307/3103155.
———. 1983. "Ionists in Industry: Physical Chemistry at General Electric, 1900–1915." *Isis* 74 (1): 7–21.

CHAPTER 11

PICTURES AND PARTS

REPRESENTATION OF FORM AND THE EPISTEMIC STRATEGY OF CELL BIOLOGY

Karl S. Matlin

The physiology of the cell cannot be fully understood unless we determine the constitution of its parts, and the relation which undoubtedly exists between its morphology and the distribution of its biochemical functions.
—*Albert Claude*

In his introduction to *General Cytology*, E. B. Wilson celebrated the broadening of cytology into a "*cellular biology* . . . in which observation and experiment, morphology and physiology, have entered into a close affiliation with one another and with biophysics and biochemistry" (Cowdry 1924, 10; italics in orig.). Wilson optimistically implied that this multidisciplinary attack on the cell would, in the near future, yield previously unimaginable insights into cell function. Indeed, the publication of *General Cytology* in 1924, edited by Edmund V. Cowdry, with chapters contributed by the leading lights of American biology, was a watershed in the development of the discipline. Chapters on cell chemistry, permeability, reactivity, and behavior not only reviewed the status of cellular biology from different perspectives, but also suggested that the way forward would be through chemical and physical analysis instead of through the morphology that had dominated the subject since development of the cell concept in the early nineteenth century.

The optimism expressed in *General Cytology* was, however, premature. While the authors were correct in their belief that chemistry and physics are key to understanding how cells function, the technologies available in 1924 were not up to the task. Light microscopes, which remained important instruments, continued to improve well into the twentieth century, but the physical limitation in resolution persisted (Bechtel 2006, 88). Methods to allow visualization of resolvable subcellular structures without the use of chemical dyes, such as phase contrast, did not appear until the early 1930s (Bradbury 1967, 292–93). Despite progress in the separate discipline of biochemistry, efforts by cytologists to understand the chemistry and physics of cells were generally applied to intact cells, not only because of technical limitations but also because of the fear among cytologists that breaking open cells would irretrievably affect the properties of interest.[1]

Attempts to probe cell chemistry were made with specific histological stains, while cell permeability was measured with other chromogenic substances (Cowdry 1924). Centrifugation was employed, but often only to see how visible cellular components redistributed within the intact cell (Beams 1943; Cowdry 1924). While it cannot be said that these methods were completely unsuccessful, they were clearly limited and offered no pathway to more in-depth analysis. In 1946, Albert Claude, who, as we shall see, was responsible for the development of radically new approaches, summarized the futility of some of these techniques:

> The intracellular topography of biochemical functions constitutes one of the major problems of cytology, and one that has benefited the least from the microscopical technique. With the successful application of staining to the study of cell morphology, the hope was entertained that specific color tests could be used under the microscope to determine the distribution of enzyme systems within the cell. Unfortunately, most of the color tests involve chemical reactions incompatible with the life of the cell and almost invariably it is found that the essential cell structures have been severely damaged or completely destroyed by the procedure. (Claude 1946a, 51)

In this essay I describe a series of discoveries made after Cowdry's book was published that finally led to the establishment of Wilson's "cellular biology." This new discipline transcended not only the limitations of traditional cytology at the turn of the century, but also the cytology envisaged by the authors of Cowdry's book. While these discoveries were most definitely spurred by the development of new technologies, the manner in which these technologies were applied to questions of cellular function was an even more significant development.

In the late 1930s and 1940s, Albert Claude and his collaborators at the Rockefeller Institute for Medical Research developed both cell fractionation and biological electron microscopy (Claude 1948). Cell fractionation consists of the mechanical disruption of cells and separation of cellular organelles and other components in a centrifugal field. While Claude was not the first to do this, the rigor of his approach paved the way for future, more refined developments. Electron microscopy was first successfully applied to whole mammalian cells in 1945 by Claude, Keith Porter, and Ernest Fullam (Porter, Claude, and Fullam 1945; Rasmussen 1997). Because of its vastly superior resolution, electron microscopy opened up what was described as "the optically empty ground substance" to examination (Porter 1953), leading to the discovery of the endoplasmic reticulum and other particulate and membrane-bound components of the cytoplasm not resolvable by light microscopy. George Palade, who joined Claude and Porter at Rockefeller just after World War II, further developed these methodologies and their application to a very high level to investigate protein synthesis and secretion (Palade 1975), particularly in the 1950s and 1960s. Then Günter Blobel, who arrived at Rockefeller in 1966, went even further to examine molecular aspects of secretion in a "cell-free" experimental system derived from isolated parts of the cell (Blobel and Dobberstein 1975a, 1975b).[2]

Based upon this case study, I argue that the way these cell biologists applied new technologies constituted a particularly effective *epistemic strategy*, that is, an experimental approach to generate mechanistic explanations of cellular phenomena. Most significantly, the strategy does not replace morphological observations with biochemical assays, but employs morphology as a constraint and guide for biochemical analysis of cell function. This heuristic use of morphology not only helps in the development of hypotheses to be experimentally tested, but also provides a context for experiments that insures that the results of these experiments are biologically meaningful (Matlin 2016). Another critical aspect of this strategy is that it is applied iteratively through multiple cycles, with each cycle leading to refinement of the hypothesis and deeper penetration into the mechanism until a molecular understanding is achieved.[3]

Finally, I use a more contemporary case study to ask if the same epistemic strategy is effective under present circumstances, when experiments are informed not only by an understanding of the cell that greatly exceeds that of the mid-twentieth century, but also by an enormous amount of information about biomolecules achieved through genomic sequencing and other "-omics" projects. Neither electron microscopy nor cell fractionation is now commonly employed in fundamental cell biology. Instead, light microscopy has been reimagined, and new, high-resolution techniques developed that overcome the limits of the past. Today living rather than fixed cells are most often observed, and parts of cells are not separated from the whole through centrifugation, but are introduced into the cell in the form of fluorescently labeled proteins, markers, and probes. Mechanistic hypotheses based upon observations of cells are still developed, but are now abstracted into computational models to both cope with complexity and extract predictions of cellular behavior based upon intricate molecular interactions.

I conclude by proposing that there is a core epistemic strategy in cell biology that facilitates investigation of cellular processes in a manner that leads to molecular explanations by defining the relationship among "pictures" of cells, their morphological representation, and the "parts" from which cells are constituted.

The Endoplasmic Reticulum

In the late 1930s Albert Claude began using centrifugation to purify and biochemically characterize particles from disrupted tumor cells (Bechtel 2006; Moberg 2012; Rheinberger 1995 and 1997). Claude soon discovered that the same particles could be isolated from normal, control cells, and gradually turned his attention away from cancer research to more fundamental cell studies as he refined his approach into an optimized method of cell fractionation using differential centrifugation (Claude 1940; Claude 1941; Claude 1943a). Claude described his approach in a pair of definitive and detailed papers published in 1946 (Claude 1946a and 1946b). Rat liver was ground up and forced through a 1mm mesh, and the resulting pulp further homogenized with a mortar and pestle. To focus on the constituents of the cytoplasm, the extract was first centrifuged at very low speed to remove unbroken whole cells and nuclei, and the resulting suspension subjected to a series of further centrifugations at successively higher speeds. From this, Claude isolated three fractions that he called the large granules, microsomes, and "supernate" (supernatant), the latter consisting of material that failed to sediment at the highest centrifugal speeds. He believed that the large particles (0.5–2.0 microns in diameter) were a mixture of mitochondria and secretory granules, but was unable to associate the microsomes with a particular cellular entity. Claude cited (and nearly quoted verbatim) E. B. Wilson's 1925 textbook (Wilson [1925] 1928) to justify his choice of the term *microsome* as appropriate because it referred to "any small granules of undefined nature" (Claude 1943a). On the basis of his biochemical and centrifugal analysis, Claude described microsomes as RNA-containing particles 50–300 nanometers in diameter, and linked the fraction to the blue-staining, "basophilic" part of the cell, observed using conventional histochemistry, by spreading isolated microsomes on a slide and applying the same stain (Claude 1943b).[4]

While Claude was not the first to perform cell fractionation (Bechtel 2006), his biochemical analysis was particularly rigorous, in that he took care to account for all the material in the original extract, something later referred to as the "balance sheet" approach (Claude 1946a).[5] In 1946, his analysis consisted primarily of determining the elemental composition of the fractions; later on, specific enzyme activities were measured (Rheinberger 1995). Claude's quantitative approach allowed him not only to say that a particular enzyme was concentrated in a specific fraction, but also that that fraction accounted for the majority of that enzyme in the cell. This ultimately permitted him and subsequent investigators to both relate functions associated with an enzyme to a distinct part of the cell and, conversely, to use that enzyme as a "marker" for further purification of the cellular component.

Claude wanted to move beyond morphology in cell studies. Nevertheless, microscopy remained a key adjunct to cell fractionation. In his study of liver cells, he knew the overall organization of the tissue from conventional light microscopy, and used this as a point of reference as fractionation proceeded. After the tissue was disrupted, the extract was examined by microscopy to determine the degree of cell breakage, and after the centrifugation steps, the isolated fractions were examined as part of their overall characterization. Claude summarized the relationship between microscopy and biochemistry in a 1948 review, stating that "it would be difficult to separate the biochemical work from the morphological observations since the microscope has constantly served as a guide or check for the chemical and biochemical studies" (Claude 1948, 123). The dilemma prior to 1945, however, was that the microsomes isolated by Claude were "submicroscopic," that is, below the resolution of light microscopy. Attempts were made to circumvent this through the use of dark-field microscopy. While this technique allowed very small cellular components to be detected, not much more could be visualized than a collection of particles jiggling with Brownian motion against a dark background. Things began to improve in 1945 when Keith Porter in Claude's group used electron microscopy to resolve previously invisible parts of cultured cells transferred intact into the microscope (Porter, Claude, and Fullam 1945).

One of Porter's signature discoveries was an intracellular structure he described as a "lace-like reticulum." Surprisingly, Claude did not mention this observation in his 1946 fractionation papers.[6] In a Harvey Lecture two years later, he described both his fractionation results and the early observations by electron microscopy, but was noncommittal about any relationship between the reticulum seen in whole cells and microsomes. He did, however, speculate that microsomes might be related to protein synthesis because of their high nucleic acid content (Claude 1948, 142).

In 1953, Porter published an in-depth follow-up to his preliminary electron microscopic observations from 1945 (Porter 1953). To convince skeptics that the lace-like reticulum (now named the *endoplasmic reticulum*, or ER), was not an artifact of the new technique, he employed phase-contrast and dark-field microscopy to look at living cells in culture, and also examined cells fixed and stained in a manner similar to those for electron microscopy, claiming that structures consistent with the reticulum were visible even by light microscopy. More significantly, Porter referred to preliminary experiments in which he had examined Claude's microsomal fraction by electron microscopy to point out morphological similarities between the isolated material and the intact ER seen in whole cells. He then pulled his findings and those of others together to conclude that "the small particles of [the microsomal] fraction . . . are morphologically similar to fragments of the endoplasmic reticulum. Even without this latter evidence of identicalness, however, the fact that both the reticulum and the small particle fraction represents the basophilic material of the cytoplasm makes it reasonable to transfer the properties of the basophilic component or the 'microsomal' fraction to the endoplasmic reticulum of the electron microscope image" (Porter 1953, 743–44).

Porter's discovery of the endoplasmic reticulum in intact cultured cells was possible because the spread periphery of the cell was thin enough to permit penetration of the electron beam. By 1953, however, techniques for embedding fixed tissue in plastic and cutting sections thin enough for electron microscopy had been developed, making it possible to look for the endoplasmic reticulum in cells "*in situ*," that is, in organs rather than in cell culture (Palade 1955b; Palade and Porter 1954; Porter and Blum 1953). In a first paper, Porter's colleague George Palade compared the appearance of the ER in primary cultures of cells grown from pieces of tissue with the same cell types observed *in situ* after thin sectioning and found that it was identical (Palade and Porter 1954). A second paper the next year looked at forty mammalian and avian cell types *in situ*. Palade concluded that the ER is a universal feature of eukaryotic cells (Palade 1955b). In a parallel paper, Palade described "a small particulate component of the cytoplasm," later known as the ribosome (Palade 1955a). Palade saw these particles not only free in the cytoplasm, but also observed that some sections revealed ER "profiles of a rough-surfaced variety," suggesting that ribosomes were attached to the ER membrane (Palade 1955b). In his second ER paper he noted that this *rough ER* was most prevalent in tissues known to engage in large amounts of protein synthesis and that the RNA concentrated in the ribosomal particles had also been "related to the process of protein synthesis" (Palade 1955b, 579). With this, the stage was set for a concerted effort to investigate the function of the ER in the cell.

Vectorial Transport

By the mid-1950s the achievements of Claude, Porter, and Palade had overcome the obstacles that had made cytology a dead-end in 1924. In the "optically empty" part of the cell, they had discovered a new structure, the ER, isolated it, and articulated a plausible function. Certainly technological advances had made these studies feasible, but even more significant is that they had not abandoned the morphological approach in favor of a purely biochemical strategy. Instead they continued to use the morphology of the cell as a guide, as outlined by Claude in 1948, to help them link isolated parts to intact structures (Claude 1948). This approach allowed them to postulate that microsomes are likely derived from the ER, and that the ER, particularly that of the "rough surfaced variety," plays some role in protein synthesis.

Before 1955, the biochemical analysis of microsomes by the Rockefeller group was largely limited to the measurement of nucleic acids, as well as nitrogen and phosphorus as stand-ins for protein and phospholipids. The activities of some metabolic enzymes, such as cytochrome oxidase, were examined, but these were all found to be concentrated in the large-granule (mitochondria) fraction (Hogeboom, Claude, and Hotchkiss 1946; Hogeboom, Schneider, and Pallade 1948). To get at the function of the ER, it was now clear that more in-depth biochemical studies were needed. Biochemistry expertise in the Rockefeller group was earlier provided by George Hogeboom, Rollin Hotchkiss, and Walter Schneider, but by 1948 Hogeboom and Schneider had relocated to the National Institutes of Health, and Hotchkiss had moved on to other things at Rockefeller (Moberg 2012).

In 1954 Palade recruited Philip Siekevitz to replace them (Bechtel 2006; Moberg 2012). The choice of Siekevitz was fortuitous if not strategic. Although he was at the time working at the University of Wisconsin, Siekevitz had previously been in Paul Zamecnik's laboratory at the Massachusetts General Hospital, one of the leading laboratories investigating the mechanism of protein synthesis. Beginning with Siekevitz in 1952, Zamecnik's group used the fractionation techniques developed at Rockefeller to isolate microsomes and ribosomes and, by incubating them in a cell-free system, showed that radioactive amino acids were incorporated into protein in the ribosomal fraction (Siekevitz 1952; Keller, Zamecnik, and Loftfield 1954; Littlefield et al. 1955).

In 1956 Palade and Siekevitz published a paper entitled "Liver Microsomes: An Integrated Morphological and Biochemical Study" that was a model for all future work on protein synthesis in the laboratory as well as the epistemic strategy that they would continue to develop (Palade and Siekevitz 1956). Using improved techniques of homogenization, centrifugation, and electron microscopy, they isolated microsomes with attached ribosomes (which they referred to as "rough microsomes") and estimated their content of nitrogen, phospholipid phosphorus, and ribonucleotides (fig. 11.1b). They also measured the activity of the enzyme diphosphopyridine nucleotide-cytochrome c reductase, which Hogeboom had shown to be concentrated in microsomes (Hogeboom 1949). Throughout the isolation procedure, Palade and Siekevitz monitored the fractionation using electron microscopy. From the order in which they presented their results in the paper, it was very clear that the study was organized around the microscopic observations. They began with a description of the morphology of the ER in the intact liver hepatocyte, and followed this with images of the total homogenate and isolated fractions as the purification proceeded: "In each experiment, small fragments of liver tissue were excised and fixed for electron microscopy before homogenizing the rest of the organ. The fate of the various cell components was followed throughout homogenization and fractionation by examining in the electron microscope samples of tissue, homogenate, and fractions derived from the same liver" (Palade and Siekevitz 1956, 172). In this manner, Palade and Siekevitz not only monitored the purification of liver microsomes during fractionation, but also made sure that the isolated microsomes were directly related to the ER seen in the intact cell.

Through their careful approach, Palade and Siekevitz were able to confirm previous suppositions that rough microsomes were fragments of the rough ER that spontaneously formed vesicles (fig. 11.1). This meant that the original, generic "microsomal fraction" of Claude was now a specific, defined part of the cell consisting primarily of membrane vesicles with attached ribosomal particles. However, their work did not just improve the procedure for isolating microsomes and provide a more detailed chemical characterization; it also began to establish that microsomes were functional cellular entities. From their images and manipulations of microsomes, they concluded that they were sealed, osmotically active vesicles with retained, electron-dense content, possibly "imprisoned" "molecules of a large size"

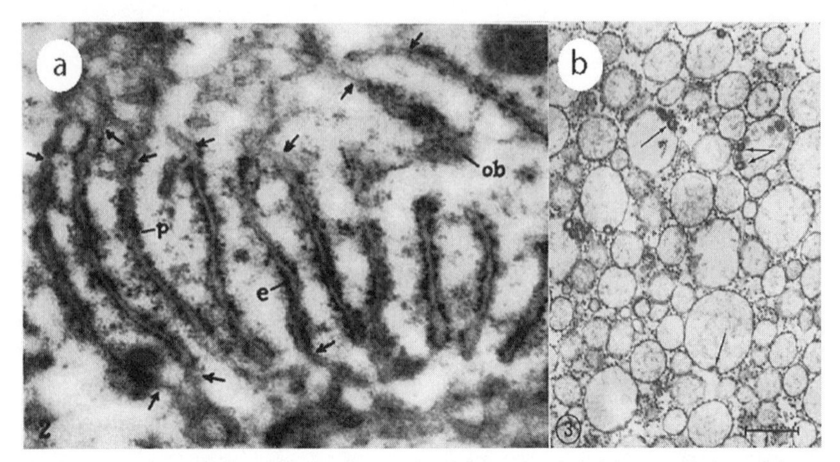

Figure 11.1. The endoplasmic reticulum and rough microsomes. (*a*) An early view of the rough endoplasmic reticulum in situ in rat liver cells (fig. 2 from Palade and Siekevitz 1956). Note the attached ribosomes appearing as poorly resolved particles (*p*). (*b*) Isolated rough microsome vesicles from a myeloma tumor cell line (fig. 3 from Blobel and Dobberstein 1975a). Due to improvements in electron microscopy, the ribosomes bound to the vesicles are clearly evident. (*a*) ©1956 Palade, G., and Siekevitz, P., *Journal of Biophysical and Biochemical Cytology* 2:171–200. (*b*) ©1975 Blobel, G., and Dobberstein, B., *Journal of Cell Biology* 67:852–62.

(Palade and Siekevitz 1956, 192). These observations, together with the fact that the ribosomes were attached to the outside of the isolated vesicles in a manner resembling their attachment on the cytoplasmic side of the ER *in situ*, suggested a clear functional organization (fig. 11.1).

In their characterization of isolated microsomes, Palade and Siekevitz went beyond straightforward measurement of chemical composition and enzymatic activity. They also began to probe the relationship between the attached particles and the microsomal vesicles by first treating the rough microsomes with versene,[7] which strongly binds divalent cations (calcium and magnesium), the detergent deoxycholate,[8] or ribonuclease, an RNA-degrading enzyme. They then recovered the treated material by centrifugation and examined it by electron microscopy. From this they concluded that most of the RNA in the fraction was in the particles, and most of the protein was in either the membrane of the vesicles or the trapped content. Ultimately, procedures like these would prove critical for defining the function of the ER at the molecular level.

After completing this study of liver microsomes, Palade and Siekevitz switched to the exocrine pancreas,[9] because the secretory cells contained abundant rough ER and secretory or *zymogen* granules. Zymogen granules, which were visible in the light microscope, were of interest because they had been shown by histochemical procedures to be likely to act as a storage site for digestive enzymes prior to their secretion (Siekevitz and Palade 1958). In embarking on their investigation of the pancreas, the goal was to broadly study the process of protein synthesis from a cellular perspective. That is, they were not particularly concerned with the biochemistry of peptide bond synthesis, but were interested in the cellular location where proteins are made and, in the case of a secretory cell like that of the exocrine pancreas, how those newly synthesized proteins are transported through and out of the cell. To accomplish this, they developed refined cell-fractionation procedures to isolate and purify rough microsomes and zymogen granules, using the guinea pig pancreas as an experimental model.

With regard to the rough microsomes, they wanted to move beyond previous studies that had demonstrated the incorporation of radioactive amino acids into mixtures of proteins, to instead track the synthesis of the specific pancreatic enzyme chymotrypsinogen shortly after injecting the animal with radioactive amino acids. This work, which culminated in a paper published in 1960, was a technical *tour de force* because it required the isolation and purification of minuscule amounts of chymotrypsinogen from cell fractions prepared at different time points (Siekevitz and Palade 1960a). Indeed, they remarked that it took five or six days to purify the enzyme from fractions resulting from a single experiment. Based upon their previous work in dissecting liver microsomes with deoxycholate, they also realized that it would be possible not only to localize newly made chymotrypsinogen to microsomes, but also to subfractionate the rough microsomes into ribosomes and content, and demonstrate that the granular material seen inside microsomes by electron microscopy was indeed chymotrypsinogen and other secretory proteins.

Following injection of radioactivity into the guinea pig, fractionation of the pancreas, and isolation of chymotrypsinogen from the fractions, they observed that one minute after isotope injection, chymotrypsinogen with the highest amount of radioactivity was associated with isolated ribosomes. By fifteen minutes, labeled chymotrypsinogen was distributed evenly among the ribosomes, microsomal content, and zymogen granules, and by forty-five minutes, it was found primarily in the zymogen granules. In a subsequent paper, they found several other digestive enzymes associated with the particles and could only release them by causing the ribosomal particles to disintegrate (Siekevitz and Palade 1960b). On the basis of these results, they concluded that synthesis commences on the ribosomes before the proteins are transferred into the "cavities" of the microsomes and then later to zymogen granules, but cautiously stated that they "realize[d] that final proof remains to be obtained by studying *in vitro* amino acid incorporation into digestive enzyme by isolated cell fractions" (Siekevitz and Palade 1960a, 626).[10]

This next step proved to be difficult. As Palade noted in a 1962 review, their attempts to incorporate amino acids into proteins using guinea pig pancreatic rough microsomes had been unsuccessful (Palade, Siekevitz, and Caro 1962). While the reason was not clear, the suspicion was that a secretory form of ribonuclease released by homogenization of the tissue was degrading key components necessary for synthesis. Palade suggested a way to circumvent this problem might be by using the pancreas from young pigeons, which contains less ribonuclease (Palade, Siekevitz, and Caro 1962).

Until a solution could be found, Siekevitz and Palade proceeded to refine the approach used in their study of chymotrypsinogen, looking at the synthesis of another enzyme, amylase, in the guinea pig pancreas (Siekevitz and Palade 1966). This work yielded significant results. As with chymotrypsinogen, they purified amylase from rough microsomes after injecting animals with radioactive amino acids.[11] This allowed them to measure not only the amounts of amylase present but also the "specific activity" of the purified enzyme—that is, the ratio of radioactivity to the mass of purified protein. Deoxycholate was again used to dissect rough microsomes into ribosomes and vesicular content, but this time they used lesser amounts of the detergent to yield residual ribosome-bound membranes without the protein content sequestered within the vesicles. The dissection was monitored both biochemically, after sedimentation of the treated microsomes, and morphologically. In a result that resembled those with chymotrypsinogen, they observed that detergent-treated rough microsomes depleted of content nevertheless were associated with amylase of the highest specific activity:

> The simplest explanation that we can offer for the above findings is that, when microsomes are gradually disrupted by detergent treatment, ribosomes begin to be detached from the membranes, and those detached most easily are those having the low specific radioactive amylase. . . . Even when disruption of the microsomes is stepped

> up by high-DOC [deoxycholate] treatment, there remains on microsomal membranes a small percentage . . . of ribosomes [that] contain an amylase of much higher specific radioactivity . . .
>
> [I]t is clear that the presence of newly synthesized, completed protein . . . on the ribosomes coincides with a firmer attachment of the ribosome to the membrane. One of the factors responsible for this situation might be that part of the amylase molecule is still firmly bound to the ribosome, while the rest of it is already anchored to the membrane of the ER. (Siekevitz and Palade 1966, 527–28)

In other words, ribosomes in the act of synthesizing secretory proteins were bound to the microsomal membrane, at least in part, by the protein they were making.

At about the same time, Colvin Redman, a postdoctoral fellow working with Siekevitz and Palade, succeeded in establishing an experimental system to study protein synthesis in a cell-free, *in vitro* system derived from pancreatic rough microsomes by following the clue that Palade mentioned in 1962. Using young pigeons, Redman removed the pancreas, which turned out to have nearly one hundred times less ribonuclease than the guinea pig pancreas, and isolated and characterized rough microsomes (Redman, Siekevitz, and Palade 1966). The microsomes were then incubated with a crude mixture of protein synthesis-stimulating factors, metabolites to supply energy, and radioactive amino acids, and were able to synthesize radioactive amylase. Redman then used deoxycholate dissection to separate membrane-bound ribosomes that had been engaged in protein synthesis from the content found within the microsomal vesicle. The results confirmed the *in vivo* findings of Siekevitz and Palade that amylase was synthesized on the bound ribosomes and then subsequently transferred through the microsomal membrane to the interior, something described by Redman as a "vectorial" process: "The system in vitro thus comprises the vectorial mechanism involved in transfer of newly synthesized secretory protein across the endoplasmic reticulum membrane into the cisternal space" (Redman, Siekevitz, and Palade 1966, 1150). This new system had many advantages, one of which was a much finer time resolution of the synthetic process than *in vivo* radioactive labeling. More importantly, it was now possible to begin to biochemically disassemble the transfer mechanism.

Gradually, investigation of the mechanism of protein synthesis and transfer across the microsomal membrane was passed along to new members of the laboratory.[12] One of these was David Sabatini, a skilled electron microscopist, who joined Palade as a graduate student to learn biochemistry. Sabatini collaborated with Redman shortly before Redman left Rockefeller to follow up on the idea that ribosomes were bound to the microsomal membrane by the *nascent chain*, the name applied to the newly synthesized secretory protein that remained associated with the ribosome while simultaneously penetrating the membrane (Redman and Sabatini 1966). If the nascent chain held the ribosomes to the membrane, then its elimination might lead to release of the ribosomes. To test this, they added the antibiotic puromycin to Redman's *in vitro* system during active synthesis of proteins. Puromycin works by prematurely terminating protein synthesis, leading to the release of any unfinished proteins from the ribosome. The results were somewhat unexpected: the ribosomes remained attached to the membrane, but the unfinished polypeptides fell off the ribosomes and continued to cross the microsomal membrane, ending up in the lumen. Clearly a continuous path existed from the site of protein synthesis in the ribosome, now known to be composed of large and small subunits, through the membrane: "The peptides being synthesized are assumed to grow within the central channel of the large ribosomal subunit (47S) in an environment which is (or can be made) continuous with the cisternal space [lumen] through a discontinuity in the ER membrane. As visualized at present, the transfer mechanism relies primarily on release from the large subunit and on structural restrictions at the ribosome-membrane junction, and hence, it is nondiscriminatory and possibly passive" (Redman and Sabatini 1966, 614).

Adopting the terminology of Redman's other paper, they referred to the phenomena they observed upon addition of puromycin as "vectorial discharge." Significantly, their model of the transfer mechanism was developed not only through biochemical analysis of *in vitro* microsomal protein synthesis, but also through Sabatini's parallel, high-resolution electron microscopy of microsome-bound ribosomes that showed attachment mediated by the large subunit (Redman and Sabatini 1966; Sabatini, Tashiro, and Palade 1966). This integrated approach was essentially the same epistemic strategy developed earlier by the Rockefeller group, only now applied to a part of the cell on which mechanistic studies were focused instead of the whole cell.

As this work was completed, another investigator joined the project. Günter Blobel, a new postdoctoral fellow from the University of Wisconsin, was a skilled biochemist whose PhD thesis had focused on an examination of both free and membrane-bound ribosomes (Blobel and Potter 1967a; 1967c; 1967b). When he first arrived at Rockefeller, he believed that Siekevitz would direct him, but soon realized that Palade was the driving force behind the secretion research. Nevertheless, he drifted while searching for his own project, purifying ribosomes and asking Palade to take pictures of his preparations in the electron microscope. This got Sabatini's attention because it appeared to overlap his interests, by now being pursued in his own laboratory within the group. Rather than compete with each other, Sabatini and Blobel began to collaborate.

The Signal Hypothesis

From the outset of the collaboration between Blobel and Sabatini, their work was informed by new information on the biochemistry of protein synthesis and the structure of ribosomes. It was now known that, while protein synthesis occurs on ribosomes, proteins are *translated* from information encoded in messenger RNAs (mRNAs) specific for each protein. Furthermore, once a ribosome becomes associated with an mRNA and begins to synthesize protein, other ribosomes can successively jump onto the mRNA to initiate another round of synthesis such that multiple ribosomes are bound at once to the mRNA like beads-on-a-string. These are called *polysomes*.

The collaboration began slowly with a pair of jointly authored papers appearing in 1970 that continued to address the attachment of ribosomes to the membrane by the nascent polypeptide chain (Blobel and Sabatini 1970; Sabatini and Blobel 1970). Isolated rough microsomes from rat liver were incubated *in vitro* to synthesize new proteins and then treated with the proteolytic enzymes trypsin and chymotrypsin. Results were as expected at this point, with ribosomes released from the membrane when the synthesizing polypeptide and some parts of the ribosomes were cleaved by the proteases, and thus were not particularly insightful. What later became valuable technically, however, was the observation that the parts of new proteins that had left the ribosome and entered the lumen of the microsome were "protected" from proteolytic degradation (Sabatini and Blobel 1970). As the collaboration continued over the next three years, other joint studies were completed with a graduate student and postdoc, but there was only incremental progress (Adelman, Sabatini, and Blobel 1973; Borgese, Blobel, and Sabatini 1973).

Up to this point, all the work had focused on what happened once ribosomes that were already membrane-bound and attached to mRNA had commenced synthesis and begun to transfer secretory proteins to the lumen of the ER. However, ribosomes were not always bound to the ER membrane. From the time that they were described in cells by Palade as a "particulate component of the cytoplasm" it was clear that there were ribosomes "free" in the cytoplasm and ribosomes bound to the ER membrane (Palade 1955a). There was a sense, not supported by hard evidence, that these two pools of ribosomes freely interchanged depending on particular physiological demands on the cell (Borgese, Blobel, and Sabatini 1973). If the latter was indeed true, then the question so far not addressed in any of the work was how ribosomes participating in the synthesis of secretory proteins actually "knew" to attach to the ER membrane and not remain free in the cytoplasm. In fact, this issue was recognized by Blobel and Sabatini and was a subject of intense discussion.

In 1971 Blobel, who was planning to attend an upcoming meeting on biomembranes, decided to present the ideas he and Sabatini had been pondering. When the resulting paper was later published as part of the symposium proceedings, it consisted of only three pages of text with one figure illustrating a speculative model (fig. 11.2a).[13] In it Blobel and Sabatini proposed that the first part of a secretory protein that is synthesized, called the amino-terminus, is a short segment of the protein (designated as *x* in the diagram) that directs or *targets* both the ribosome-mRNA complex and the new protein in the process of synthesis to the ER membrane. The segment then associates with a "factor" that mediates binding of the ribosome to the membrane, permitting the nascent polypeptide to cross the membrane as synthesis continues (fig. 11.2a). The model was consistent with what was known about both the process of protein synthesis and the attachment of ribosomes to the membrane: that is, the amino terminus of new proteins was exposed first from the large ribosomal subunit during synthesis, and the large subunit binds to the membrane, but otherwise it was unsupported by any direct data.

The following year, evidence for the model came from an unexpected source. Tim Harrison, a graduate student with George Brownlee and Cesar Milstein at Cambridge University, was attempting to purify mRNA for the small subunit of a protein produced by the immune system called the immunoglobulin light chain. At the time, there was no easy way to identify specific mRNA's except to add them to an *in vitro* system capable of protein synthesis, and then look for the appearance of a protein of the correct size, using the relatively new technique of SDS gel electrophoresis (fig. 11.3).[14] Harrison's *in vitro* system was crude by Rockefeller standards, sometimes

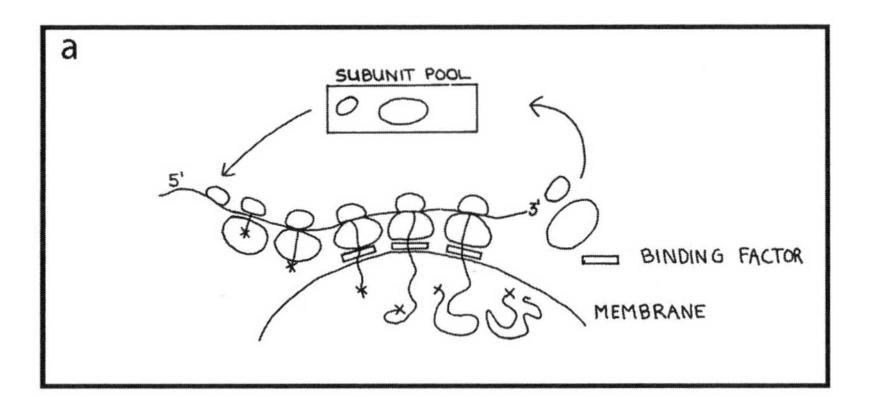

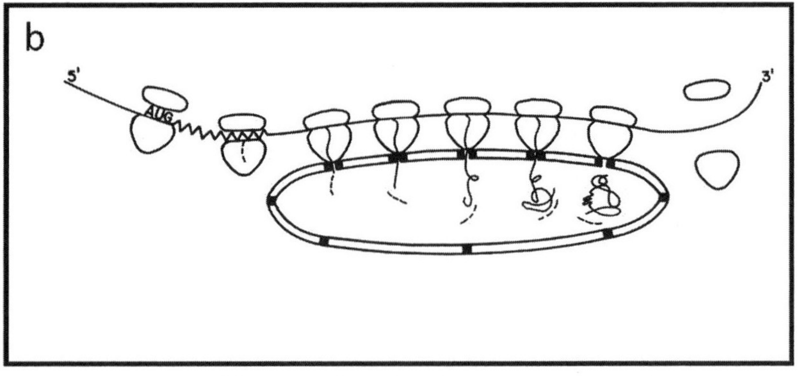

Figure 11.2. The 1971 (*a*) and 1975 (*b*) models of the signal hypothesis, from Blobel and Sabatini 1971 and Blobel and Dobberstein 1975a. In the earlier model, the signal sequence is depicted as an *x* and is not removed from the growing polypeptide chain after it reaches the ER lumen. In the 1975 model, the signal sequence appears as a dotted line that is proteolytically cleaved after translocation across the membrane. The diagram in (*a*) is used with permission of Günter Blobel. (*b*) ©1975 Blobel, G., and Dobberstein, B., *Journal of Cell Biology* 67:852–62.

containing microsomal membranes and sometimes not (Milstein et al. 1972; Matlin 2011). In a first series of experiments, the synthesized immunoglobulin chain was observed to move in the electric field on the gel to a position corresponding to the same molecular size as the authentic light chain included as a control. However, Harrison also noticed the synthesis of another protein very slightly larger than the authentic light chain. In subsequent experiments, he determined that synthesis of the smaller, authentic protein required the presence of membranes. Together with Milstein,

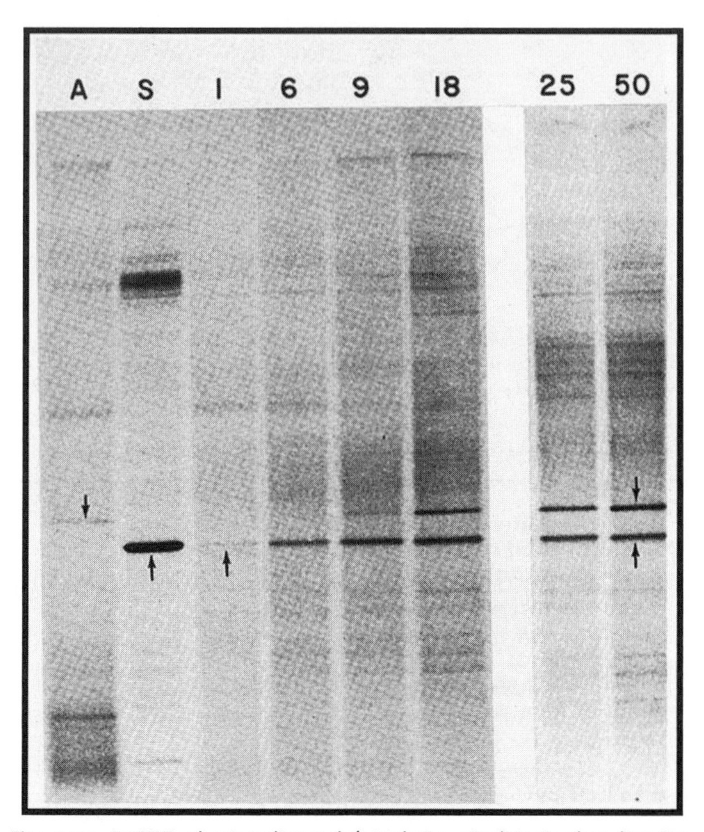

Figure 11.3. An SDS gel autoradiograph (a technique to detect only radioactive proteins) illustrating the small size difference between "bands" corresponding to the precursor that retains the signal sequence (downward-pointing arrows) and the processed or mature form (upward-pointing arrows). This is figure 8 from Blobel and Dobberstein 1975a. Harrison and Milstein used a similar technique in their work (Milstein et al. 1972). In Blobel's study the numbers from 1 to 50 above certain vertical "lanes" correspond to successive samples taken during a "read-out" experiment (from 1–50 minutes) as described in the text. ©1975 Blobel, G., and Dobberstein, B., *Journal of Cell Biology* 67:852–62.

he then mapped fragments of both the larger and smaller proteins, showing that they were identical except for one part at the amino-terminus, the first part of the protein that was synthesized. Because immunoglobulins are secretory proteins, they proposed that the extra segment was a "signal" that directed the protein to the ER membrane, where it was then removed (Milstein et al. 1972). At the time, Harrison and Milstein knew nothing of the Blobel and Sabatini model (Matlin 2011).

The publication of Harrison's findings galvanized the Blobel and Sabatini labs when they realized that their model might indeed be true. However, Sabatini soon left Rockefeller to become the chairman of cell biology at New York University, and his added duties distracted him from a full-time focus on the problem. Blobel, however, continued the work vigorously. Harrison and Milstein had only postulated that the larger protein containing the putative signal was a precursor of the smaller protein, but had not shown it definitively. Blobel believed that rigorous proof required an almost completely defined experimental system, and set about purifying the most critical individual components. By the end of 1974, he began to get results, and one year later published a pair of definitive papers (Blobel and Dobberstein 1975a; 1975b).

As if he wanted to dramatically demonstrate the universality of the mechanism that he hoped to prove, Blobel purified components of his system from diverse biological sources. Rough microsomes were isolated from both mouse myeloma tumor cells and dog pancreas,[15] and characterized by electron microscopy (fig. 11.1b). The myeloma microsomes were chosen because myeloma cells synthesized the same protein studied by Harrison and Milstein, the immunoglobulin light chain. Dog pancreas was not only a rich source of rough microsomes synthesizing a variety of secretory proteins, but, like Redman's pigeon pancreas, had significantly less ribonuclease than pancreas from other sources. Small ribosomal subunits were purified from rabbit reticulocytes (immature red blood cells), large subunits from rat liver rough microsomes, a mixture of factors needed to initiate and continue protein synthesis from another mouse tumor, mRNA for light chain from the same myeloma, and mRNA for globin (the core protein of hemoglobin), also from reticulocytes (Blobel and Dobberstein 1975a; 1975b).

Blobel used these components to create two separate systems for *in vitro* protein synthesis. The first he called the "readout system." This contained either rough microsomes from myeloma cells or something he called "detached ribosomes." The rough microsomes consisted of membrane-bound *polysomes*, the ribosome beads on the mRNA string, in the process of synthesizing light chain. In the polysomes, the ribosomes nearest the distal end of the mRNA at the time of isolation were assumed to have almost finished synthesizing light chain, while those at the beginning of the mRNA were just getting started. Blobel's "detached ribosomes," were really detached polysomes, that is, the same polysomes present in the rough microsomes but released intact from the membrane by treatment with the detergent deoxycholate. To determine if the smaller and larger light chains were synthesized, Blobel incubated both the rough microsomes and the detached polysomes with other factors required for continued protein synthesis to "read out" the light chain mRNAs—that is, to complete the synthesis of the light chain proteins that had started synthesis before the myeloma cells were disrupted. He compared the protein products of the readout by SDS gel electrophoresis. Because the readout system did not contain factors to start new protein synthesis, the synthesis of light chain from the beginning in the readout system was impossible. When Blobel examined light chains made by incubating the rough microsomes, most were of the smaller size identical to authentic light chain. Light chain made by detached polysomes, however, consisted of both the smaller and larger proteins previously seen by Harrison and Milstein (fig. 11.3). Blobel then repeated readout of the detached polysomes, but this time he took samples at different times after synthesis had been reactivated (fig. 11.3). Under these conditions, he saw that the first proteins completed were of the authentic size, and the last were larger chains (Blobel and Dobberstein 1975a; 1975b; Matlin 2011). Although the synthesis of both sizes of the light chain had begun before isolation of the rough microsomes, Blobel reasoned that the proteins furthest along at the time the polysomes were released from the membrane by detergent had already reached and perhaps begun traversing the microsomal membrane on the way to the lumen. In contrast, the proteins seen at later time points had only just begun synthesis before isolation. These, he reasoned, were not sufficiently long to have exited each ribosome in the polysomes and encountered the membrane before it was dissolved by detergent. To Blobel, these results supported a precursor-product relationship between the larger and smaller, authentic-sized protein, as well as confirming the observation of Harrison and Milstein that "processing" of the precursor by removal of the extra segment was dependent on the membrane (Blobel and Dobberstein 1975a; 1975b; Matlin 2011).

Blobel called the second system the "initiation system." This was reconstituted from a mixture of protein synthesis factors, including those needed to initiate new protein synthesis, purified ribosomal subunits, and mRNA. Dog pancreas microsomal membranes that had been "stripped" clean of bound ribosomes were also added. When light chain was synthesized in this system from the beginning in the absence of the stripped microsomes, only the larger, putative precursor protein was detected. However, in the presence of the membranes, mostly the smaller protein was seen, suggesting that the precursor had been "processed" by the membranes. When Blobel added mRNA for the nonsecretory protein globin to the initiation system instead of the light chain message, the protein produced was the same size, regardless of the presence of membranes. Most importantly, when Blobel treated samples from each condition with proteolytic enzymes after synthesis was complete, only the light chain synthesized in the presence of membranes was protected from degradation. In the system charged with globin mRNA, proteolysis degraded the globin protein with or without the addition of membranes. Apparently light chain made in the presence of membranes was not only processed to the correct, authentic size, but had also crossed into the interior of the microsomal membrane vesicles and was inaccessible to proteases (Blobel and Dobberstein 1975a; 1975b; Matlin 2011).

Separately, Blobel used mRNAs purified from dog pancreas rough microsomes to synthesize a mixture of pancreatic secretory proteins in the initiation system. In the presence of stripped microsomal membranes, many of the synthesized proteins "shifted" to smaller sizes in comparison to those made in the absence of membranes. Protein sequencing of the beginning, amino-terminal parts of the larger, putative precursor proteins revealed that all were about twenty amino acids larger than the smaller versions. The individual sequences of the extensions were not identical, but showed significant homology (Devillers-Thiery et al. 1975). These extensions corresponded to the signals postulated by Harrison and Milstein (Milstein et al. 1972).

On the basis of these results, Blobel reformulated his previous speculative model to include removal of the short extension, or signal sequence as he now called it, on the amino-terminus of the precursor proteins during the process of protein transfer across the membrane (Blobel and Dobberstein 1975a). He also made an explicit prediction about how proteins crossed the membrane. Instead of suggesting a vague "discontinuity" (Redman and Sabatini 1966), he proposed that the additional segment targeting the protein synthesis machinery to the membrane stimulated the assembly of a "transient tunnel" that facilitated the transfer process. Blobel called this revised model the "signal hypothesis." According to this hypothesis, the signal sequence was the "address" that directed the nascent secretory protein together with the ribosome/mRNA complex to its destination, the cytoplasmic surface of the ER membrane.

Over the next twenty-five years, the refined *in vitro* system developed by Blobel enabled the almost complete molecular description of the transfer (or translocation, as it is now called) process. Components predicted in the 1975 model, including the protease ("signal peptidase") that clipped off the signal sequence and the tunnel, were identified and characterized, along with other elements that had not been predicted, such as the signal-recognition particle, SRP (Walter and Blobel 1980; 1982; Matlin 2011; Matlin 2002). If one reflects, however, only on events up to 1975 instead of the later achievements, it is astounding to consider that the epistemic strategy begun by Albert Claude, using morphological examination of the cell and its parts to guide biochemical investigation, would ultimately yield the molecular details of Blobel's signal hypothesis. Although dramatically refined as new knowledge and techniques became available, there was no essential deviation from this strategy—even Blobel's landmark 1975 papers contained electron micrographs of rough microsomes (Blobel and Dobberstein 1975a). While the focus moved away from cells to smaller and smaller parts, the context of the cell, the big picture, was never abandoned.

The Epistemic Cycle and Modern Cell Biology

In the previous sections I described efforts by cell biologists to explain cellular phenomena through the application of an epistemic strategy developed in the latter half of the twentieth century. This strategy, illustrated in figure 11.4, is based upon the microscopic examination of whole cells leading to a representation of cellular morphology. Cellular form depicted in this manner plays essential roles in both development of a mechanistic hypothesis and testing of that hypothesis. In the former case, cellular form is used heuristically to constrain the scope of mechanistic possibilities to those consistent with the represented form, providing a framework for the incorporation of preexisting biological knowledge. For the latter, form provides a point of reference and biological context to guide physical decomposition of cells into parts, enabling their dynamic and molecular characterization (Matlin 2016).[16] The ensuing experimental findings are then compared with predictions of the mechanistic hypothesis. The degree of agreement between results and predictions then determines if modifications to either the hypothesis or the experimental setup are warranted. I call this process of hypothesizing and testing through guided decomposition an epistemic cycle because its success is dependent on iteration, with each new round of the cycle leading to a further perturbation of the biological system that ultimately provides a more detailed and accurate explanation of the cellular phenomenon under investigation.[17]

Although Albert Claude initiated this approach, it reached a more refined level with the work of George Palade in the 1950s, who referred to it explicitly as an integrated strategy when he and Philip Siekevitz used it to

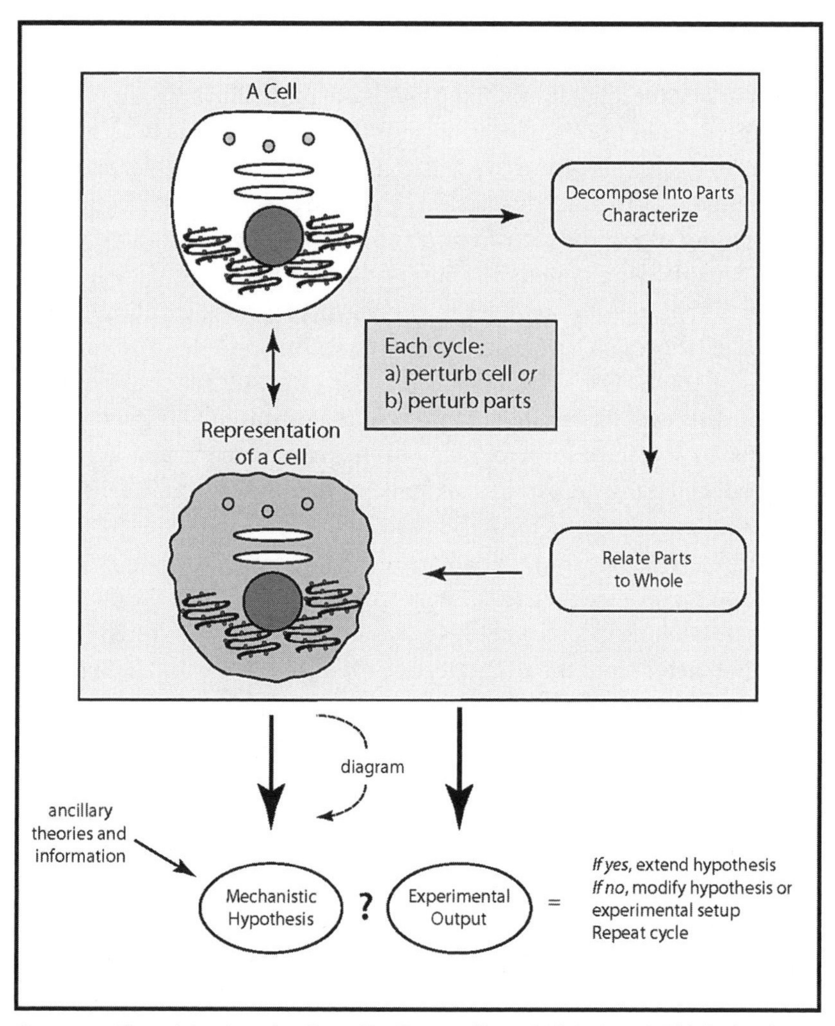

Figure 11.4. The epistemic cycle of investigation used by cell biologists. Cell biologists begin with a representation of a cell in the form of a microscopic image. The representation, sometimes converted into a simplified diagram, heuristically constrains possible mechanisms to those consistent with cellular morphology. Theories and information from biochemical and genetic approaches contribute to development of an initial mechanistic hypothesis. Then, through a cycle of decomposition, localization, and recomposition guided by the representation of the cell, experimental results are generated that either support the hypothesis or require its modification. Future iterations of the cycle in which the cell or its parts are experimentally perturbed further refine the mechanistic hypothesis.

investigate the function of the ER (Palade and Siekevitz 1956). In its original form, it reached an apotheosis with the work of Günter Blobel, who provided molecular explanations of the biological process of secretory protein targeting, membrane translocation, and segregation within the lumen of the ER. To get to this ultimate point, Palade and his colleagues prepared representations of whole cells (within organized tissues) using electron microscopy, and noted the distribution and amount of rough ER in the cytoplasm. They then integrated their observations with existing histochemical data on the "basophilic" part of the cytoplasm, physiological correlations, RNA localization data, and biochemical results linking RNA to protein synthesis to yield the initial hypothesis that the rough ER is key to secretory protein synthesis. To test this, they then homogenized and fractionated the cells, took electron micrographs of the fractions, and compared the images of isolated microsomes to similar structures in whole cells. They also biochemically characterized the isolated parts. After iterating this investigative cycle a few times, they concluded that rough microsomes were equivalent to the rough ER. In the next cycles they perturbed the biological system by labeling cells with radioactive amino acids for different times, providing insight into the dynamics of protein synthesis and its relationship to the next stages of secretion.

Once it was evident from rough microsomal morphology—ribosomes on the outside, closed vesicles, and trapped content—that this isolated part retained essential characteristics of the ER *in situ*, Blobel was justified in perturbing isolated microsomes instead of whole cells. He treated them with detergents and chemicals, gaining insights into the nature of the membrane barrier and the attachment of ribosomes. In subsequent experiments, he stimulated protein synthesis in the presence of isolated microsomes and eventually reconstituted the entire process of secretory protein synthesis and translocation in the ER. Ultimately, Blobel and others identified and characterized the molecular parts responsible for protein synthesis and transfer into the ER. Each investigative step was another turn on the epistemic cycle, yielding more and more detailed revisions of the initial hypothesis. Vague postulates about the relationship of rough microsomes to secretory protein synthesis became specific ideas about "vectorial transfer" and ribosome attachment; membrane "discontinuities" evolved into membrane "channels"; and previously unimagined concepts, such as signal-mediated protein targeting, emerged along with the evidence to support them.

Even if one accepts this characterization of the epistemic strategy used by twentieth-century cell biologists as accurate, it is reasonable to ask if it remains valid under current circumstances. Modern biology is said to be transformed not only by new instruments, as occurred in the past, but also by the "big data" provided by those instruments (Callebaut 2012). The sequencing of the human genome was followed by the development of many so-called "high throughput" approaches that created other "-omes" beyond the genome. These are essentially databases of molecular structures and interactions. In theory, if not in actual practice, all the molecular parts of living systems are now accessible, and their molecular activities predictable. These achievements made concrete the complexity of living systems and stimulated a move to different forms of systems biology whose goals are to decipher higher-level processes in living systems by exploiting -omic data. While these data may be of little explanatory value by themselves (Krohs and Callebaut 2007), they are undeniably useful. If so, then how might they be best incorporated into the investigation of biological phenomena? I propose that the same epistemic strategy used by cell biologists in the pre-omic period still informs modern cell biology.

As one example, I describe recent attempts to understand the onset of asymmetric organization, or *polarization*, of the *Caenorhabditis elegans* single-celled embryo (Dawes and Munro 2011; Munro, Nance, and Priess 2004; Nance, Munro, and Priess 2003). These studies have certain features that, superficially at least, seem unrelated to the cell biology of Palade and Blobel. Following fertilization, live embryos are observed over time by light microscopy. Specific proteins, whose involvement in the processes under investigation has been inferred from prior data and detailed characterization of the molecules themselves, are expressed in the live embryos as fluorescent "fusion" proteins, so that their movements and, to a certain extent, interactions with other similarly tagged proteins can be observed and quantitated in real time. RNA-interference, a technology that permits the amount of expression of specific proteins to be modulated, is used to perturb the system. Finally, mechanistic hypotheses, while still expressed in diagrams like those of Blobel, are also formalized mathematically to make specific predictions, often depicted as animations of cellular events. Using this approach, it has been possible to determine the mechanism by which a set of "Par" proteins become asymmetrically localized in the one-cell embryo through their dynamic oligomerization and mutual inhibition in the absence of any stable intracellular boundaries (Dawes and Munro 2011).[18]

To understand how this combination of novel technologies, imaging, and computational modeling might fit the previously described epistemic cycle, I outline events leading up to the most recent experiments.[19] Polarization of the embryo was inferred from the asymmetric redistribution of cytoplasmic granules visualized in simple light microscopic observations of live embryos. At the same time, movements of cytoplasmic material in the cortex of the embryo revealed contractions that, by analogy, suggested the involvement of actin and myosin, proteins responsible for muscle contraction (Hird and White 1993). Par proteins were discovered in genetic mutants that failed to polarize, and biochemical as well as genetic studies suggested that they both physically interacted and were capable of reciprocal inhibition. Despite this, it was clear that the process by which these molecular components came together to generate and maintain polarity was too dynamically complex to be intuited by conventional approaches. Based upon the accumulated prior knowledge about Par proteins and the known functional characteristics of actin and myosin, an initial hypothesis proposed that actomyosin contractility initiated by fertilization was responsible for asymmetrically distributing the Par proteins and maintaining their distribution. While initial studies supported this general model, it became evident from observed diffusion of the Par proteins in the embryo that the maintenance of their polarized localization was independent of the contractile apparatus (Dawes and Munro 2011; Munro, Nance, and Priess 2004). A revised hypothesis based on interactions among and between the Par proteins was then developed and formalized into a mathematical model. Predictions from this model were experimentally tested through depletion of a specific Par protein. This experiment validated the model, which was then manipulated to make new, experimentally testable predictions.[20]

This strategy conforms to the epistemic cycle described in figure 11.4 once certain steps are updated to correspond to modern technical approaches. As with the investigation of secretion, study of polarization begins with representation of the cell. Instead of static images, however, the live embryos are visualized by light microscopy. Initial observations are combined with extensive ancillary information, in this case in the form of detailed biochemical and genetic data about specific molecules and their interactions, to yield a mechanistic hypothesis (fig. 11.4). In addition to its diagrammatic depiction, the hypothesis is eventually mathematically formulated to enable the proposed complex relationships between molecules to yield specific predictions. Decomposition into parts is not accomplished physically, but instead is carried out by expressing fluorescent proteins (the parts) and observing their behavior in the living cell. In this manner, relating the parts to the whole—a key step in the epistemic strategy—is intrinsic to the experimental setup because the expressed proteins are always embedded within the whole cell.[21] Experimental observations, which now yield dynamic and quantitative data generated in the context of the whole cell, are compared with the predictions of the mechanistic hypothesis. Based upon this comparison, the cycle is iterated with perturbation, this time mediated by RNA interference modulation of the molecular parts.[22] Gradually, with further iterations, a molecular, dynamic process occurring in the cell comes into focus.

Strategies of Investigation

In this essay I have used a historical case study to describe an epistemic strategy successfully exploited by cell biologists studying protein secretion in the latter half of the twentieth century. What is remarkable about this work is that it extended from the first description of the ER found in the "optically empty" part of the cytoplasm to, ultimately, a detailed molecular description of one primary function of the ER, the segregation of secretory proteins within membranes of the secretory pathway. I then attempted to update this strategy by applying it to a more current case study in systems biology that uniquely integrates microscopic observations of cells with molecular manipulations and computational approaches. This new combination has the potential to penetrate the intuitively inaccessible complexity of living systems to explain biological phenomena at the molecular level.

Neither case study establishes the generality of this epistemic strategy in the investigation of cell biological phenomena. I propose, nevertheless, that features of this strategy are generally applicable, a view that I believe is supported by recent scholarship in the philosophy of biology. On the basis of the work of William Wimsatt, I have separately argued that representation of form by cell biologists, as in micrographs, has essential heuristic qualities that constrain mechanistic hypotheses to those more likely to be biologically meaningful, and provide a point of reference for decomposition (Matlin 2016). Both these features are prominent elements of the proposed epistemic cycle. The idea that investigation of cellular phenomena requires a cellular context has been emphasized by Wagner and Laubichler, who argue that cells have ontological priority, and that investigative strategies that fail to take this into account may not be able to link molecular events with cellular functions (Wagner and Laubichler 2000). Similarly,

Richard Burian argues that studies incorporating biological form in a manner similar to my proposal played an essential part in discoveries credited to molecular biology, stating that "*in many cases, it is only in the context of higher level structures, functioning properly, that molecular mechanisms operate 'correctly'*" (Burian 1996, 81; italics in original).

Krohs and Callebaut make a related point in their critique of "top-down" and "bottom-up" strategies in systems biology, stating that "'unbiased modularization' [selection of modules based, for example, on networks] . . . gives up the established link between physiology and function" (Krohs and Callebaut 2007, 208). Thus, I argue, relying on representations of whole cells and relating decompositions to the entire cellular context are key aspects of cell biologists' epistemic strategy. In a later paper, Callebaut argued for a "multiperspectival" approach in response to an era of big-data biology that some feel, as Callebaut notes, threatens to make "the search for underlying mechanisms to distinguish causation from spurious correlation . . . obsolete" (Callebaut 2012, 74). Multiperspectivism is built into the epistemic strategy of cell biology: a biochemical and molecular perspective on the parts is linked to the perspective of the whole cell through the process of guided decomposition. No perspective is definitive or even completely accurate, but the cyclic *process of iteration* moves investigation closer to a detailed explanation of the biological phenomenon. Maureen O'Malley also invokes iterativity as a key feature of "more pluralistic, pragmatic accounts of scientific inquiry" (O'Malley, Elliott, and Burian 2010, 414). On the basis of work by Hasok Chang, O'Malley describes "epistemic iterativity" as a means by which "each step of understanding proceeds in a locally satisficing way to others. 'Imperfect ingredients' are thrown together in order to make something just a little bit better while remaining imperfect in the light of subsequent inquiry" (O'Malley, Elliott, and Burian 2010, 414; Chang 2004, 226).[23] Among cell biologists, each mechanistic hypothesis is considered incomplete or inadequate, a false model, but one that is a *means* to a truer theory as cycles are iterated on the way to molecular understanding (Wimsatt 2007). O'Malley believes that systems biologists discovered this iterative strategy. Instead, I argue, the epistemic cycle is fundamental, discovered and exploited but not created by cell biologists.

Acknowledgments

I am most grateful to all participants in the two "Updating Cowdry" workshops, students and faculty on the Committee on Conceptual and Historical Studies of Science at the University of Chicago, and, particularly, to Jane Maienschein, Bill Wimsatt, and Garland Allen for their constructive comments and insights. This work was supported in part by a grant from the National Library of Medicine and funds from the Edwin S. Webster Foundation.

Notes

1 In 1935 Frederick Gowland Hopkins, the great Cambridge biochemist, noted in an address that "indeed most biologists had felt [until the twentieth century] that [the study of dynamic chemical events in the cell] was outside the scope of chemical science. It was their faith that the subtle molecular events which must underlie the visible functions displayed by living organisms are initiated and controlled by the inherent but elusive potency of protoplasm in its integrity. They felt, therefore, with apparent reason, that the methods of the chemist, which must destroy that integrity, could not illuminate this field of biological reality" (Hopkins 1949).

2 Throughout most of this essay I focus on discoveries made at the Rockefeller Institute of Medical Research, which later became the Rockefeller University. While this is somewhat artificial because scientific progress does not depend solely on the contributions from one institution, in this particular case many substantive developments that built systematically upon each other did occur in the same laboratory over a period of 30–40 years.

3 The concept of mechanism relevant to my discussion is that of Bechtel: "A mechanism is a structure performing a function in virtue of its component parts, component operations, and their organization. The orchestrated functioning of the mechanism is responsible for one or more phenomena" (Bechtel 2006, 26). Some of the ground that I cover here has been previously trod by others (Bechtel 2006; Rasmussen 1997; Rheinberger 1995). Both Rheinberger and Bechtel focus, as I do, on experimental processes driving discoveries of cellular functions, and my analysis owes considerable debt to both. Bechtel's history of cell biology stops in 1970, while Rheinberger turns from the origins of cell biology to deal primarily with the mechanism of protein synthesis as studied in the laboratory of Paul Zamecnik (Rheinberger 1997). Rasmussen concentrates almost exclusively on the history of biological electron microscopy.

4 As I discuss subsequently, the term *microsomes* was eventually and specifically defined as "vesicular fragments of the endoplasmic reticulum."

5 Claude remarked, "In the experiments here discussed, and in future papers dealing with the biochemistry of the various fractions, special emphasis has been attached to the quantitative aspect of the results. Efforts have been made to determine the yield of each fraction and to express, whenever possible, the activity exhibited by a particular fraction, in terms of the total activity possessed by the unfractionated liver extract" (Claude 1946a, 53).

6 Claude did reference Porter's paper in the second of his cell fractionation publications, but only to make a minor point about mitochondria (Claude 1946b, 70).

7 Versene is now called EDTA (ethylene-diamine tetra-acetate). The technical term for the association of EDTA with divalent cations is *chelation*.

8 At about the same time, Zamecnik's laboratory also used deoxycholate to separate ribosomes from the microsomal membrane (Littlefield et al. 1955). While this work was not cited in the 1956 Palade and Siekevitz paper, Siekevitz was likely aware of it (Moberg 2012, 142). Deoxycholate is a mild detergent related to bile salts that was also used by biochemists at the time to free enzymes from cellular particulate matter (Matlin 2016).

9 The exocrine pancreas synthesizes and releases (secretes) digestive enzymes into the gut. The endocrine pancreas, which is part of the same organ in mammals, produces insulin and glucagon, and secretes them into the bloodstream.
10 *In vitro*: literally "in glass," generally synonymous with "cell-free" experiments conducted with purified cellular material in test tubes.
11 Fortuitously, amylase could be more easily purified than chymotrypsinogen by precipitating it with large amounts of glycogen (a polymeric sugar).
12 Some of what follows in this and the subsequent section is derived from the author's interviews.
13 While interviews consistently attribute the model to joint discussions between Blobel and Sabatini, the actual cartoon was drawn by Blobel and appeared in his grant application submitted late in 1970.
14 Protein electrophoresis in polyacrylamide gels containing the detergent SDS (sodium dodecyl sulfate) had only recently begun to be widely used at the time of Harrison's studies. A revolutionary technique, it permitted complex mixtures of proteins to be separated according to their molecular size at very high resolution and made it easy to compare many different experimental samples at once.
15 A myeloma is a tumor of antibody-secreting cells. Because much of the protein synthesis in myeloma cells is devoted to a single antibody molecule, microsomes isolated from the cells are enriched in ribosomes and mRNA-making light chain.
16 The strategy described here obviously owes much to the approach of decomposition and localization articulated by Bechtel and Richardson and subsequently refined (Bechtel 2006; Bechtel and Abrahamsen 2013; Bechtel and Richardson 1993). It also resembles in certain respects Giere's model of scientific reasoning as extended by Griesemer (Griesemer 2000). Space limitations preclude, however, appropriate discussion of these relationships.
17 Rheinberger hints at such an iterative process in his look at the early days of cell fractionation and electron microscopy at the Rockefeller Institute, stating, "It was a technique of manipulating whole cells that made it possible to shuttle back and forth between representations of material recovered from the test tube and structures that could be obtained from in situ preparation" (Rheinberger 1995, 73).
18 Par proteins are the products of genes identified in *C. elegans* as *Par*tition-deficient mutations that disrupted the asymmetric first division of the zygote. They are now known to be important in the asymmetric organization of most cells.
19 It was very unusual that the work on the secretory mechanism occurred in a continuum in a single department at the same institution (although see my previous note about focusing on the Rockefeller [Institute] University). I propose that in most cases, successive epistemic cycles leading to more and more detailed explanations of a particular biological phenomenon take place in multiple laboratories over time, building upon each other but not departing from the essential cyclic strategy.
20 The work on this topic is ongoing in the laboratory profiled and in other laboratories. Some of what is described here is based upon an interview with Ed Munro.
21 While selection of the parts on which to focus is clearly biased by the mechanistic hypothesis, this is no different than in the secretion studies when the conditions of physical decomposition and fractionation were adjusted for the isolation of specific

compartments (i.e., the ER) that featured in the mechanistic hypothesis. One reason why I refer to the use of form to help develop a mechanistic hypothesis as a heuristic process is that it suggests educated guesses, such as which organelles or proteins might be involved in the phenomena under investigation. Of course, one other characteristic of a heuristic process is that the guesses might be wrong—that is, heuristics do not guarantee success (Matlin 2016).

22 It is important to note that this implementation of the epistemic cycle can take full advantage of -omic data sets. Mechanistic models for polarization were based upon microscopic observations of whole cells but were informed by molecular details developed through modern techniques of cloning and sequencing that have made -omics possible. However, the causal burden of explaining polarization at the molecular level was not placed solely on the molecules, but was shaped by the inherently biological constraints provided by morphological representation of the embryo.

23 The term *satisficing* originated with Herbert Simon, who defined it as "decisions that are 'good enough'" in the context of heuristic strategies in artificial intelligence (Simon 1996, 27).

References

Adelman, M. R., D. D. Sabatini, and G. Blobel. 1973. "Ribosome-Membrane Interaction: Nondestructive Disassembly of Rat Liver Rough Microsomes into Ribosomal and Membranous Components." *Journal of Cell Biology* 56 (1): 206–29.

Beams, H. W. 1943. "Ultracentrifugal Studies on Cytoplasmic Components and Inclusions." In *Biological Symposia: Frontiers in Cytochemistry*, edited by N. L. Hoerr, 71–90. Lancaster, PA: Jacques Cattell Press.

Bechtel, W. 2006. *Discovering Cell Mechanisms: The Creation of Modern Cell Biology*. Cambridge: Cambridge University Press.

———, and A. Abrahamsen. 2013. "Diagrams as Vehicles for Scientific Reasoning." Manuscript.

———, and R. C. Richardson. 1993. *Discovering Complexity*. Princeton, NJ: Princeton University Press.

Blobel, G., and B. Dobberstein. 1975a. "Transfer of Proteins Across Membranes. I. Presence of Proteolytically Processed and Unprocessed Nascent Immunoglobulin Light Chains on Membrane-Bound Ribosomes of Murine Myeloma." *Journal of Cell Biology* 67 (3): 835–51.

———, and B. Dobberstein. 1975b. "Transfer of Proteins Across Membranes. II. Reconstitution of Functional Rough Microsomes from Heterologous Components." *Journal of Cell Biology* 67 (3): 852–62.

———, and V. R. Potter. 1967a. "Ribosomes in Rat Liver: An Estimate of the Percentage of Free and Membrane-Bound Ribosomes Interacting with Messenger RNA *in Vivo*." *Journal of Molecular Biology* 28 (3): 539–42.

———, and V. R. Potter. 1967b. "Studies on Free and Membrane-Bound Ribosomes in Rat Liver II. Interaction of Ribosomes and Membranes." *Journal of Molecular Biology* 26 (2): 293–301. doi: 10.1016/0022-2836(67)90298-7.

———, and V. R. Potter. 1967c. "Studies on Free and Membrane-Bound Ribosomes in Rat Liver. I. Distribution as Related to Total Cellular RNA." *Journal of Molecular Biology* 26 (2): 279–92. doi: 0022-2836(67)90297-5 [pii].

———, and D. Sabatini. 1970. "Controlled Proteolysis of Nascent Polypeptides in Rat Liver Cell Fractions. I. Location of the Polypeptides within Ribosomes." *Journal of Cell Biology* 45 (1): 130–45.

Borgese, D., G. Blobel, and D. D. Sabatini. 1973. "In Vitro Exchange of Ribosomal Subunits between Free and Membrane-Bound Ribosomes." *Journal of Molecular Biology* 74 (4): 415–38.

Bradbury, S. 1967. *The Evolution of the Microscope*. Oxford: Pergamon Press.

Burian, R. M. 1996. "Underappreciated Pathways toward Molecular Genetics as Illustrated by Jean Brachet's Cytochemical Embryology." In *The Philosophy and History of Molecular Biology: New Perspectives*, edited by S. Sarkar, 67–85. Dordrecht: Kluwer Academic Publishers.

Callebaut, W. 2012. "Scientific Perspectivism: A Philosopher of Science's Response to the Challenge of Big Data Biology." *Studies in History and Philosophy of Biological and Biomedical Sciences* 43 (1): 69–80. doi: 10.1016/j.shpsc.2011.10.007.

Chang, H. 2004. *Inventing Temperature*. Oxford: Oxford University Press.

Claude, A. 1940. "Particulate Components of Normal and Tumor Cells." *Science* 91 (2351): 77–78. doi: 10.1126/science.91.2351.77.

———. 1941. "Particulate Components of Cytoplasm." *Cold Spring Harbor Symposia on Quantitative Biology* 9: 263–71. doi: 10.1101/SQB.1941.009.01.030.

———. 1943a. "The Constitution of Cytoplasm." *Science* 97 (2525): 451–56. doi: 10.1126/science.97.2525.451.

———. 1943b. "Distribution of Nucleic Acids in the Cell and the Morphological Constitution of the Cytoplasm." In *Biological Symposia: Frontiers in Cytochemistry*, edited by N. L. Hoerr, 111–29. Lancaster, PA: Jacques Cattell Press.

———. 1946a. "Fractionation of Mammalian Liver Cells By Differential Centrifugation: I. Problems, Methods, and Preparation of Extract." *Journal of Experimental Medicine* 84 (1): 51–59.

———. 1946b. "Fractionation of Mammalian Liver Cells By Differential Centrifugation: II. Experimental Procedures and Results." *Journal of Experimental Medicine* 84 (1):61–89.

———. 1948. "Studies on Cells: Morphology, Chemical Constitution and Distribution of Biochemical Functions." *Harvey Lectures* 43: 121–64.

Cowdry, Edmund V., ed. 1924. *General Cytology: A Textbook of Cellular Structure and Function for Students of Biology and Medicine*. Chicago: University of Chicago Press.

Dawes, A. T., and E. M. Munro. 2011. "PAR-3 Oligomerization May Provide an Actin-Independent Mechanism to Maintain Distinct Par Protein Domains in the Early *Caenorhabditis elegans* Embryo." *Biophysical Journal* 101 (6): 1412–22. doi: 10.1016/j.bpj.2011.07.030.

Devillers-Thiery, A., T. Kindt, G. Scheele, and G. Blobel. 1975. "Homology in Amino-Terminal Sequence of Precursors to Pancreatic Secretory Proteins." *Proceedings of the National Academy of Sciences* 72 (12): 5016–20.

Griesemer, J. 2000. "Development, Culture, and the Units of Inheritance." *Philosophy of Science*. 67: S348–68. doi: 10.2307/188680.

Hird, S. N., and J. G. White. 1993. "Cortical and Cytoplasmic Flow Polarity in Early Embryonic Cells of *Caenorhabditis elegans*." *Journal of Cell Biology* 121 (6): 1343–55.

Hogeboom, G. H. 1949. "Cytochemical Studies of Mammalian Tissues. II. The Distribution of Diphosphopyridine Nucleotide-Cytochrome c Reductase in Rat Liver Fractions." *Journal of Biological Chemistry* 177 (2): 847–58.

———, A. Claude, and R. Hotchkiss. 1946. "The Distribution of Cytochrome Oxidase and Succinoxidase in the Cytoplasm of the Mammalian Liver Cell." *Journal of Biological Chemistry* 165 (2): 615–29.

———, W. C. Schneider, and G. E. Pallade. 1948. "Cytochemical Studies of Mammalian Tissues I. Isolation of Intact Mitochondria from Rat Liver: Some Biochemical Properties of Mitochondria and Submicroscopic Particulate Material." *Journal of Biological Chemistry* 172 (2): 619–35.

Hopkins, F. G. 1949. "The Spirit of Modern Biochemistry." In *Hopkins and Biochemistry*, edited by J. Needham and E. Baldwin, 264–68. Cambridge: W. Heffer and Sons.

Keller, E. B., P. C. Zamecnik, and R. B. Loftfield. 1954. "The Role of Microsomes in the Incorporation of Amino Acids into Proteins." *Journal of Histochemistry and Cytochemistry* 2 (5): 378–86.

Krohs, U., and W. Callebaut. 2007. "Data without Models Merging with Models without Data." In *Systems Biology: Philosophical Foundations*, edited by F. C. Boogerd, F. J. Bruggeman, J.-H. S. Hofmeyr, and H. V. Westerhoff, 181–213. Amsterdam: Elsevier.

Littlefield, J .W., E. B. Keller, J. Gross, and P. C. Zamecnik. 1955. "Studies on Cytoplasmic Ribonucleoprotein Particles from the Liver of the Rat." *Journal of Biological Chemistry* 217 (1): 111–23.

Matlin, K. S. 2002. "The Strange Case of the Signal Recognition Particle." *Nature Reviews Molecular Cell Biology* 3 (7): 538–42. doi: 10.1038/nrm857.

———. 2011. "Spatial Expression of the Genome: The Signal Hypothesis at Forty." *Nature Reviews Molecular Cell Biology* 12 (5): 333–40. doi: 10.1038/nrm3105.

———. 2016. "The Heuristic of Form: Mitochondrial Morphology and the Explanation of Oxidative Phosphorylation." *Journal of the History of Biology* 49 (1): 37–94. doi: 10.1007/s10739-015-9418-3.

Milstein, C., G. Brownlee, T. Harrison, and M. Mathews. 1972. "A Possible Precursor of Immunoglobulin Light Chains." *Nature: New Biology* 239 (91): 117–20.

Moberg, C. L. 2012. *Entering an Unseen World*. New York: Rockefeller University Press.

Munro, E. M., J. Nance, and J. R. Priess. 2004. "Cortical Flows Powered by Asymmetrical Contraction Transport PAR Proteins to Establish and Maintain Anterior-Posterior Polarity in the Early *C. elegans* Embryo." *Developmental Cell* 7 (3): 413–24. doi: 10.1016/j.devcel.2004.08.001.

Nance, J., E. M. Munro, and J. R. Priess. 2003. "*C. elegans* PAR-3 and PAR-6 Are Required for Apicobasal Asymmetries Associated with Cell Adhesion and Gastrulation." *Development* 130 (22): 5339–50. doi: 10.1242/dev.00735.

O'Malley, M. A., K. C. Elliott, and R. M. Burian. 2010. "From Genetic to Genomic Regulation: Iterativity in MicroRNA Research." *Studies in History and Philosophy of Science, Part C: Studies in History and Philosophy of Biological and Biomedical Sciences* 41 (4): 407–17. doi: 10.1016/j.shpsc.2010.10.011.

Palade, G. 1955a. "A Small Particulate Component of the Cytoplasm." *Journal of Biophysical and Biochemical Cytology* 1 (1): 59.

———. 1955b. "Studies on the Endoplasmic Reticulum. II. Simple Dispositions in Cells in Situ." *Journal of Biophysical and Biochemical Cytology* 1 (6): 567–82.

———. 1975. "Intracellular Aspects of the Process of Protein Synthesis." *Science* 189 (4200): 347–58.

———, and K. R. Porter. 1954. "Studies on the Endoplasmic Reticulum. I. Its Identification in Cells in Situ." *Journal of Experimental Medicine* 100 (6): 641–56.

———, and P. Siekevitz. 1956. "Liver Microsomes: An Integrated Morphological and Biochemical Study." *Journal of Cell Biology* 2 (2): 171–200.
———, P. Siekevitz, and L. G. Caro. 1962. *Structure, Chemistry and Function of the Pancreatic Exocrine Cell.* Chichester, UK: John Wiley & Sons.
Porter, K. R. 1953. "Observations on a Submicroscopic Component of Cytoplasm." *Journal of Experimental Medicine* 97 (5): 727–50.
———, and J. Blum. 1953. "A Study in Microtomy for Electron Microscopy." *Anatomical Record* 117 (4): 685–709. doi: 10.1002/ar.1091170403.
———, A. Claude, and E. F. Fullam. 1945. "A Study of Tissue Culture Cells by Electron Microscopy: Methods and Preliminary Observations." *Journal of Experimental Medicine* 81 (3): 233–46.
Rasmussen, N. 1997. *Picture Control.* Stanford, CA: Stanford University Press.
Redman, C. M., and D. D. Sabatini. 1966. "Vectorial Discharge of Peptides Released By Puromycin from Attached Ribosomes." *Proceedings of the National Academy of Sciences* 56 (2): 608–15.
———, P. Siekevitz, and G. E. Palade. 1966. "Synthesis and Transfer of Amylase in Pigeon Pancreatic Microsomes." *Journal of Biological Chemistry* 241 (5): 1150–58.
Rheinberger, H.-J. 1995. "From Microsomes to Ribosomes: 'Strategies' of 'Representation.'" *Journal of the History of Biology* 28 (1): 49–89.
———. 1997. *Toward a History of Epistemic Things.* Stanford, CA: Stanford University Press.
Sabatini, D. D., and G. Blobel. 1970. "Controlled Proteolysis of Nascent Polypeptides In Rat Liver Cell Fractions. II. Location of The Polypeptides in Rough Microsomes." *Journal of Cell Biology* 45 (1): 146–57.
———, Y. Tashiro, and G. E. Palade. 1966. "On the Attachment of Ribosomes to Microsomal Membranes." *Journal of Molecular Biology* 19 (2): 503–24.
Siekevitz, P. 1952. "Uptake of Radioactive Alanine in Vitro into the Proteins of Rat Liver Fractions." *Journal of Biological Chemistry* 195 (2): 549–65.
———, and G. E. Palade. 1958. "A Cytochemical Study on the Pancreas of the Guinea Pig. I. Isolation and Enzymatic Activities of Cell Fractions." *Journal of Biophysical and Biochemical Cytology* 4 (2): 203–18.
———, and G. E. Palade. 1960a. "A Cytochemical Study on the Pancreas of the Guinea Pig. V. In Vivo Incorporation of Leucine-1-C14 into the Chymotrypsinogen of Various Cell Fractions." *Journal of Cell Biology* 7 (4): 619–30. doi: 10.1083/jcb.7.4.619.
———, and G. E. Palade. 1960b. "A Cytochemical Study on the Pancreas of the Guinea Pig. VI. Release of Enzymes and Ribonucleic Acid from Ribonucleoprotein Particles." *Journal of Biophysical and Biochemical Cytology* 7 (4): 631–44.
———, and G. E. Palade. 1966. "Distribution of Newly Synthesized Amylase in Microsomal Subfractions of Guinea Pig Pancreas." *Journal of Cell Biology* 30 (3): 519–30.
Simon, H. A. 1996. *The Sciences of the Artificial.* 3rd ed. Cambridge, MA: MIT Press.
Wagner, G. P., and M. D. Laubichler. 2000. "Character Identification in Evolutionary Biology: The Role of the Organism." *Theory in Biosciences* 119 (1): 20–40.
Walter, P., and G. Blobel. 1980. "Purification of a Membrane-Associated Protein Complex Required for Protein Translocation across the Endoplasmic Reticulum." *Proceedings of the National Academy of Sciences* 77 (12): 7112–16.

———, and G. Blobel. 1982. "Signal Recognition Particle Contains a 7S RNA Essential for Protein Translocation across the Endoplasmic Reticulum." *Nature* 299 (5885): 691–98. doi: 10.1038/299691a0.

Wilson, E. B. (1925) 1928. *The Cell in Development and Heredity*. 3rd ed., with corrections. New York: Macmillan. Orig. pub. as *The Cell in Development and Inheritance* in 1896; 2nd ed., 1900.

Wimsatt, W. C. 2007. *Re-Engineering Philosophy for Limited Beings*. Cambridge, MA: Harvard University Press.

CHAPTER 12

OBSERVING THE LIVING CELL

SHINYA INOUÉ AND THE REEMERGENCE OF LIGHT MICROSCOPY

Rudolf Oldenbourg

Many of the observations reported in Cowdry's book *General Cytology* from 1924 were made using the light microscope, the primary tool of the time to study cells, either living or fixed. Staining was used to enhance contrast and highlight specific structures, revealing details down to a fraction of a micron. With the invention of the electron microscope in the 1930s, and during its heyday in the 50s and 60s, much smaller details emerged in cells that were fixed, embedded, and stained for this new and captivating imaging tool. Results, however, were dependent on the fixation, embedding, and staining procedures, a fact that electron and light microscopy seemed to have in common at the time.

Against this backdrop, cell biologists Shinya Inoué and his mentor Katsuma Dan insisted on making observations in living cells, using the light microscope and noninvasive optical means to enhance contrast and highlight specific cell structures. Polarized light microscopy was particularly interesting, as it revealed order at a molecular scale that is usually hidden to the light microscope and offered a way to bridge the resolution gap between light and electron microscopy, while enabling observations on living cells. In the early 1950s, in the Marine Biological Laboratory's Lillie Auditorium, Inoué demonstrated the existence of parallel submicroscopic fibers in the mitotic spindle by showing movies taken in polarized light of actively dividing cells. Until then, many had questioned the reality of those fibers, as well as their roles in the anaphase separation of chromosomes and the division of the cytoplasm itself. A decade later, in a landmark study, Inoué revealed the packing of DNA in living sperm heads by interpreting his polarized light observations in terms of the structure of DNA, a feat that linked light microscopy with the then new discipline of molecular biology.

Thus, the seeds for the reemergence of light microscopy were sown, with a further dramatic advance in the 1980s when Inoué systematically exploited the combination of light microscopy, electronic imaging, and digital image processing to reveal ever finer and subtler detail inside living cells, tissues, functioning model systems, and whole organisms. Since then, a confluence of spectacular new optical techniques and a plethora of ways to highlight and label cell structures and functions have set a new stage on which light microscopy advances our understanding of living cells and life itself.

This article is necessarily selective and cannot present the breadth and depth of microscope developments and the many contributions by Shinya Inoué. For a more complete treatment, we refer to Inoué's insightful articles and books, some of which tell his own story about microscopes, living cells, and dynamic molecules (Inoué 2008; Inoué 2012).

The Light Microscope as Primary Observation Tool for Cytology Leading Up to Cowdry's Book

The term *cell* was first coined by Robert Hooke, who used it in his book *Micrographia* of 1665, describing observations of thin slices of cork that he prepared and examined under the microscope. He described the pores in a thin cork slice as cells that had walls, "much like a Honey-comb, but that the pores of it were not regular; yet it was not unlike a Honey-comb in these particulars" (fig. 12.1; Hooke 1665). It is interesting to note that the *cyto* in cytology derives from the Greek word *κύτος* (*kytos*), "a hollow," and relates to the first description of cells by Robert Hooke as hollow or empty pores separated by walls (for the history of the cell concept, see Harris 1999).

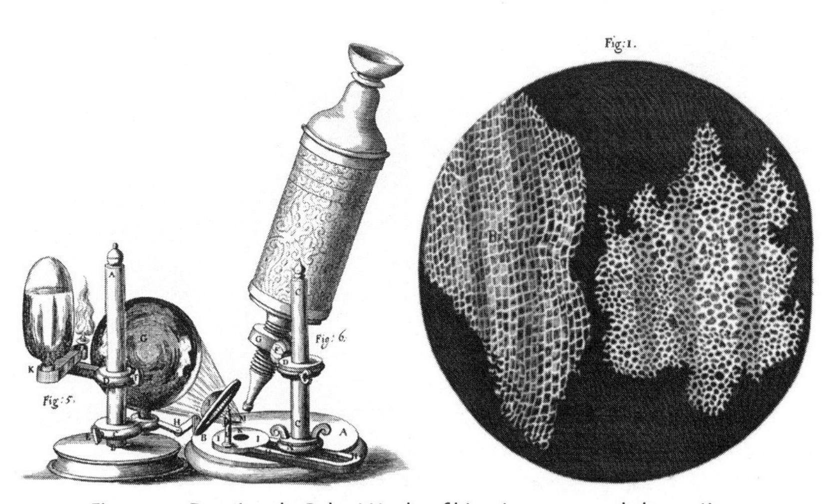

Figure 12.1. Drawings by Robert Hooke of his microscope and observations of cells or pores in thin slices of cork published in *Micrographia* in 1665. Hooke 1665 Schemes 1 and 11; from the Gutenberg Project.

Well into the 1800s, building good microscopes was a craft rather than a science, and performance of the instruments varied widely. In the second half of the nineteenth century, the instrument makers Carl Zeiss and Ernst Leitz, the physicist Ernst Abbe, and the glass chemist Otto Schott developed the first practical and theoretical means to build reliable microscopes that performed near the limit of resolution. The standards that they developed are still valid today (Abbe 1873).

What was lacking, however, was contrast in images of living cells, which were mostly transparent. Cells were full of organelles and structures that were nearly invisible yet tantalizing to those biologists who were eager to explore life based on the laws of physics and chemistry. Edmund Beecher Wilson writes in the introduction to *General Cytology* in 1924,

> It may seem strange that the subject should so long have been dominated by morphological studies, especially on fixed and stained cells, when we recall those illuminating researches on living cells by Dujardin, Max Schultze, DeBary, Kühne, and other pioneers, which led to a general recognition of protoplasm as the physical basis of life. The explanation lies in part in the failure of the earlier microscopes to make visible in living cells an organization in any degree adequate to explain

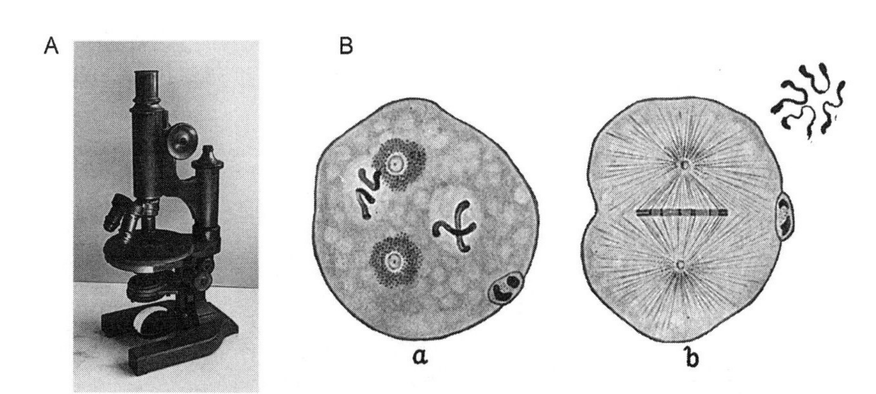

Figure 12.2. Light microscope and observational drawings from the period before 1924. *A*. Photograph of a Leitz microscope in the MBL's collection. (Photo by the author, with support from Louie Kerr and William Haskins.) *B*. Fertilized *Ascaris megalocephala* egg fixed in alcohol and picric acid. (*a*) Germ nuclei approaching between the attraction spheres of the first cleavage spindle, each containing two chromosomes. (*b*) The first cleavage spindle fully formed; it contains four chromosomes, which are shown in a polar view of the same spindle in the small figure to the right above. After Boveri 1888; reproduced from Lillie 1924, figure 10, with permission of the University of Chicago Press.

> the vital activities, while the fixation of cells by certain coagulating agents, such as dilute acetic, osmic, or chromic acids, often makes visible a definite and complex structure brought still more clearly into view by the use of certain dyestuffs, such as carmine or hematoxylin. (Cowdry 1924, 4)

Wilson seems to acknowledge here that observations of the dynamic protoplasm in living cells were instrumental in the initial recognition that life has a physical basis. And yet, the poor rendering of the constituents of protoplasm in the light microscopes made fixing and staining samples an indispensable part of cytology at the time. Their careful application and the recognition of the many sources of errors that these methods can introduce have nevertheless made possible many of cytology's most fundamental discoveries (fig. 12.2).

The Impact of the Electron Microscope

Developments in physics during the first half of the twentieth century, including the concept of the particle-wave duality of light and even material particles, made it possible to envision a different type of microscope, one that uses electrons as radiation. An electron beam, even after moderate acceleration by an electric potential of 1V, corresponds to radiation of wavelength 1.23 nm, which is about five hundred times smaller than the wavelength of visible light. Hence, diffraction of electron beams probe much smaller distances in a sample compared to the diffraction of visible light, and images formed by diffracting electrons instead of light can potentially have much higher resolution. Intrigued by this potential, Ernst Ruska pursued fundamental work in electron optics and designed the first electron microscope in 1932 (Knoll and Ruska 1932; Ruska 1980; Ruska 1993).

Initially, electron microscopes were used in material science, since samples had to be prepared as very thin sheets and kept in a vacuum for imaging. In the 1950s, George Palade and Keith Porter developed specialized fixation, embedding, sectioning, and staining methods for biological samples, leading to sudden advances in observation of biological tissues and cell fine structure in never-before-seen detail (fig. 12.3*B*; Palade 1952; Porter and Blum 1953; Porter and Kallman 1953; Palade and Porter 1954).

The improved resolution and specific contrast in an electron microscope were, however, gained by stopping all life processes in a specimen, removing or permeabilizing the cell membrane, adding fixative and staining agents, embedding the resultant structures in a solid matrix, then

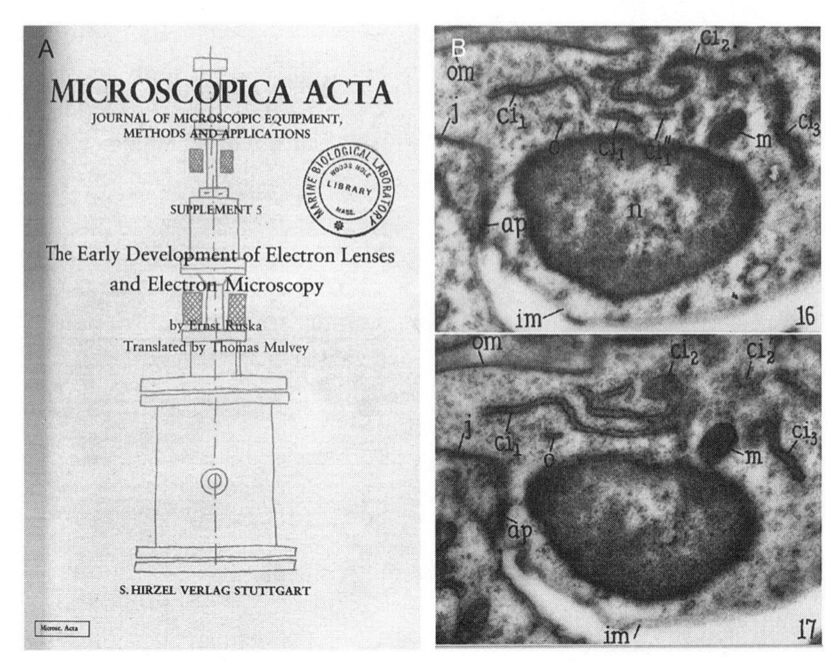
A

MICROSCOPICA ACTA

JOURNAL OF MICROSCOPIC EQUIPMENT, METHODS AND APPLICATIONS

SUPPLEMENT 5

The Early Development of Electron Lenses and Electron Microscopy

by Ernst Ruska

Translated by Thomas Mulvey

S. HIRZEL VERLAG STUTTGART

Figure 12.3. Electron microscope. *A*, cover page of comprehensive article by Ernst Ruska on the early history of the electron microscope. The background shows a sketch of the first electron microscope drawn by Ruska in 1931. *B*, electron micrographs by George Palade and Keith Porter representing two serial sections through an endothelial cell. The nucleus, sectioned close to its surface, appears at *n* and a mitochondrial profile at *m*. The cell membrane faces the lumen at *im* and the pericapillary spaces at *om*. The profiles marked *o*, ci_1, ci_2, ci_3 are taken to represent the endoplasmic reticulum. *A*, from Ruska 1980. *B*, from Palade and Porter 1954, plate 61.

sectioning the block and revealing its molecular architecture in an often destructive imaging process. Because of the many opportunities to introduce artifacts due to the extensive manipulation of the sample, it is mandatory to use several, complementary preparative methods, compare their results, and extrapolate them to the living state. After all this, you still have only a snapshot, though in great detail, of structures whose dynamics are a critical part of their function.

Phase and Polarized Light Microscopy Techniques for Imaging Living Cells

To this day, light microscopy remains the primary tool to study living cells, tissues, and whole organisms that function in their physiological environments. However, until the early 1950s, most of the molecular architecture of living cells and tissues remained hidden because of a lack of contrast, while stains and labels compromised cellular health and function and could not be relied upon on living material.

Technical developments that started in the 1930s, though, introduced optical "tricks" that enhanced the weak contrast and detectability of cellular architecture and allowed the observation of healthy living cells and their architectural dynamics in the light microscope. In the 1930s, Ernst Zernike discovered the principles of phase-contrast imaging and applied it to microscopy (Zernike 1942). Commercial instruments became available after World War II and invigorated the observation of living cells, which tend to display little contrast in standard, wide-field microscopes (fig. 12.4).

In the early 1950s, Francis H. Smith (Smith 1952) invented the differential interference contrast (DIC) microscope, with a similar aim of creating contrast in transparent specimens. A DIC microscope is a beam-shearing interferometer in which the two interfering beams are sheared by only a small amount, generally by less than the diameter of the Airy disk of the imaging optics. The technique produces a shadow-cast image that displays the difference between the optical path-length of the two beams (fig. 12.5*B*).

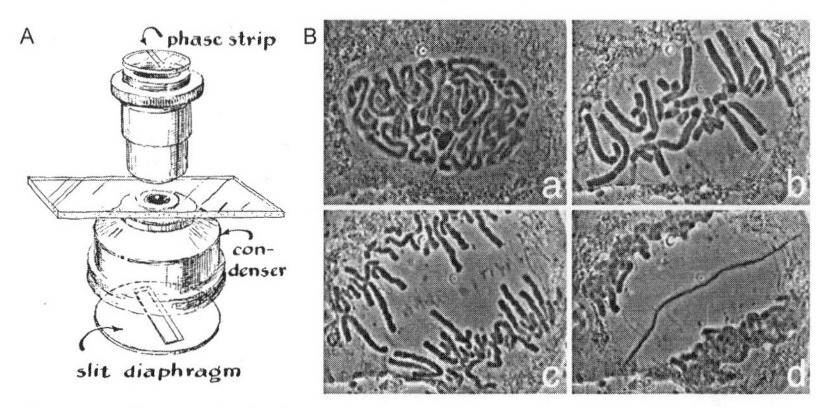

Figure 12.4. Phase contrast microscope: *A*, arrangement of condenser and objective with slit diaphragm in the front focal plane of the condenser and the phase strip in the back focal plane of the objective. In modern microscopes, the slit and phase mask are circular instead of linear. *B*, mitosis and cell plate formation in a flattened endosperm cell of the African blood lily, *Haemanthus katherinae*, observed in phase contrast with cine-micrography. *A*., from Zernike 1958, figure K-7. *B*, from Bajer and Molebajer 1956, figure 8; available as a video from the Cell Image Library, online at http://www.cellimagelibrary.org/images/11952. With permission of Springer.

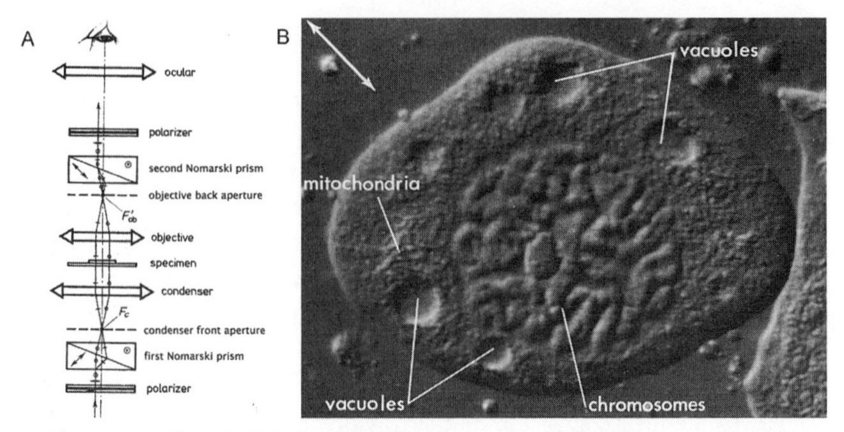

Figure 12.5. Differential interference microscope: *A*, schematic of DIC microscope with Nomarski prisms. In contrast to Wollaston prisms in the original design by F. H. Smith, Nomarski prisms are designed to be placed outside the objective and condenser lens. *B*, endosperm cell of *Haemanthus katherinae*, viewed with Zeiss/Nomarski DIC equipment. The arrow indicates the optical shear direction. *A*, schematic after Pluta 1989, figure 7.14. With permission from Elsevier. *B*, from Allen, David, and Nomarski 1969, figure 11.

Smith placed a Wollaston prism at the front focal plane of the condenser and a second one in the back focal plane of the objective lens. In 1952, Georges Nomarski proposed an improved design based on a special type of Wollaston prism, called the Nomarski prism, which simultaneously introduced spatial displacement and angular deviation of two orthogonally polarized beams (Nomarski 1955). The Nomarski prism can be placed outside the objective lens (fig. 12.5*A*) and is therefore compatible with high-resolution and highly corrected modern microscope optics.

Another way of enhancing contrast of native biological structures relies on the polarization of light. Polarization, like the phase of a light wave, cannot be discerned by the human eye. Therefore, to recognize changes in the polarization of light that has passed through biological material, additional components are added to the microscope optical path to make these subtle changes visible. In addition to polarizers that modulate the intensity of light according to its polarization, a polarized light microscope also employs a compensator, which is used to compare and thereby measure the optical anisotropy of the sample with the known birefringence of the compensator.

Polarized light microscopy was first used more than one hundred and fifty years ago to study plant and animal tissues (see Valentin 1861, which also includes the early history of polarized light microscopy). The term *striated muscle*, for example, was coined from observations with the polarizing microscope, which revealed birefringent (anisotropic) A-bands and isotropic I-bands in muscle tissue. The I-band is not strictly isotropic, but less birefringent than the A-band, which is now known as the domain where actin and myosin filaments overlap in the sarcomere of striated muscle. Already in 1875, Engelmann reported that in striated muscle, the birefringent A-band was the origin of contractility (Engelmann 1875).

In the 1920s, Wilhelm J. Schmidt observed animal cells and tissues in the polarizing microscope and published his findings in two celebrated monographs (as discussed by Daniel Liu, chapter 10 in this volume; Schmidt 1924; Schmidt 1937). One included the first micrographs of the mitotic spindle in polarized light, recorded in actively dividing cells (Schmidt 1937). At first, Schmidt reported the spindle birefringence as originating from a combination of spindle fibers and chromosomes; two years later he corrected himself in a more comprehensive study, and assigned the spindle birefringence to the fibers alone (Schmidt 1939). Schmidt recognized that the positive spindle birefringence was compatible with dense arrays of parallel filaments spanning the distance between the spindle poles and the chromosomes. The parallel filaments are responsible for the birefringence of the spindle and make it appear with high contrast in a polarizing microscope, even without any stains or labels (fig. 12.6).

Inspired by the findings of W. J. Schmidt and prompted by his mentor Katsuma Dan, Shinya Inoué built his own polarizing microscope and went on to successfully reproduce Schmidt's observations. During World War II, Inoué studied biology in Japan and had the great fortune to be taught by Katsuma Dan, an exceptional scientist and Japanese citizen who had traveled the world, had lived and studied in the United States, and was married to Jean Clark Dan, an American and a scientist in her own right. The Dans were particularly interested in cell division and knew how to light the passion for this subject in a bright and curious student. After a first failed attempt to observe the spindle birefringence in fertilized sea urchin eggs in a room darkened by air raid shutters during the war, Shinya Inoué did not let up, but following the end of the war built his own polarizing microscope at the Misaki lab near Tokyo, calling it Shinya-Scope Nr. 1, which clearly revealed the birefringent spindles in fertilized sand dollar eggs (Inoué 2012).

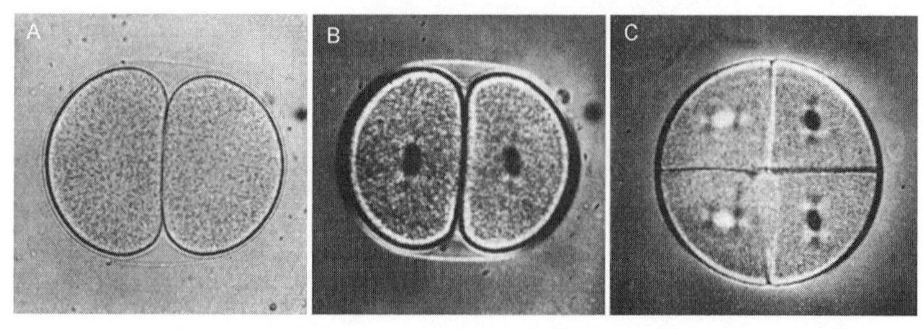

Figure 12.6. Microscope observations of a living and dividing green sea urchin egg, as reported by W. J. Schmidt in 1939. *A*, two-cell state in regular bright field in which the mitotic spindle remains nearly invisible. *B*, same as *A*, viewed between crossed polarizers and a thin mica plate that enhances the contrast of the mitotic spindle birefringence. *C*, four-cell state with four spindles in two groups that exhibit reverse contrast due to their change in spindle orientation with respect to the compensator axes; egg diameter ~150 μm. From Schmidt 1939, figures 8 and 9. With permission of Springer.

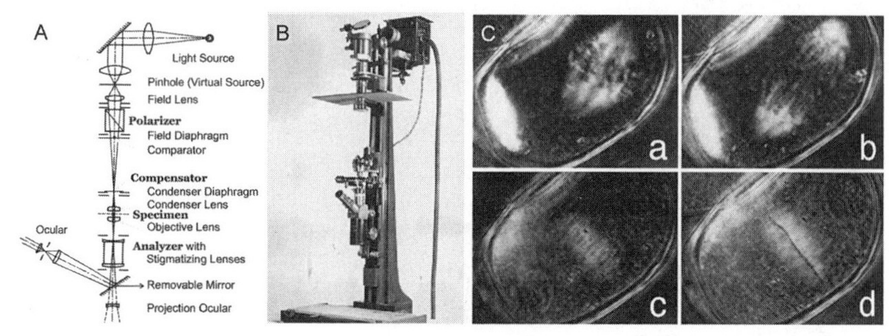

Figure 12.7. Polarized light microscope. *A*, schematic. *B*, photograph of Shinya-Scope Nr. 2. This might well have been the first inverted microscope used in biology. *C*. Mitosis and cell plate formation in centrifuged pollen mother cell of the Easter lily, *Lilium longiflorum*, observed with polarization microscopy and recorded on 16 mm celluloid film in 1952. The film was transferred to video (available at the Cell Image Library, http://www.cellimagelibrary.org/images/11957), and four still frames were selected from the video. *A* and *B* are from Inoué 1951, figures 1 and 3.

In 1948, Shinya Inoué arrived at Princeton University, New Jersey, for graduate studies in biology, supported by a scholarship arranged by the Dans. A year later, Inoué spent his first summer at the Marine Biological Laboratory in Woods Hole, starting a long tradition of summer research at MBL, which quickly became Inoué's scientific home. At Princeton, Inoué built the second version of the Shinya-Scope (fig. 12.7*A*), which he used to record the spindle dynamics during cell division in cinematographic time lapse, clearly demonstrating that the mitotic/meiotic spindle in living cells consists of many birefringent fibers and their component fibrils (later identified as microtubules; see fig. 12.7) (Inoué 1953; Inoué and Oldenbourg 1998). When comparing the observations of dividing plant cells in phase-contrast (fig. 12.4*B*) versus polarized light (fig. 12.7*C*), it is striking to note the complementary information gained from both techniques: phase contrast reveals the differences in optical density, highlighting chromosomes and the phragmoplast, a molecular scaffold assembling in plant cells during late cytokinesis and separating the two daughter cells, while polarized light observations reveal the birefringence of the fibrous arrays that surround these structures and guide the chromosome separation and assembly of the phragmoplast.

In subsequent studies, Inoué and collaborators established the reversible assembly and disassembly of spindle fibers and fibrils in living cells exposed to antimitotic drugs (Inoué 1952), in cold temperatures (Inoué 1964), under hydrostatic pressure (Salmon 1975), and when regular water was exchanged with heavy water (Inoué and Sato 1967). In each of these studies, polarized light microscopy was the key in observing the submicroscopic fine structure of the spindle. Its dynamic architecture could be followed in real time, at high temporal and spatial resolution, with no staining or labeling required.

After further refinements in the microscope optics, Inoué and Sato revealed the packing of DNA in living sperm heads by interpreting their polarized light observations in terms of the structure of DNA (Inoué and Sato 1966). Extensive modeling was required to link the molecular structure of the DNA polymer, whose optical anisotropy is dominated by the carbon rings in the stacked base pairs, to the recorded birefringence patterns in the sperm heads (fig. 12.8). This feat was enabled in part by the invention of rectified optics for the polarizing microscope by Inoué and Hyde (Inoué and Hyde 1957). Rectified objective and condenser optics enabled the simultaneous realization of high resolution and high sensitivity in polarized light microscopy.

In the early 1980s, Shinya Inoué and (independently) Robert D. Allen introduced a further important advance for light microscopy by adding a high-quality video camera to the microscope. The initial intention was simply to display the microscope field of view to a larger group of viewers who attended either of the two MBL microscopy courses, one organized by Allen, and the other by Inoué. The video equipment turned out to have unexpected advantages that resulted in the video camera becoming a

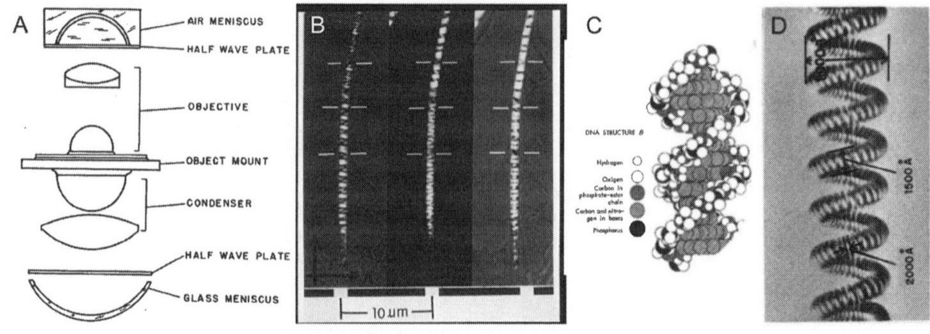

Figure 12.8. Rectified polarized light microscopy of cave cricket sperm head. *A*, Inoué and Hyde introduced polarization rectifiers consisting of an appropriately shaped glass-air interface and a half wave plate to counteract the polarization aberrations that occur on lens surfaces of high NA condenser and objective lenses. *B*, cave cricket sperm head observed in polarizing microscope equipped with rectified optics. *C*, B-DNA according to the Watson-Crick model. *D*. Preliminary model of a coiled coil arrangement of DNA strands in the sperm head proposed by Inoué and Sato, based on the birefringence patterns and the molecular DNA model by Watson and Crick. *A*, from Inoué and Hyde 1957, text-figure 2. *B*, *C*, *D*, from Inoué and Sato 1966, figures 7, 26, and 29.

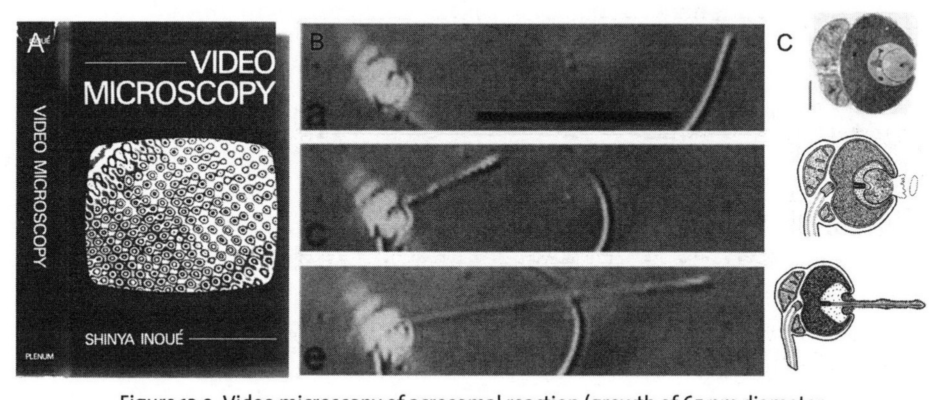

Figure 12.9. Video microscopy of acrosomal reaction (growth of 65 nm diameter acrosomal process by actin polymerization) in Thyone sperm. *A*, book cover of the first comprehensive treatise on video microscopy, merging the fields of microscopy, electronic imaging, and digital image processing and analysis. *B*, three stages, 1.5 seconds apart, in the elongation of the acrosomal reaction; scale bar 20 μm. (Video available from the Cell Image Library, http://www.cellimagelibrary.org /images/11973.) *C*. Electron micrograph (*top*) and schematics of structural changes in Thyone sperm undergoing the acrosomal reaction. *A*, from Inoué 1986. With permission of Springer. *B*, Tilney and Inoué 1982; *C*. From Inoué and Tilney 1982, figures 1 and 6.

permanent fixture on a research-grade microscope. For example, by matching the camera's resolution to the optical resolution of the microscope and taking advantage of electronic controls in the video circuitry, Inoué (Inoué 1981) and Allen (Allen et al. 1981) were able to greatly improve the visibility of the finest details near the resolution limit of light microscopy (fig. 12.9). The combination of video recording with digital image processing and analysis further enhanced the utility of the light microscope as a quantitative tool for studying living cells and tissues in their physiological environments. Quickly, the new technique was adopted in most every mode of light microscopy used in biology today (Inoué 1986).

Enhancing Polarized Light Microscopy Using Liquid Crystal Devices and Digital Image Processing

Electronic image capture and digital image processing brought new opportunities to exploit the quantitative nature of polarized light microscopy. For the first time, instantaneous measurements of intensities in every resolved image point became possible; images could be converted into digital format, stored, and immediately processed. In addition, liquid crystal devices suitable for imaging and instantly manipulating the polarization of light became available. Recognizing the ongoing revolution in light microscopy, I joined Shinya Inoué at MBL in 1989, and with Inoué's support developed the liquid crystal polarizing microscope, the LC-PolScope, which significantly advanced the utility of the polarized light microscope in biology and many other fields (Oldenbourg 1996).

The optical design of the LC-PolScope builds on the traditional polarizing microscope, introducing two essential modifications: the specimen is illuminated with nearly circularly polarized light, and the traditional compensator is replaced by a universal compensator built from liquid crystal devices (fig. 12.10*A*, plate 1). Image acquisition and processing algorithms are used to compute images that represent the retardance and slow axis orientation in each resolved image point (fig. 12.10*B*, 12.10*C*, plate 1; see Oldenbourg and Mei 1995; Shribak and Oldenbourg 2003). Slow axis refers to the polarization orientation of light that experiences the highest refractive index, hence the slowest speed, when passing through the specimen in a given direction.

In figure 12.10*A*, the retardance and slow axis information that was calculated based on the raw PolScope images was used to generate the false color image at the bottom right of panel *A*. Red represents the horizontal

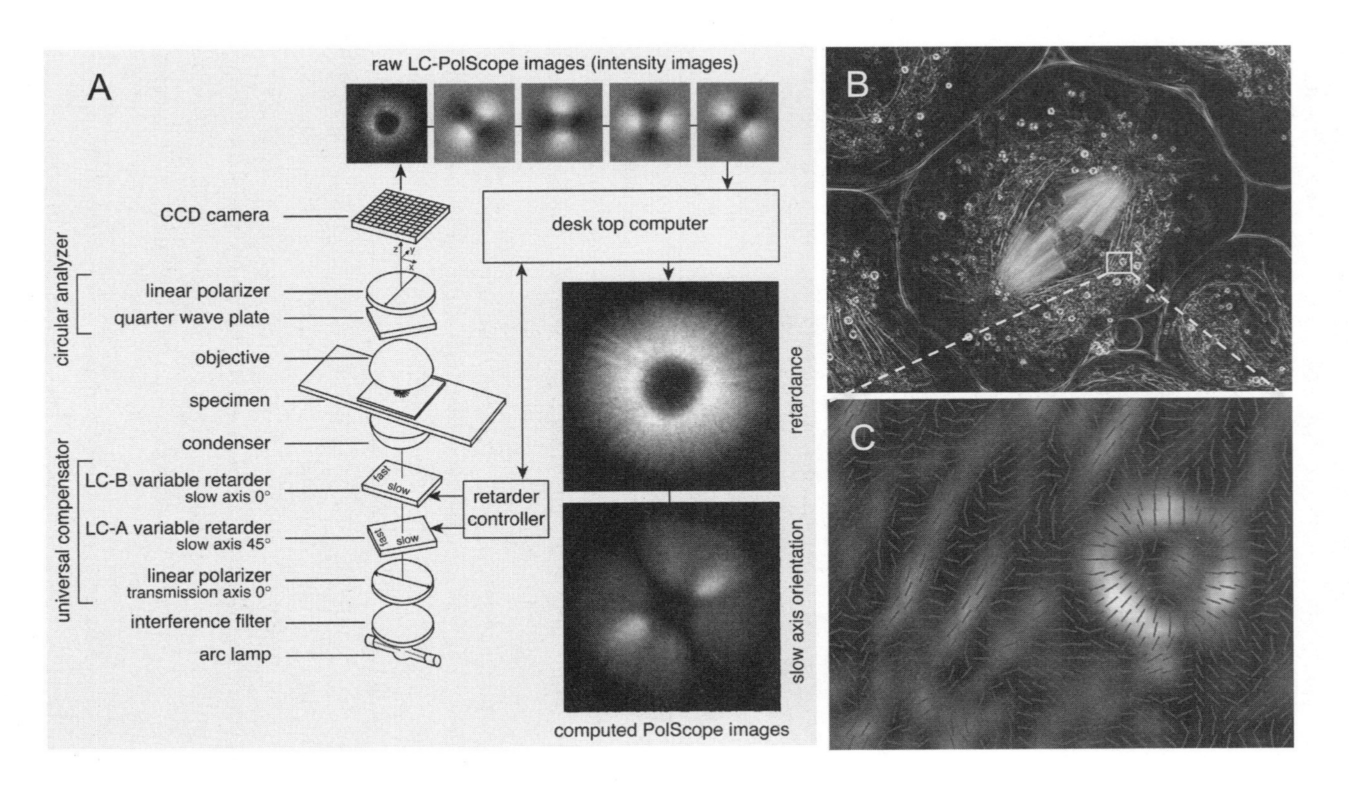
A
raw LC-PolScope images (intensity images)
CCD camera
circular analyzer
linear polarizer
quarter wave plate
objective
specimen
condenser
universal compensator
LC-B variable retarder
slow axis 0°
LC-A variable retarder
slow axis 45°
linear polarizer
transmission axis 0°
interference filter
arc lamp
fast
slow
fast
slow
desk top computer
retarder
controller
retardance
slow axis orientation
computed PolScope images
B
C

slow axis, light blue, the vertical, and so on. Michael Shribak has devised an ingenious way to create this combination image of retardance and slow axis orientation optically for direct viewing in the polychromatic polscope (Shribak 2015). In collaboration with Shinya Inoué, Shribak also invented the orientation-independent differential interference contrast (OI-DIC) microscope (Shribak and Inoué 2006) and combined it with the LC-PolScope (Shribak et al. 2008). The combined OI-DIC and LC-PolScope microscope produces two complementary images: the first displays the optical density, while the second displays the optical anisotropy, combining the effects seen in figures 12.4*B* and 12.7*C*.

Fluorescence Microscopy

The fluorescence microscope has become the most widely used imaging tool among cell biologists today. The popularity of the technique is due to several factors, including the development of fluorescent probes that are compatible with the living state to highlight specific cell structures and physiological conditions. Equally important were developments of highly efficient optical, electronic, and digital methods for acquiring, processing, and analyzing the weak fluorescent signals.

Figure 12.10. LC-PolScope instrument. *A*, schematic of the optical design (*left*) that builds on the traditional polarized light microscope with the conventional compensator replaced by two liquid crystal variable retarders, LC-A and LC-B. The combination of microscope optics, electro-optic components, electronic imaging, and digital image processing is used to generate computed PolScope images of an aster consisting of microtubule fibers radiating from a centrosome. *B*, LC-PolScope image of a live primary spermatocyte from the crane fly, *Nephrotoma suturalis*. Brightness is directly proportional to the retardance of the specimen, independent of the orientation of the birefringent structure. The flattened cell reveals with great clarity the birefringence of spindle microtubules extending from the chromosomes to the spindle poles. The birefringence of other cell organelles, such as elongated mitochondria, which surround the spindle like a mantle, and small spherical lipid droplets, are also evident against the dark background of the cytoplasm. The pole-to-pole distance of the spindle is approx. 25 μm. *C*, magnified cytoplasmic region of *B* with red lines indicating the slow axis direction for each pixel. The lipid droplet to the right has a highly birefringent shell with slow axes perpendicular to the droplet's surface, while mitochondria have slow axes that are parallel to the mitochondria's long axes. Figure adapted from Mehta, Shribak, and Oldenbourg 2013.

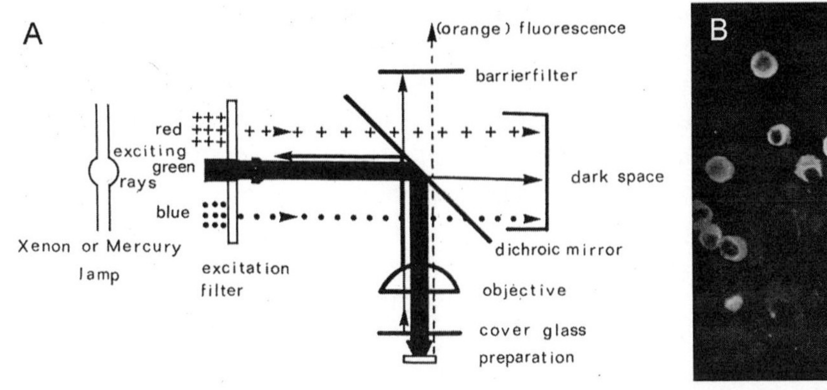

Figure 12.11. *A*, fluorescence microscope with the epi-illuminator developed by Johan Ploem (Ploem 1967). The development of a dichroic mirror (45°) that reflects short-wavelength light and transmits longer-wavelength light enabled the merging of the illumination with the imaging path before the objective lens. Hence, the objective is used for both, focusing the excitation light into the specimen and collecting the emitted fluorescence for projecting it as an image onto the eyepiece or camera. Additional excitation and emission/barrier filters improved the contrast of the weak fluorescence against the intense excitation light. *B*, bone marrow cells stained with anti-IgG fluorescein isothiocyanate conjugate (FITC), and an anti-kappa tetramethylrhodamine isothiocyanate conjugate (TRITC). Excitation with blue incident light resulted in green fluorescence of cells containing mainly FITC, orange fluorescence of cells containing mainly TRITC, and yellow and yellow-green fluorescence of cells containing both FITC and TRITC. From Ploem 1967, figures 3 and 6a.

In the 1960s, an important step forward in separating the weak fluorescence from the strong excitation light was taken by Johan Ploem when he proposed an optical arrangement of filters and a dichroic mirror, which allowed the merging of the excitation and imaging path (fig. 12.11, plate 2; Ploem 1967).

To this day, the compact assembly of filters and a dichroic mirror into a "fluorescence filter cube," as proposed by Ploem and first implemented by the Leitz/Leica company, is the distinctive feature of nearly every fluorescence microscope. Several features of the cube, combined with epi-illumination, allowed for the efficient rejection of the bright excitation light from the weak fluorescence, which typically is five orders of magnitude weaker than the excitation light. This advance in instrumentation paved the way for the many biochemical developments of fluorescent probes and labeling technologies that are chemically specific to cellular structures and are compatible with the living state, even to the point that the cells themselves express the fluorescent molecules, such as fluorescent proteins.

The discovery of fluorescent proteins and the development of methods to genetically encode their expression in most every cell type and to co-express and link them with other proteins of interest were major breakthroughs that expanded the use of fluorescence microscopy for live cell imaging (Chalfie 2009; Shimomura 2009; Tsien 2009). It also attracted the interest of Shinya Inoué, who discovered, together with his MBL colleague Osamu Shimomura, that crystals of the green fluorescent protein (GFP) emit highly polarized fluorescence (fig. 12.12*A*, plate 3; Inoué et al. 2002), suggesting that individual GFP molecules do the same. Inspired by this discovery, Vrabioiu and Mitchison devised an orientationally constrained

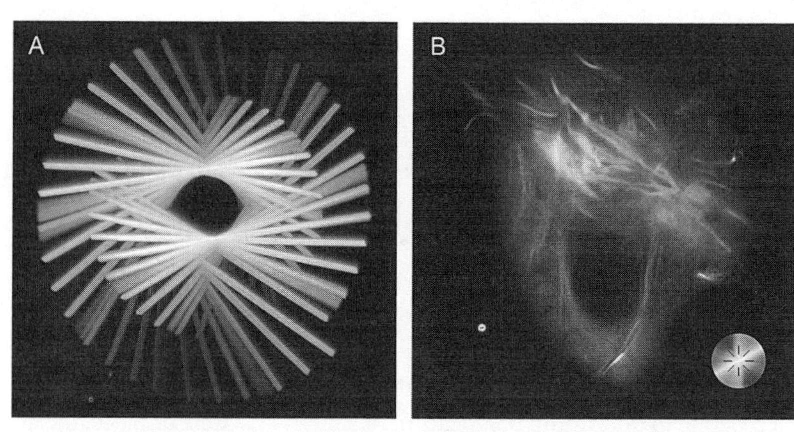

Figure 12.12. *A*, fluorescence of a GFP crystal illuminated with 460 nm wavelength plane polarized light and observed through 527/15 nm band-pass filter. The panel is a collage, superimposing images of the same crystal rotated in 10° steps. During the specimen rotation, the transmission axis of the polarizer in the illumination path remained horizontally oriented. No analyzer was present. The crystal fluorescence is brightest when the crystal axis is oriented horizontally, while it is dimmest when the axis is oriented vertically. Image width is 100 μm. *B*, mammalian epithelial cell (MDCK) expressing a septin-GFP construct, which incorporates into septin fibers. The polarized fluorescence of the constrained GFP dipoles in septin fibers is presented in color, which encodes the orientation of the polarization axis according to the color wheel inset on the bottom right. The white color of fluorescence from the cytosol reveals the lack of common alignment of septin-GFP molecules suspended in the cytosol. The image was recorded with the Fluorescence LC-PolScope. Image width is 50 μm. *A*, image by Shinya Inoué, see Inoué et al. 2002 for a description of their work. *B*, from DeMay et al. 2011a, figure 5.

GFP-septin construct and used its polarized fluorescence to determine septin filament organization and dynamics in living yeast (Vrabioiu and Mitchison 2006; Vrabioiu and Mitchison 2007).

Motivated by these breakthroughs, Shinya Inoué's collaborators at the MBL developed the liquid crystal–based fluorescence polarizing microscope, Fluorescence LC-PolScope, that sequentially excites fluorophores with linearly polarized light of four orientations: 0°, 45°, 90° and 135° (DeMay et al. 2011b; McQuilken et al. 2015). The use of four linear polarizations, rotated in 45° steps, enables the measurement of polarized fluorescence that is oriented at any angle in the image plane, unlike the approach that uses only two orientations that are parallel or perpendicular to a previously identified symmetry axis, as described by Axelrod (Axelrod 1989). The Fluorescence LC-PolScope was used to analyze the 3D architectural rearrangement of constrained, GFP-tagged septin assemblies during cytokinesis of budding yeast, the filamentous fungus *Ashbya*, and mammalian epithelial cells (fig. 12.12*B*, plate 3; DeMay et al. 2011a). With the development of polarized fluorescence microscopy, the seemingly disparate goals of structural specificity by analyzing the polarization of light and molecular specificity by employing chemical markers have been combined in a single imaging tool.

Fluorescence microscopy is the basis of many new developments in light microscopy that now have breached the resolution barrier, formerly thought insurmountable, and achieved so-called super-resolution. With the ability to turn fluorescent emitters inside cells on and off at will (or stochastically), it has become possible to exploit the ability of the light microscope to determine the location of a single particle or molecule at a much higher precision than it is possible to tell two of them apart. While the image of a submicroscopic particle, broadened by diffraction, is two hundred or more nanometers wide, depending on the microscope optics used, its exact location can be determined within a few nanometers, with the precision only depending on the number of collected photons emitted by the particle. Several techniques identified by specific acronyms, such as PALM (photo-activated localization microscopy), STORM (stochastic optical reconstruction microscopy), and STED (stimulated emission depletion), are making use of this fundamental property and create images of fluorescently labeled structures at resolutions that are near those achieved by electron microscopy. Alas, the images represent the labels and not the native structures themselves and more often than not, collecting the images is not compatible with the living state.

Conclusion

Over the years, the pace of change in light microscopy has quickened through an ever-closer interaction between the engineering fields of optics, mechanics, electronics, and digital processing and analysis, as well as between physics and chemistry, all in the service of the biological sciences and their quest to understand life and its smallest unit, the cell. Cell biology was transformed by new approaches to light microscopy, partly developed under the mantra, "Demand healthy living cells, the technology will follow"; and by reciprocity: biological questions drive technical developments, and technical developments drive new biological questions.

Shinya Inoué looms large among his colleagues who transformed light microscopy from an observational tool, as it was used in the time before Cowdry, to a precise tool that can be used to measure and reveal the molecular interactions and architectural dynamics that create the exacting and conserved framework for living matter and life itself.

Acknowledgment

I am immensely grateful to Shinya Inoué, who was a demanding and exacting, yet patient and always generous mentor, who taught by example, combining a passion both for creating tools and applying them to life's persistent question: What is life? I thank my MBL colleagues Tomomi Tani and Michael Shribak for their contributions to an article we wrote together on the legacy of Shinya Inoué, to appear in the *Biological Bulletin*, that set the stage for my writing this book contribution. I am grateful to Karl Matlin and Jane Maienschein for their patience and guidance in the preparation of this manuscript, which was supported by a grant from the National Institutes of Health (GM114274).

References

Abbe, E. 1873. "Beiträge zur Theorie des Mikroskops und der mikroskopischen Wahrnehmung." *M. Schultz's Archiv für mikroskopische Anatomie* 9: 413–68.

Allen, R. D., G. B. David. and G. Nomarski. 1969. "The Zeiss-Nomarski Differential Interference Equipment for Transmitted-Light Microscopy." *Zeitschrift für wissenschaftliche Mikroskopie* 69 (4): 193–221.

———, J. L. Travis, N. S. Allen, and H. Yilmaz. 1981. "Video-Enhanced Contrast Polarization (AVEC-POL) Microscopy: A New Method Applied to the Detection of Birefringence in the Motile Reticulopodial Network of *Allogromia Laticollaris*." *Cell Motility* 1 (3): 275–89.

Axelrod, D. 1989. "Fluorescence Polarization Microscopy." Methods in Cell Biology, 30: 333–52.

Bajer, A., and J. Molebajer. 1956. "Cine-Micrographic Studies on Mitosis in Endosperm. 2: Chromosome, Cytoplasmic and Brownian Movements." *Chromosoma* 7 (6–7): 558–607.

Boveri, T. 1888. "Zellen-Studien II." *Jenaische Zeitschrift für Naturwissenschaft* 22: 683–882.

Chalfie, M. 2009. "GFP: Lighting Up Life." In *The Nobel Prizes, 2008*, edited by K. Grandin. Stockholm: Nobel Foundation.

Cowdry, Edmund V. 1924. *General Cytology: A Textbook of Cellular Structure and Function for Students of Biology and Medicine*. Chicago: University of Chicago Press.

DeMay, B. S., X. Bai, L. Howard, P. Occhipinti, R. A. Meseroll, E. T. Spiliotis, R. Oldenbourg, and A. S. Gladfelter. 2011a. "Septin Filaments Exhibit a Dynamic, Paired Organization That Is Conserved from Yeast to Mammals." *Journal of Cell Biology* 193 (6): 1065–81.

———, N. Noda, A. S. Gladfelter, and R. Oldenbourg. 2011b. "Rapid and Quantitative Imaging of Excitation Polarized Fluorescence Reveals Ordered Septin Dynamics in Live Yeast." *Biophysical Journal* 101(4): 985–94.

Engelmann, T. W. 1875. "Contractilität und Doppelbrechung." *Pflüger's Archiv für die gesamte Physiologie des Menschen und der Tiere* 11 (1): 432–64.

Harris, H. 1999. *The Birth of the Cell*, New Haven, CT: Yale University Press.

Hooke, R. 1665. *Micrographia: Some Physiological Descriptions of Minute Bodies Made by Magnifying Glasses with Observations and Inquiries Thereupon*, London: Royal Society.

Inoué, S. 1951. "Studies of the Structure of the Mitotic Spindle in Living Cells with an Improved Polarization Microscope." PhD thesis, Princeton NJ: Princeton University.

———. 1952. "The Effect of Colchicine on the Microscopic and Submicroscopic Structure of the Mitotic Spindle." *Experimental Cell Research*, supp. 2:305–18.

———. 1953. "Polarization Optical Studies of the Mitotic Spindle. I. The Demonstration of Spindle Fibers in Living Cells." *Chromosoma* 5:487–500.

———. 1964. "Organization and Function of the Mitotic Spindle." In *Primitive Motile Systems in Cell Biology*, edited by R. H. Allen and N. Kamiya, 549–98. New York: Academic Press.

———. 1981. "Video Image Processing Greatly Enhances Contrast, Quality, and Speed in Polarization-Based Microscopy." *Journal of Cell Biology* 89 (2): 346–56.

———. 1986. *Video Microscopy*. New York: Plenum Press.

———. 2008. *Collected Works of Shinya Inoué: Microscopes, Living Cells, and Dynamic Molecules*. Singapore: World Scientific.

———. 2012. *Through Yet Another Eye: Personal Recollections*. Falmouth MA: Self-published.

———, and W. L. Hyde. 1957. "Studies on Depolarization of Light at Microscope Lens Surfaces. II. The Simultaneous Realization of High Resolution and High Sensitivity with the Polarizing Microscope." *Journal of Biophysical and Biochemical Cytology* 3 (6): 831–38.

———, and R. Oldenbourg. 1998. "Microtubule Dynamics in Mitotic Spindle Displayed By Polarized Light Microscopy." *Molecular Biology of the Cell* 9 (7): 1603–7.

———, and H. Sato. 1966. "Deoxyribonucleic Acid Arrangement in Living Sperm." In *Molecular Architecture in Cell Physiology*, edited by T. Hayashi and A. G. Szent-Gyorgyi, 209–48. Englewood Cliffs, NJ: Prentice Hall.

———, and H. Sato, H. 1967. "Cell Motility By Labile Association of Molecules: The Nature of Mitotic Spindle Fibers and Their Role in Chromosome Movement." *Journal of General Physiology*, supp. 50 (6): 259–92.

———, O. Shimomura, M. Goda, M. Shribak, and P. T. Tran. 2002. "Fluorescence Polarization of Green Fluorescence Protein." *Proceedings of the National Academy of Sciences* 99 (7): 4272–77.

———, and L. G. Tilney. 1982. "Acrosomal Reaction of Thyone Sperm. I. Changes in the Sperm Head Visualized By High Resolution Video Microscopy." *Journal of Cell Biology* 93 (3): 812–19.

Knoll, M., and E. Ruska. 1932. "Beitrag zur geometrischen Elekronenoptik, I." *Annalen der Physik* 404 (5): 607–40.
Lillie, F. R. 1924. "Fertilization." In Cowdry 1924, 668–793.
McQuilken, M., S. B. Mehta, A. Verma, G. Harris, R. Oldenbourg, and A. S. Gladfelter. 2015. "Polarized Fluorescence Microscopy to Study Cytoskeleton Assembly and Organization in Live Cells." *Current Protocols in Cell Biology* 67: 4.29.1–4.29.13.
Mehta, S. B., M. Shribak, and R. Oldenbourg. 2013. "Polarized Light Imaging of Birefringence and Diattenuation at High Resolution and High Sensitivity." *Journal of Optics* 15: 094007–094020.
Nomarski, G. 1955. "Nouveau dispositif pour l'observation en contraste de phase differentiel." *Journal de Physique et le Radium*, supp. 16: S88.
Oldenbourg, R. 1996. "A New View on Polarization Microscopy." *Nature* 381 (6585): 811–12.
———, and G. Mei. 1995. "New Polarized Light Microscope with Precision Universal Compensator." *Journal of Microscopy* 180 (2): 140–47.
Palade, G. E. 1952. "A Study of Fixation for Electron Microscopy." *Journal of Experimental Medicine* 95 (3): 285–98.
———, and Porter, K. R. 1954. "Studies on the Endoplasmic Reticulum. I. Its Identification in Cells in Situ." *Journal of Experimental Medicine* 100 (6): 641–56.
Ploem, J. S. 1967. "The Use of a Vertical Illuminator with Interchangeable Dichroic Mirrors for Fluorescence Microscopy with Incidental Light." *Zeitschrift für wissenschaftliche Mikroskopie* 68 (3): 129–42.
Pluta, M. 1989. *Advanced Light Microscopy*. Vol. 2. *Specialized Methods*. Amsterdam: Elsevier Science.
Porter, K. R., and J. Blum. 1953. "A Study in Microtomy for Electron Microscopy." *Anatomical Record* 117 (4): 685–710.
———, and F. Kallman. 1953. "The Properties and Effects of Osmium Tetroxide as a Tissue Fixative, with Special Reference to Its Use for Electron Microscopy." *Experimental Cell Research* 4 (1): 127–41.
Ruska, E. 1980. "The Early Development of Electron Lenses and Electron Microscopy." *Microscopica Acta*, supp. 5: 1–140.
———. 1993. "The Development of the Electron Microscope and of Electron Microscopy." Nobel Media AB 2014. Online at http://www.nobelprize.org/nobel_prizes/physics/laureates/1986/ruska-lecture.html.
Salmon, E. D. 1975. "Spindle Microtubules: Thermodynamics of In Vivo Assembly and Role in Chromosome Movement." *Annals of the New York Academy of Sciences* 253:383–406.
Schmidt, W. J. 1924. *Die Bausteine des Tierkörpers in polarisiertem Lichte*, Bonn: Cohen.
———. 1937. *Die Doppelbrechung von Karyoplasma, Zytoplasma und Metaplasma*, Berlin: Bornträger.
———. 1939. "Doppelbrechung der Kernspindel und Zugfasertheorie der Chromosomenbewegung." *Zeitschrift für Zellforschung und mikroskopische Anatomie Abt. B. Chromosoma* 1 (1): 253–64.
Shimomura, O. 2009. "Discovery of Green Fluorescent Protein, GFP." Nobel Media AB 2014. Online at http://www.nobelprize.org/nobel_prizes/chemistry/laureates/2008/shimomura-lecture.html.
Shribak, M. 2015. "Polychromatic Polarization Microscope: Bringing Colors to a Colorless World." *Scientific Reports* 5:17340.

———, and S. Inoué. 2006. "Orientation-Independent Differential Interference Contrast Microscopy." *Applied Optics* 45 (3): 460–69.
———, J. LaFountain, D. Biggs, and S. Inoué. 2008. "Orientation-Independent Differential Interference Contrast Microscopy and Its Combination with an Orientation-Independent Polarization System." *Journal of Biomedical Optics* 13 (1): 014011.
———, and R. Oldenbourg. 2003. "Techniques for Fast and Sensitive Measurements of Two-Dimensional Birefringence Distributions." *Applied Optics* 42 (16): 3009–17.
Smith, F. H. 1952. "Interference Microscope." *US Patent, Number 2,601,175*. USA: Francis Hugh Smith.
Tilney, L. G., and S. Inoué. 1982. "Acrosomal Reaction of Thyone Sperm. II. The Kinetics and Possible Mechanism of Acrosomal Process Elongation." *Journal of Cell Biology* 93 (3): 820–7.
———, and S. Inoué. 1985. "Acrosomal Reaction of the Thyone Sperm. III. The Relationship between Actin Assembly and Water Influx During the Extension of the Acrosomal Process." *Journal of Cell Biology* 100 (4): 1273–83.
Tsien, R. Y. 2009. "Constructing and Exploiting the Fluorescent Protein Paintbox." Nobel Media AB 2014. Online at http://www.nobelprize.org/nobel_prizes/chemistry/laureates/2008/tsien-lecture.html.
Valentin, G. 1861. *Die Untersuchung der Pflanzen- und der Thiergewebe in polarisiertem Lichte*. Leipzig: Verlag von Wilhelm Engelmann.
Vrabioiu, A. M., and T. J. Mitchison. 2006. "Structural Insights into Yeast Septin Organization from Polarized Fluorescence Microscopy." *Nature* 443 (7110): 466–69.
———. 2007. "Symmetry of Septin Hourglass and Ring Structures." *Journal of Molecular Biology* 372 (1): 37–49.
Zernike, F. 1942. "Phase Contrast: A New Method for the Microscopic Observation of Transparent Objects, Part II." *Physica* 9:974–86.
———. 1958. "The Wave Theory of Microscopic Image formation." In *Concepts of Classical Optics*, edited by J. Strong, 525–36. San Francisco: W. H. Freeman.

CHAPTER 13

ENRICHING THE STRATEGIES FOR CREATING MECHANISTIC EXPLANATIONS IN BIOLOGY

William Bechtel

The approach of explaining phenomena by identifying and characterizing responsible mechanisms has a long history in biology (for historical reviews, see Coleman 1971; Allen 1979). After Theodor Schwann (1839) identified cells as the basic units in which metabolic processes such as fermentation occur, a host of researchers developed and deployed a very impressive variety of strategies for structurally decomposing cells and functionally characterizing what their components did. The discovery of chromosomes and the characterization of the operations of mitosis and meiosis were prominent nineteenth-century successes. Vitalist critics persisted throughout the century in arguing that phenomena such as fermentation could not be explained mechanistically, but Eduard Buchner's (1897) demonstration of fermentation in a cell-free extract inspired the quest for explanations of cell activities in terms of chemical reactions catalyzed by enzymes.

As reflected in Edmund Cowdry's (1924) *General Cytology*, mechanistically inspired biologists in the first decades of the twentieth century were developing new techniques for decomposing mechanisms into their parts (e.g., staining cell preparations to identify organelles) and operations (e.g., inhibiting enzymes to identify steps in reactions). As reflected in Cowdry's own chapter on the mitochondrion and the Golgi apparatus, a major aspiration was to localize different cellular activities in specific organelles. The development of new techniques continued in the decades after Cowdry's book. Cell fractionation and electron microscopy played central roles, facilitating what George Palade (1987) described as a bridge between morphology (providing increased detail about cell structure) and biochemistry (characterizing cell reactions). Within the newly constituted discipline of cell biology, researchers combined these techniques to offer new, mechanistic explanations of phenomena such as oxidative metabolism and protein synthesis (Bechtel 2006).

Only much more recently did philosophers of science develop detailed accounts of mechanistic explanations (Machamer, Darden, and Craver 2000; Bechtel and Abrahamsen 2005), and they have now become a central focus of philosophical analysis. Much of the philosophical inquiry focuses on the processes by which scientists develop mechanistic explanations by (1) linking phenomena to be explained to mechanisms and (2) decomposing these mechanisms into responsible parts and operations (Bechtel and Richardson 1993/2010; Craver and Darden 2013). All accounts of mechanisms also acknowledge the importance of how these components are organized in generating the phenomena. Organization becomes a focus for biologists as they attempt to recompose a mechanism, at least conceptually, to show that it can generate the phenomenon. But philosophers have paid less attention to how biologists recompose mechanisms. In contemporary research papers, in which most of the text and figures are devoted to presenting new experimental findings about parts or operations, the researchers may conclude with a description of a mechanism, often accompanied by a diagram. The diagram typically does much of the work of conveying how the researchers conceive of the parts and operations fitting together. The emphasis on recomposing mechanisms, often relying on mechanism diagrams, is more frequent in commentaries on research papers and in review articles.

Despite playing a central role in biologists' attempts to recompose mechanisms, philosophers, including those who discuss mechanistic explanations, have offered little analysis of how diagrams play that role (but see Abrahamsen, Sheredos, and Bechtel 2018). Here I note two features of diagrams that make them particularly well suited for representing the organization of a mechanism. First, abstract shapes or icons are used to represent parts, and arrows (often of varied formats) are employed to show how an operation performed by one part affects other parts (Tversky [2011] refers to these as "glyphs"). Second, researchers use the two spatial dimensions of a diagram[1] to represent the relations between parts, sometimes indicating how they are spatially related (as in fig. 13.1 below), and sometimes how they are functionally related (as in fig. 13.2). A diagram can better present the often very complicated relations between parts and operations of a mechanism than text (often, if no diagram is presented, readers construct diagrams for themselves). One important advantage of diagrams is that they allow viewers to direct their attention to different components of the mechanism as desired. A diagram, however, is static; it cannot, by itself, show how the parts, working together, are able to produce the phenomenon. To understand this, viewers must mentally animate the diagram (Hegarty 1992) by rehearsing the operations specified by the arrows in their imagination. An explanatory text often includes a narration of

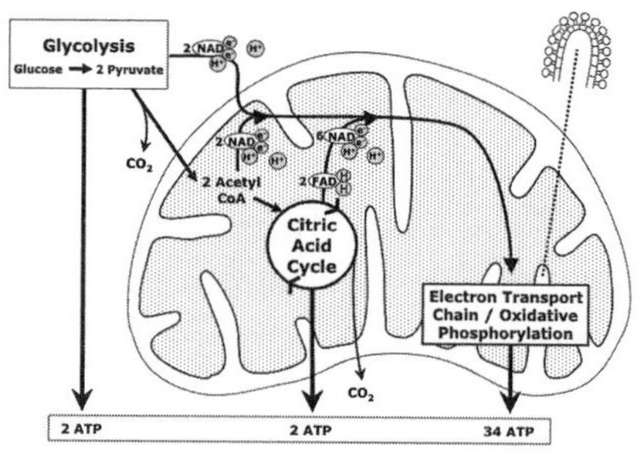

Figure 13.1. Conception of the mechanisms involved in basic energy metabolism as arrived at by the 1960s using the tools of cell fractionation and electron microscopy.

the operations of the mechanism and serves to guide the viewer in mentally animating the diagram.

Mental animation of a diagram, guided by narration, usually suffices for establishing understanding when the operations of the proposed mechanism are construed as executed sequentially, "from start or set-up to finish or termination conditions" (to borrow the language used in Machamer, Darden, and Craver's [2000] definition of mechanism) whenever the starting conditions are present. Assuming regular sequential operation is a good heuristic starting point and the resulting explanations, which I refer to as *basic mechanistic explanations* (Bechtel 2011), have often seemed to suffice to explain biological phenomena. Researchers, for example, typically view biochemical pathways as sequences of individual reactions.

However, the strategies of decomposing mechanisms to identify parts and operations have continued unabated. In the case of almost all biological mechanisms, including those thought to be well understood, researchers are discovering more and more parts. Researchers often confront three problems in recomposing these parts into a mechanistic explanation that they can mentally simulate. First, the parts do not operate just in a single sequence but in many parallel streams that proceed at different rates yet interact with each other at numerous points. It is challenging to rehearse mentally multiple interacting sequences of activity at the same time. Second, researchers frequently discover that operations they view as later in the sequence feed back to those they view as earlier. In such circumstances, their mental simulations have to take into account the effects of later operations when rehearsing the earlier one. Third, many of the additional parts that researchers identify for one mechanism turn out to be parts that also belong to other mechanisms. Mechanisms cease to have well-delineated boundaries but are embedded in large, interactive networks in which long-range connections modulate the behavior of what were taken to be independent mechanisms (Bechtel 2015).

Diagrams can represent multiple pathways, nonsequential organization, and the embeddedness of mechanisms in larger networks. But the challenge of understanding them—understanding how the mechanism portrayed generates the phenomenon one is trying to explain—begins to stress human cognitive capacities. To appreciate the problems, one can look ahead to figures 13.2 and 13.3, or one can think about the challenges in understanding human agents that are engaged in complex systems—for example, an orchestra player who not only must coordinate with her instrument but also with those around her, whose behavior is in part affected by her own behavior. Humans cannot mentally simulate all the operations and interactions proposed in the account of the mechanism to determine whether the proposed mechanism would generate the phenomenon. To make progress, researchers have had to enrich their strategies for investigating mechanisms beyond those that worked in developing basic mechanistic explanations.

Seeking to address these challenges is a major motivation of biologists who have adopted the name *systems biology*[2] (see, for example, Alon 2007; Ideker, Galitski, and Hood 2001; Kitano 2002). I focus here on two major strategies that systems biologists have adopted to address these problems and how these complement the more traditional strategies of mechanistic research. The first involves mathematically characterizing multiple interacting components and simulating their collective behavior computationally; the second involves representing components as networks of nodes and edges and deploying graph-theoretic and other tools for analyzing these graphs. I introduce these explanatory strategies and show how they differ from and yet complement those that gave rise to the development of basic mechanistic explanations.

In the next section I introduce the strategy of basic mechanistic explanation by describing its application to cells and show how it contributed to the growth of biological knowledge. In the subsequent section I explore how this strategy approached its limits as researchers encountered feedback loops that engender oscillatory behavior, such as the daily rise and fall of human body temperature or of the concentrations of various proteins. Body temperature and protein synthesis are circadian phenomena, in that they are controlled by an internal mechanism that generates oscillations of approximately 24 hours. The strategies of basic mechanistic explanation could not determine whether the proposed mechanism would produce the observed sustained oscillations or ones that would dampen over time. In the following two sections, I show how circadian researchers are supplementing basic mechanistic approaches with the two additional explanatory strategies introduced above.

As I briefly discuss in the concluding section, the introduction of new strategies for advancing explanations is not novel. The strategies for developing basic mechanistic explanations of cell phenomena themselves developed over the nineteenth and twentieth centuries, and the mechanistic explanations of cell behavior that figure prominently in contemporary textbooks are products of those strategies. Researchers have turned to new strategies of computational modeling and network analysis to cope with the success of those strategies in discovering more and more components organized and interacting in more complex ways. Almost certainly, as biological research continues, additional strategies will be developed. But computational modeling and network analyses are already generating explanations that are more dynamic and integrated than the basic mechanistic explanations advanced in the twentieth century.

The Quest for Basic Mechanistic Explanations

The idea of putting different types of components together to accomplish what individually they could not do has deep roots in engineering design. Ancient Greeks combined simple machines such as the wheel and axle and the pulley into compound machines such as the crane. Descartes vigorously advanced the idea that phenomena in the natural world, including those associated with living organisms, also result from machines. As was true of the mechanistic explanations espoused by Descartes, many of the early proposals of biological mechanisms were highly speculative. By the nineteenth century, however, researchers began to develop techniques that enabled them to pursue mechanistic accounts grounded in empirical knowledge of the parts and operations. These techniques involve identifying candidate mechanisms and then decomposing them into their parts and operations. For example, in the wake of the chemical revolution, chemists analyzed the constitution of different molecules in living organisms, identified reactions that transformed one into another, and discovered that biological tissues often contained catalysts (later termed *enzymes*) that facilitated these chemical reactions (Berzelius 1836). It was in this same period that Schwann (1839), utilizing newly improved microscopes, identified cells as the basic units of living systems and went on to associate them with processes that transform food into new tissue and energy, which he named *metabolism*.

Schwann could go no further in explaining the metabolic power of cells than to appeal to their distinctive chemical composition, which he proposed resulted from a process of cell formation through an iterative depositing of different materials around a core, as in crystal formation. Further advances awaited the development of new tools of biochemistry and cell biology. Starting around the beginning of the twentieth century, biochemists identified biochemical groups that are transferred between substrates in reactions, and by the 1930s Gustav Embden, Otto Meyerhof, and numerous other investigators had pieced together an account of the glycolytic pathway, a sequence of reactions transforming glucose to pyruvate that yielded modest production of ATP, the molecule that provides energy for other cellular functions. The introduction of the new techniques of cell fractionation and electron microscopy, beginning in the 1940s, was pivotal in generating a basic understanding of the mechanisms responsible for the subsequent oxidation of pyruvate to carbon dioxide and water, coupled with much more synthesis of ATP than glycolysis (Bechtel 2006). Cell fractionation enabled researchers to differentiate fractions with different enzymatic composition that originate in different organelles of the cell, while electron microscopy permitted visual identification of these organelles and their structure. In particular, cell fractionation allowed the localization of the citric acid (Krebs) cycle and electron transport to the mitochondrion while electron microscopy revealed the distinctive organization involving the inner membrane protruding into the cell interior. As shown in figure 13.1, the citric acid cycle was localized to the inner matrix, while electron transport and the coordinated phosphorylation reactions were localized to the cristae.

As discussed above, decomposition is only part of the strategy for constructing mechanistic explanations. The behavior of a mechanism depends on how components are organized and how the operation performed by one part affects others. Important graphical components of figure 13.1 are the arrows that link the different operations, establishing what Machamer, Darden, and Craver (2000) refer to as *productive continuity*. To understand the activity of the mechanism, researchers mentally animate the operations. A researcher familiar with the different operations in figure 13.1 can envisage a molecule of glucose being oxidized to yield two molecules of pyruvate, which, after being transformed into acetyl-CoA, enter the citric acid cycle, and so on. Once research has succeeded in identifying a mechanism, decomposing it into its parts and operations, and recomposing these into an organized set of productively continuous operations that researchers can mentally animate, the search for a mechanistic explanation seems to have reached a successful conclusion.[3]

Pushing Basic Mechanistic Explanation to Its Limits

The account of metabolism in the previous section followed the strategy of basic mechanistic explanation in offering a largely feed-forward account from start to termination conditions. The one exception is that the citric acid cycle involves a feedback loop in which the initial component of the cycle is regenerated from a product and new incoming acetyl CoA. As typically approached, however, this does not present any special challenges, as someone animating the diagram need only follow the Krebs cycle to the point where it outputs to other operations. In this section I consider how further pursuit of mechanistic research did push the basic mechanistic strategy to its limits when research on glycolysis revealed the type of complex dynamics involved (the periodic increase and decrease in concentrations of intermediary metabolites) and linked these to feedback loops. I then turn to a different phenomenon, circadian rhythms, where the complex dynamics were known from the outset. In both cases, mechanistic strategies could generate part but not all of what was needed for an explanation.

The reason the discovery of complex dynamics posed a problem for basic mechanistic explanation is that in pursuing such accounts, researchers assume that mechanisms function in a regular manner. Any change in how the mechanism responds is assumed to be due to the external input, not processes endogenous to the mechanism. Whenever glucose is available to the first reaction in the glycolytic pathway, the subsequent reactions are thought to occur seriatim. To a first approximation, many biological mechanisms do operate in this manner, presenting ideal conditions for the successful application of mechanistic research strategies. However, the data biologists record frequently manifests substantial variability. Researchers often attribute this to measurement errors or dismiss it as noise. In some cases, however, a pattern is found that reveals underlying dynamical behavior that is fundamental to the phenomenon. This happened in research on glycolysis. While measuring concentrations of glycolytic intermediates in yeast, using spectrophotometric techniques, Amal Ghosh and Britton Chance (1964) discovered that the concentration of NADH oscillated with a period of about a minute. Benno Hess, Arnold Boiteux and J. Krüger (1969) subsequently demonstrated periodic oscillations in the concentrations of other reactants. Moreover, neighboring reactants in the glycolytic pathway oscillated in phase with each other, whereas those on opposite sides of two major reactions were phase-reversed (i.e., 180° out of phase). One of these is the reaction in which fructose-6-phosphate (F6P) is phosphorylated to fructose-diphosphate (FDP) at the expense of transforming ATP to ADP through the action of the enzyme phosphofructokinase (PFK).

By focusing on PFK, researchers were able to extend the mechanistic explanation to partly account for this oscillation. PFK is an allosteric enzyme that contains binding sites for multiple molecules, and the binding at one site causes conformation changes at other sites, altering reactivity at those sites. In particular, PFK binds with three of the products of the main reaction, namely FDP and ADP, as well as AMP (a product generated by removing another phosphate group from ADP). When PFK is bound to FDP, ADP, and AMP, the reaction from F6P to FDP runs faster. The downstream effect of this positive feedback loop is to increase the production of NADH and, even further downstream, the synthesis of ATP from ADP or AMP. This further effect realizes a long-range negative feedback loop that counters the short-term positive feedback loop; their joint action is first to speed up NADH synthesis and then to slow it down. This verbal narrative suggests how the parts of the glycolytic system combine to produce oscillations, but it is important to note that there is an alternative possibility—that the system reaches an equilibrium at which NADH concentrations stop oscillating. This presents a limit to the basic mechanistic strategy: it is not able to determine which outcome will be realized.

Glycolytic oscillation was discovered in the context of an already worked out mechanism, and it remains unclear whether it plays a functional role in the metabolism of yeast. But in the case of many other physiological functions, oscillations clearly play a functional role. For example, rhythmic contraction of muscles is crucial for the circulation of blood. Increasingly, neuroscientists are discovering that subthreshold oscillations of ions in neurons, detectable by EEG or in resting state fMRI, figure in the coordination of processing in different regions of the brain. From here out I focus on a system whose primary function is to maintain an oscillation, circadian rhythms. These are oscillations with a period of approximately 24 hours that are generated endogenously in many living organisms, can be entrained to the light-dark cycle in the environment, and regulate many physiological and behavioral activities.

One of the first challenges circadian researchers faced was to establish that these daily rhythms, observed in physiological measures such as body temperature or in physical behaviors such as running, are generated within the organism and are not simply responses to cues from the environment. The crucial evidence was provided by the fact that the period found through studies of behavior in conditions in which cues have been removed (referred to as *free-running* conditions) varied slightly from 24 hours. After establishing that these rhythms were endogenous, research turned to figuring out the mechanism (which, early on, was referred to as a *clock*).

Initially, progress in identifying and decomposing the clock was slow. Research on fruit flies resulted in one of the first breakthroughs. Fruit flies exhibit circadian oscillations in locomotive behavior as well as in timing of their eclosion from their pupae. Ronald Konopka and Seymour Benzer (1971) identified a gene, *period* (*per*), which, when mutated in different ways, resulted in slow or fast rhythms or arrhythmic behavior. Once cloning became available, Paul Hardin, Jeffrey Hall, and Michael Rosbash (1990) demonstrated that both *per* mRNA and the protein PER oscillate with a period of 24 hours, with the concentration of the mRNA peaking about four hours before the concentration of the protein peaks. Knowing that negative feedback is a design principle that can generate oscillations, they proposed that the mechanism had the form of a transcription-translation feedback loop (TTFL), according to which PER feeds back to inhibit the transcription of *per*.

To see how such a mechanism could generate oscillations, one can try to simulate its operation mentally. Start in the state in which the concentration of PER is low. Since there is little inhibition on the rate of *per* transcription and translation, the concentration of PER gradually increases. But as it does so, it increasingly inhibits *per* transcription, stopping the increase in its concentration. Since PER gradually degrades, its concentration will now start to decline. As it declines, the inhibition is reduced, and the concentration of PER begins to rise again. As with glycolytic oscillation, basic mechanistic explanation reaches a limit, in that one could also narrate a scenario in which the mechanism approaches a steady state and stops oscillating. I return to this in the next section, but first follow the history a bit more to see how the conception of the circadian clock expanded.

One shortcoming of the TTFL proposed by Hardin and colleagues (1990) was that researchers could not find a DNA-binding site on the PER protein. Such a site is required if PER is to bind to the promoter of its own gene and block its own transcription. This gap in the account was only filled

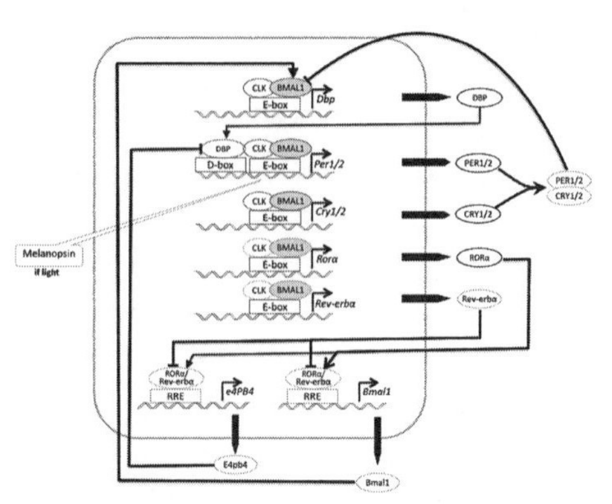

Figure 13.2. A mechanism diagram of the mammalian circadian clock, showing the principal genes and proteins that figure in feedback loops as understood circa 2005.

when Joseph Takahashi's group (Vitaterna et al. 1994) undertook a comparable search for mutants in mice and identified a gene they named *Clock* (Circadian locomotor output cycles kaput).[4] Mutations to *Clock* resulted in altered circadian rhythms, but, more importantly, CLOCK possessed the needed region to bind to the promoter (known as an E-box) on *per*. A homolog of *Clock* was soon identified in fruit flies, and within a few years, three homologs of *per* were found in mammals. Together these genes and proteins constitute the major negative feedback loop shown in the top portion of figure 13.2 (the figure shows PERs as dimerizing with CRYs, and CLOCK with BMAL1). Figure 13.2 presents many but not all the parts and operations that had been identified and fit into an account of the circadian mechanism by 2005. As the figure makes clear, these additional parts figure in various feedback loops, both positive and negative.

As noted above, although this diagram shows many of the crucial parts and operations, it is static. Viewers must supply the dynamics as they try to mentally simulate the operations. The additional feedback loops make this task even more difficult, and they have not alleviated the problem that one could equally simulate sustained oscillations or ones that dampen to a steady state.

In this section I have considered two cases in which the strategies of basic mechanistic research were pushed to their limits. With the discovery of the occurrence of oscillation in glycolysis, researchers were able to localize the responsible enzyme and provide a qualitative narrative of how it operated. In the case of circadian rhythms, researchers likewise began with a single feedback loop and could offer a narrative as to how it generated oscillations. But in neither case could they differentiate this narrative from one in which the system approached a steady state at which oscillations ceased. A further limit was reached when researchers identified multiple feedback loops in a diagram such as figure 13.2. Mentally simulating more than one feedback loop operating at once is challenging. Moving forward to show how these mechanisms generated the phenomena required supplementing the strategies for developing mechanistic explanations with new explanatory strategies.

Explaining Dynamic Behavior Through Computational Simulation

The previous section revealed one respect in which oscillatory phenomena stretch basic mechanistic explanation to its limits. Feedback systems generate oscillations, but many dampen over time as the system reaches a steady state. From mentally simulating a mechanism such as that shown in figure 13.2, or even the initial proposal for a feedback loop involving *per* alone, one cannot tell whether the oscillation will be maintained or whether it will reach a point at which the increase of PER from transcription and translation is perfectly compensated by the rate at which it degrades.

The main alternative to which scientists turn when mental simulation fails is computational simulation. When this method is applied in cell and molecular biology, modelers create differential equations to describe how the concentrations of individual components are affected by the concentrations of one or more other components. What one realizes quickly with feedback systems is that if one assumes that all the equations are linear, oscillations resulting from feedback will quickly dampen. One or more nonlinear equation is required. To determine how the TTFL proposed by Hardin et al. (1990) would behave, Albert Goldbeter (1995) constructed a computational model consisting of five differential equations.[5] The first equation represents how the concentration of *per* mRNA (*M*) changes as a result of being increased by the rate of transcription of the gene *per* (first term after the equal sign) and decreased by the rate of decay of the protein PER (second term):

$$\frac{dM}{dT} = v_s \frac{K_I^n}{K_I^n + P_N^n} - v_m \frac{M}{K_m + M}$$

The symbols v_s, v_m, K_I, and K_m, as well as n, are parameters, and the choice of values for parameters has a significant effect on how the simulation behaves. I will, however, only comment on n, which appears as the coefficient of the variable P_N in the denominator of the first term. P_N represents the concentration of PER in the nucleus. Putting it in the denominator has the effect of reducing the increase of M as P_N increases, thereby capturing the role of PER in the nucleus in inhibiting its own transcription. How much of an inhibitory effect P_N has is critically affected by the coefficient n, known as the Hill coefficient, which reflects how many molecules of PER are required to inhibit *per* transcription. In the second term, M appears in the numerator as well as the denominator, where a parameter is added to it. This has the effect of increasing the rate of decay in a nonlinear manner as the amount of M increases. Using what he claimed were biologically plausible values for parameters and applying each of the five equations iteratively, Goldbeter (1995) showed that they generated stable oscillations in variables such as M.

The introduction of more components into the mechanistic account raised the possibility that the new system would yield dampened oscillations. Accordingly, Jean-Christophe Leloup and Goldbeter (2008) added equations and used simulations to show that the expanded computational model would, with what again were assumed to be biologically plausible parameters, generate sustained oscillations. Leloup and Goldbeter also incorporated components in this model enabling them to account for the entrainment of the oscillator by light and for known circadian pathologies, such as delayed sleep-phase syndrome.

Goldbeter deployed these computational models to understand how an already proposed mechanism would behave. They provide what Adele Abrahamsen and I have referred to as *dynamic mechanistic explanations* (Bechtel and Abrahamsen 2010). By using that term, we sought to emphasize the complementary roles of computational modeling in ascertaining the dynamics of a system and the mechanistic analysis of its composition. Although the roles are complementary, computational models contribute to explanations in a very different manner than does the identification of the parts and operations of the mechanism.

Mechanistic explanations have often been presented as an alternative to the type of deductive-nomological (DN) explanations that figured in earlier philosophy of science. According to the DN model, explanation involved deriving a description of a phenomenon from statements of one or more laws and initial conditions. Basic mechanistic explanations do not invoke laws or perform derivations. Computational simulations are more closely aligned with DN explanations than mechanistic explanations (Krakauer et al. 2011). First, while the equations used in computational simulations are typically not what one would characterize as laws, but rather mathematically characterized regularities, they are a basis for deriving results. Second, equations, like laws, are general. They specify relations between values of variables without specifying the parts whose properties are varying. Different parts with varying properties may be characterized by the same generalization. Third, invoking Craver's distinction between *how-possibly* and *how-actually* accounts, computational models provide *how-possibly* accounts—whether they characterize a putative or an actual mechanism, what they show is that such a mechanism *could* generate the phenomenon.

The explanatory import of computational modeling in biology extends beyond contexts in which it is used to understand the behavior of a hypothesized mechanism. I briefly highlight four additional roles. Sometimes they are the object of experiments that are designed to better understand how the modeled mechanism functions. This involves intervening on the computational model by, for example, investigating the effects of other parameter values or of removing or adding components to the model. As experimental researchers discovered more components of the mammalian circadian clock, modelers tried strategies such as fixing the values of variables for some of the components in their models to see if that affected the ability of the model to generate sustained oscillations. It is much easier to intervene on a computational model than on the biological tissue. The results of such experiments on computational models only provide information about what would happen in a real biological preparation to the extent that the model correctly describes the actual biological system. Yet, by drawing attention to what is possible under the current hypothesis, they can play an important role in the interactive engagement of modeling and experimentation.

A second additional use of computational models is to identify and characterize design principles (Green, Levy, and Bechtel 2014). Increasingly, biologists are approaching biological systems with the mindset of engineers. When engineers design systems, they often put together modules that are themselves composed of components, but organized according to principles from which their behavior in various contexts can be determined. Biologists can make use of such principles to understand how mechanisms they encounter operate. Uri Alon (2007) pioneered the investigation of motifs: small networks of two, three, or four nodes that are organized in a particular way (e.g., two units negatively feeding back on each other). Through computational simulations (typically models using either Boolean or differential equations), he and others have determined how motifs will behave in any system in which parameter values fall within a specified range. When researchers identify an instance of such a motif in a biological system, they can immediately infer its behavior. Design principles are not tied to any particular mechanism but abstract from them; nonetheless, they can be applied in understanding the behavior of actual mechanisms in which components are organized in the manner specified.

A third use of computational models is closely related to the search for design principles. While one can construct detailed computational models that adhere closely to the details of a particular mechanism, one can also relax those constraints to develop models that generalize across a broader range of phenomena. This involves abstracting (Levy and Bechtel 2013) or coarse graining (Krakauer et al. 2011) by, for example, relaxing constraints of the range of variables or considering parameters in an extended range. When successful, this approach can reveal general principles. Biology is often contrasted with physics insofar as there do not seem to be a small set of basic principles (laws) that can be applied universally. But that does not mean one cannot achieve varying degrees of generality, and computational models provide one vehicle for doing so.

A final use of computational models is to provide an understanding of global states of systems and how systems might evolve. A useful way to represent the behavior of a complex system with many components, each of which changes its state over time, is in terms of a state space in which each dimension corresponds to a variable characterizing the system. The current state of the system will correspond to a point in the space, and change in the system will correspond to a trajectory through the possible states of the system. By studying trajectories in state space, investigators can identify the structure in the state space—for example, discovering that it contains attractor states to which the system will evolve from a variety of other states (the basin of the attractor). This structure can be productively represented as a landscape in which the points at which the system will stabilize are located at the bottom of valleys.

Although one can develop a state-space representation of an empirically studied system, it is much easier to run multiple simulations with a computational model of a dynamical system. Using the simple computational model of the TTFL that he advanced in 1995, Goldbeter was able to map a landscape that contained a limit cycle attractor—a closed loop of states in state space corresponding to the oscillation of *per* mRNA and PER concentrations. The cycle is called a limit cycle due to the fact that, from a variety of starting points not on the cycle, the system will evolve toward the cycle. In more complex dynamical systems, there are multiple attractors in the landscape. Moreover, one can represent alterations to the system as changes in the identity and location of attractors. By constructing computational models of hypothetical complex systems and studying the resulting landscapes and how they can change, one can acquire intuitive ideas about what is happening in natural systems (e.g., that the perturbation that leads to cancerous growth creates a new attractor) (Huang and Kauffman 2012).

Computation modeling invokes a very different explanatory strategy than the mental animation of mechanism diagrams that figured in basic mechanistic explanation. It emphasizes the abstract, possible system, not the concrete, actual mechanism. When directed at particular mechanisms, it can provide information about how the mechanism will behave that cannot be generated from the mechanistic account itself. Through its extended uses, it can facilitate the discovery of generalized principles and enable researchers to address additional questions about possible systems that lie beyond the scope of basic mechanistic accounts.

Explaining the Integration of Mechanisms Through Graph-Theoretic Analyses

In this section I address a second limitation that often confronts basic mechanistic explanations—that the same strategies used for initially identifying parts of a mechanism end up identifying a large-holistic system implicated in multiple phenomena, not just the one under investigation. Mechanistic research often assumes that biological mechanisms are independent entities that, when provided the right inputs, depend only on their inner workings to generate the phenomenon they are invoked to explain. That is, it assumes there is a natural boundary to a mechanism and that the parts one identifies reside within it. In some cases, research on a mechanism begins with a delineated structure (e.g., a cell organelle or a brain region) and a characterization of what it does. But in many cases, as in the investigations of circadian rhythms discussed above, research begins with a part of the mechanism (e.g., a gene), and the account of the mechanism is further developed by determining which other parts (1) interact with that part in generating the phenomenon and (2) change the phenomenon when they are altered. When researchers turn to recomposing the mechanism, they seek to identify the place of these entities within the mechanism.

The problem with the strategy just outlined is that it is extremely sensitive to the techniques available at the time to identify components that have an effect on the phenomenon being explained and to measure those effects. Traditional strategies such as inhibiting or stimulating parts were limited to investigating a small number of potential parts. Often this would yield on the order of ten parts, as seen in figure 13.2 in the case of circadian rhythms. With the development of new techniques in the 2000s, however, researchers have identified many more genes that have effects on circadian oscillations. For example, using small interfering RNAs to knock down 17,631 known and 4,837 predicted human genes in U2OS (human osteosarcoma) cells containing a luciferase reporter attached to the known clock gene *Bmal1*, John Hogenesch, Steve Kay, and their collaborators identified nearly 1,000 genes that resulted in low-amplitude circadian oscillations (Zhang et al. 2009). Due to challenges in analyzing period in these cases, they were not further analyzed. They focused instead on 343 genes that clearly increased the amplitude or altered the period of circadian rhythms (they only counted the gene if it produced deviations more than three standard deviations from the mean).

The researchers selected seventeen genes on which to perform a dose-dependent knockdown, and in sixteen cases established dose-dependent effects comparable to those that previous research had found with genes already regarded as clock genes. In addition, the researchers analyzed protein interactions and determined that some of the proteins synthesized from these genes interacted directly with known clock genes, whereas others were further removed. Many of them are part of pathways such as those for insulin and hedgehog signaling, cell cycle, and folate metabolism. For example, down-regulating several components of the insulin pathway (*JNK, IKK, MTOR, APKC*, and *PYK*) results in longer-period oscillations, while down-regulating another, *PFK*, results in shorter-period oscillations. Since these are also pathways that previously had been shown to be regulated by the circadian clock, Zhang et al. conclude collectively that "the clock is massively interconnected and functionally intertwined with many biological pathways" (Zhang et al. 2009, 207).

These results present a new challenge in conceptualizing the clock mechanism. It would not be productive to simply add all of these genes and proteins to what is regarded as the clock mechanism. As noted, many are part of other mechanisms; such a move would quickly lead to treating the whole cell or organism as the mechanism for all phenomena. The great success of mechanistic research has stemmed from its ability to decompose systems into relevant mechanisms and their parts, and to show how these contribute to the phenomena under investigation. If biology is to continue to build upon this success, researchers need ways to draw boundaries around mechanisms in order to generate recognizably, if more complex, mechanistic explanations. How can they do so?

As in many other fields, biologists are increasingly invoking tools to analyze networks to understand biological mechanisms (Barabasi and Oltvai 2004; Mitra et al. 2013; Prokop and Csukás 2013). Most fundamentally, network approaches provide new tools for representing biological organization. They also provide new tools with which to reason about organization and its consequences for the behavior of mechanisms. To illustrate the potential of this approach, I start with the network diagram (reproduced in fig. 13.3 and plate 4) that Zhang et al. used to present their results. Shown in light and dark blue are the proteins that are normally construed as constituting the circadian clock (ARNTL and ARNTL2 are alternate names for BMAL1 and BMAL2; NR1D2 and NR1D2 are alternate names for Rev-Erbα and Rev-Erbβ). In purple, red, and green, they show proteins that, when knocked down, increased the amplitude or altered the period of the clock. In pink are proteins that link those proteins that affect the clock when knocked down and the core components of the clock.

The first important role of network representations in biology is to provide new perspectives on organization both within and between mechanisms. These perspectives draw from graph theory, which provides a number of measures for analyzing network organization. In the vocabulary of graph theory, networks consist of nodes (the circles in the above diagram) and edges (the lines connecting the circles). One graph-theoretic measure, cluster analysis, identifies as modules nodes that are highly interconnected (clustered). As a result of these connections, modules represent candidate mechanisms. Sometimes the mechanisms identified in this manner correspond roughly to those identified by the classical procedure of starting with the phenomenon and finding parts that affect it. The highly connected nodes in the center, colored in light or dark blue, correspond to the traditionally construed clock mechanism.

Network analysis, however, is most useful when it offers accounts that differ from those advanced directly from mechanistic research. One such role is to identify additional components beyond those differentiated by classical mechanistic research (see, e.g., Ravasz et al. 2002; Kelley and Ideker

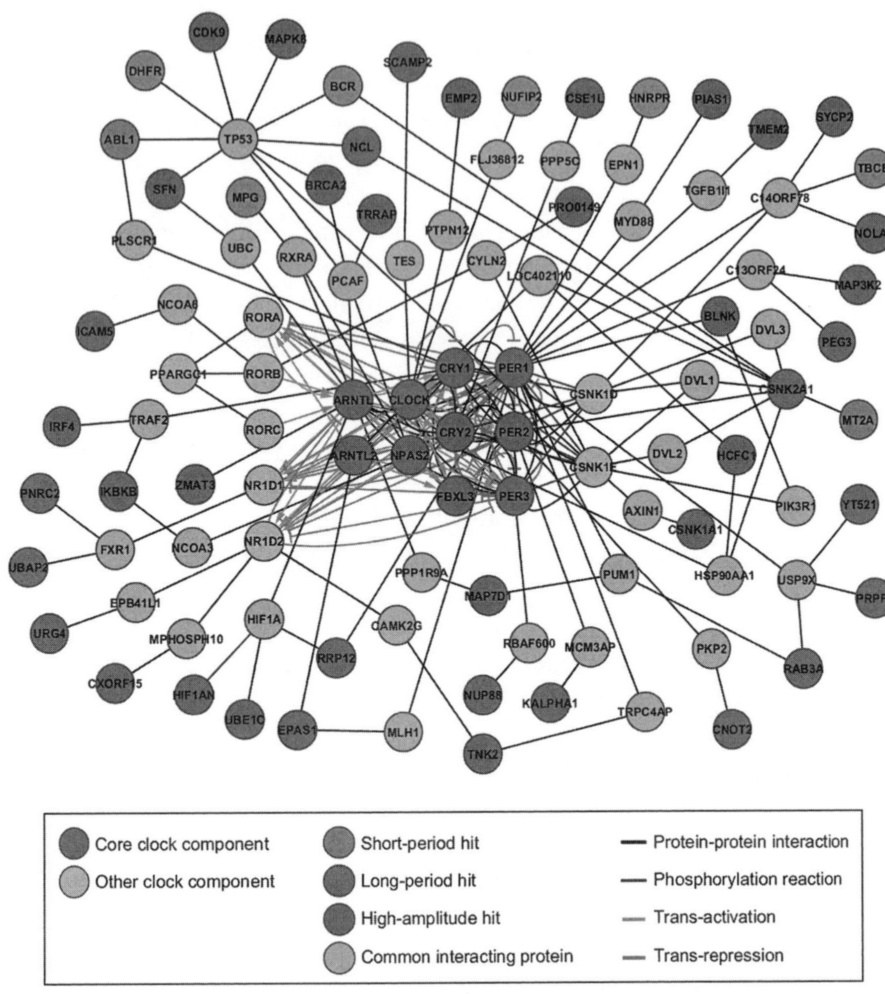

Figure 13.3. Zhang et al.'s (2009) representation of the various proteins that, when knocked down, have effects on circadian rhythms. See text and legend for details. Reprinted from *Cell*, vol. 139, Eric E. Zhang, Andrew C. Liu, Tsuyoshi Hirota, Loren J. Miraglia, Genevieve Welch, Pagkapol Y. Pongsawakul, Xianzhong Liu, Ann Atwood, Jon W. Huss, Jeff Janes, Andrew I. Su, John B. Hogenesch, Steve A. Kay, "A Genome-wide RNAi Screen for Modifiers of the Circadian Clock in Human Cells," pages 199–210. Copyright (2009), with permission from Elsevier.

2005). At other times, modules may point to the existence of functional organization within and between mechanisms that may not have been discovered by traditional approaches (Costanzo et al. 2010).

Another important measure used in graph analysis that provides insights to the organization within and between mechanisms is degree distribution—the distribution of the number of edges from a given node. Early graph theorists assumed that degree would be distributed normally, but in many networks it is not normally distributed. Rather, a few nodes have unusually high degree. These nodes are referred to as hubs and they may serve either to integrate a module/mechanism (TP53 in the upper left in fig. 13.3 and plate 4) or to facilitate integration between modules/mechanisms (CSNK2A1 on the right). Some components of the core clock, such as Per1, have extensive connections to other components of the clock and also to units elsewhere, suggesting an important role in integrating clock components with components of other mechanisms that are both regulated by and regulate clock function.

Second, networks are not only the product of numerous experimental inquiries, but they can also serve as a guide both to further experimentation and to modeling. A network representation reveals many unsuspected, indirect connections between nodes and so can guide inferences about how the effects of perturbing one node will spread to others. In many cases, modeling of the network provides a guide to what sorts of effects one should expect. Researchers often annotate network representations with information from the Gene Ontology project about where in the cell genes are expressed, the functions the proteins perform, and the larger biological processes in which they figure. Particularly valuable from the point of view of understanding mechanisms is that Gene Ontology represents parts and functions hierarchically, which then supports predictions as to the effects of perturbing individual genes (Yu et al. 2016). Given that Gene Ontology involves curated information about the entities and their functions that have been identified in experimental research, it is not surprising that the hierarchical units more or less correspond to traditional mechanisms. What is less to be expected is that new data-driven approaches to generating ontologies, such as NeXO (Dutkowski et al. 2013), also yield hierarchical structures corresponding to mechanisms. When combined with information from these ontologies, network representations enable researchers to make new inferences about the effects of perturbations (e.g., knocking out of two genes) not only within traditionally characterized mechanisms but also across mechanisms.

The challenge I raised at the beginning of this section was where to draw the boundaries around a mechanism once we recognize how the components of a mechanism are interconnected with other entities in large networks. In the above paragraphs I noted that modules in graph representations often correspond to mechanisms initially identified in more traditional ways. But given that there are a number of entities outside the modules that affect activity within them, researchers must often exercise discretion as to where to draw the boundaries. Recent research, for example, has identified several instances in which an operation within the circadian mechanism depends directly on a component usually treated as part of the metabolic mechanism. One involves the binding of the dimer of Clock and Bmal1 to the E-box on *per* and other genes, as shown in the center of figure 13.2. Clock has been identified as a histone acetyl transferase—by adding acetyl groups to the chromatin, it affects how tightly the chromatin is bound and hence whether genes such as *Per* can be transcribed. This function of Clock is regulated by SIRT1 binding to it. The concentration of SIRT1 is itself dependent on levels of NAD^+, a central component in glycolysis and other metabolic pathways (Bellet et al. 2011). If one focuses on the way in which NAD^+ modulates SIRT1 activity and thereby modulates circadian rhythms, then one might include it in the circadian mechanism. If the research question involves how transcription of circadian genes is modulated, a researcher might include not just SIRT1 but even the metabolic processes, such as glycolysis, that oxidize and reduce NAD^+ in the relevant mechanism. Decisions as where to locate the boundaries of mechanisms are constrained by both the interconnectivity of components in larger networks and the questions researchers are investigating.

The characterization and analysis of network organization constitutes a different strategy than is traditionally employed in mechanistic research. As in the case of computational modeling, in network analysis, researchers abstract from the specific composition of the mechanism and focus on the ways edges connect nodes. The goal of network analysis is to identify organizational principles at an abstract or coarse-grained level, and to determine the behaviors they make possible. The application of these results to specific actual networks that realize a pattern of organization involves a reasoning process much like derivation. The resulting abstract analysis plays a different explanatory role than the specification of the parts and operations. By identifying a mechanism with a module in a network, one can use the network representation to identify ways in which the module is affected by other modules. To investigate how such modulation is actually achieved, researchers need to engage further with mechanistic strategies that identify specific parts and operations. Network analysis performs a function complementary to traditional mechanistic inquiry, revealing multiple ways in which a mechanism is situated among other mechanisms and creating conditions in which researchers can select which entities to treat as mechanisms and investigate further.

Expanding Mechanism's Explanatory Strategies

The strategies scientists pursue to explain the phenomena of interest to them change over time. Cytology and cell biology through the nineteenth and twentieth centuries witnessed the cultivation of new techniques to identify mechanisms and decompose them into their parts and operations. This is evident in *General Cytology* and again with the introduction of electron microscopy and cell fractionation. These techniques enabled cell biologists to identify organelles and localize specific chemical reactions in each. On the basis of these results, researchers set about recomposing mechanisms, often using diagrams and mentally rehearsing the operations depicted (e.g., fig. 13.1). They offered accounts of how cells perform a wide range of activities that fit the pattern of basic mechanistic explanations. Even as contemporary cell researchers move beyond these accounts, they remain success stories.

But as a result of developing new techniques for decomposing cells to identify genes and proteins, especially those that operate on a mass scale, cell biologists found that the parts and operations could no longer be recomposed into basic mechanistic explanations. The discovery of multiple operations occurring in parallel, feedback relations between operations regarded as later in the process and those regarded as earlier, and interactions with components regarded as parts of other mechanisms rendered the project of invoking mental simulation to animate static diagrams insufficient. One could not settle whether the mechanism would generate, for example, sustained circadian rhythms or dampened oscillations.

To address these questions, biologists availed themselves of new strategies that could complement those of basic mechanistic explanation. I have focused on two—computational simulation and network analysis—that are playing important roles in contemporary systems biological approaches to explaining cellular phenomena such as circadian rhythms. I have emphasized the differences between these strategies and mechanistic strategies of the past. Both computational simulations and network analyses abstract from the details of the composition of mechanisms and appeal to general principles. Researchers apply the results of such abstraction to understand the operation of specific mechanisms through processes like derivation. From a computational analysis, researchers can find out how a proposed mechanism will behave and explore other possibilities. From a network analysis, researchers can make inferences based on characteristic modes of organization and propose plausible boundaries for mechanisms, while recognizing that they are situated in an environment that affects their operation.

Deploying new strategies to advance explanatory objectives is a recurring theme in biology. What is important to recognize is that the conception of explanation is also being extended. Recomposition achieved through computational modeling and network analysis differs from that achieved by mentally animating a mechanism. Nonetheless, contemporary cell biology is still mechanistic. In particular, biologists are still decomposing biological systems into parts and operations (with high-throughput procedures, this process is accelerated). And the resulting explanations still appeal to these parts and operations. And when possible, biologists still attempt to mentally rehearse the operations in mechanism diagrams. The new techniques complement existing mechanistic strategies. The resulting perspective is a pluralistic one in which different explanatory strategies each make a complementary contribution to the pursuit of mechanistic explanation in contemporary cell biology.

Acknowledgments

I thank Garland Allen, the audiences at the two Updating Cowdry Workshops at Woods Hole, and the editors of this volume, Jane Maienschein, Karl Matlin, and Manfred Laubichler, for their very helpful and constructive comments on earlier versions of this chapter.

Notes

1 Drawn or printed diagrams are limited to two dimensions (although on occasion overlays are used to reflect additional dimensions); sometimes researchers find it important to represent a mechanism in three dimensions and build physical models.

2 Systems biology has intellectual roots in the cybernetics and general systems theory movements of the twentieth century, and these traditions have informed some of the modeling approaches adopted in systems biology. But far more fundamental was the development of techniques for collecting massive data about genes, proteins, and metabolites.

3 Biochemists since the beginning of the twentieth century have invoked mathematical representations to characterize operations within a mechanism. For example, the Michaelis-Menten equation is employed to determine how the concentration of the

substrate affects the rate of production of the product. But this mathematical analysis is not required for understanding the behavior of the mechanism, characterized qualitatively.

4 The naming convention for genes in fruit flies is to use lowercase italics. In mammals, gene names are also in italics, but they begin with an initial capital. Protein names conventionally are all in roman capital letters.

5 The term *model* is used in a wide variety of ways. Explanatory accounts, such as those presented in a mechanism diagram, are often referred to as (mechanistic or explanatory) *models*. Sets of equations that describe a mechanistic model and are used to generate a simulation are referred to as a *computational model*, which is often abbreviated as just *model*. It is usually clear from context what is meant by the term. In this section, *model* refers to a computational model.

References

Abrahamsen, A., B. Sheredos, and W. Bechtel. 2018. "Explaining Visually: Mechanism Diagrams." In *Routledge Handbook of Mechanisms*, edited by S. Glennan and P. Illari. London: Routledge.

Allen, G. E. 1979. *Life Science in the Twentieth Century*. London: Cambridge University Press.

Alon, U. 2007. *An Introduction to Systems Biology: Design Principles of Biological Circuits*. Boca Raton, FL: Chapman and Hall/CRC.

Barabasi, A. L., and Oltvai, Z. N. 2004. "Network Biology: Understanding the Cell's Functional Organization." *Nature Reviews Genetics* 5:101–13.

Bechtel, W. 2006. *Discovering Cell Mechanisms: The Creation of Modern Cell Biology*. Cambridge: Cambridge University Press.

———. 2011. "Mechanism and Biological Explanation." *Philosophy of Science* 78: 533–57.

———. 2015. "Can Mechanistic Explanation Be Reconciled with Scale-Free Constitution and Dynamics?" *Studies in History and Philosophy of Science, Part C: Studies in History and Philosophy of Biological and Biomedical Sciences* 53:84–93.

———, and A. Abrahamsen. 2005. "Explanation: A Mechanist Alternative." *Studies in History and Philosophy of Science, Part C: Studies in History and Philosophy of Biological and Biomedical Sciences* 36:421–41.

———, and A. Abrahamsen. 2010. "Dynamic Mechanistic Explanation: Computational Modeling of Circadian Rhythms as an Exemplar for Cognitive Science." *Studies in History and Philosophy of Science, Part A* 41: 321–33.

———, and R. C. Richardson. (1993) 2010. *Discovering Complexity: Decomposition and Localization as Strategies in Scientific Research*. Cambridge, MA: MIT Press. Originally published in 1993. Princeton, NJ: Princeton University Press.

Bellet, M. M., R. Orozco-Solis, S. Sahar, K. Eckel-Mahan, and P. Sassone-Corsi. 2011. "The Time of Metabolism: NAD+, SIRT1, and the Circadian Clock." *Cold Spring Harbor Symposia on Quantitative Biology* 76:31–38.

Berzelius, J. J. 1836. "Einige Ideen über eine bei der Bildung organischer Verbindungen in der lebenden Naturwirksame, aber bisher noch nicht bemerkte Kraft." *Jahres-Bericht über die Fortschritte der Chemie* 15:237–45.

Buchner, E. 1897. "Alkoholische Gärung ohne Hefezellen (Vorläufige Mittheilung)." *Berichte der deutschen chemischen Gesellschaft* 30:117–24.

Coleman, W. 1971. *Biology in the Nineteenth Century: Problems of Form, Function, and Transformation*. New York: John Wiley.

Costanzo, M., A. Baryshnikova, J. Bellay, Y. Kim, E. D. Spear, C. S. Sevier, H. Ding, et al. 2010. "The Genetic Landscape of a Cell." *Science* 327:425–31.
Cowdry, Edmund V. 1924. *General Cytology: A Textbook of Cellular Structure and Function for Students of Biology and Medicine*. Chicago: University of Chicago Press.
Craver, C. F., and L. Darden. 2013. *In Search of Mechanisms: Discoveries across the Life Sciences*. Chicago: University of Chicago Press.
Dutkowski, J., M. Kramer, M. A. Surma, R. Balakrishnan, J. M. Cherry, N. J. Krogan, and T. Ideker. 2013. "A Gene Ontology Inferred from Molecular Networks." *Nature Biotechnology* 31:38–45.
Ghosh, A. K., and B. Chance. 1964. "Oscillations of Glycolytic Intermediates in Yeast Cells." *Biochemical and Biophysical Research Communications* 16:174–81.
Goldbeter, A. 1995. "A Model for Circadian Oscillations in the *Drosophila* Period Protein (PER)." *Proceedings of the Royal Society of London. B: Biological Sciences* 261:319–24.
Green, S., A. Levy, and W. Bechtel. 2014. "Design sans Adaptation." *European Journal for Philosophy of Science* 5:15–29.
Hardin, P. E., J. C. Hall, and M. Rosbash. 1990. "Feedback of the *Drosophila period* Gene Product on Circadian Cycling of Its Messenger RNA Levels." *Nature* 343: 536–40.
Hegarty, M. 1992. Mental Animation: Inferring Motion from Static Displays of Mechanical Systems. *Journal of Experimental Psychology: Learning, Memory, and Cognition* 18:1084–1102.
Hess, B., A. Boiteux, and J. Krüger. 1969. "Cooperation of Glycolytic Enzymes." *Advances in Enzyme Regulation* 7:149–67.
Huang, S., and S. A. Kauffman. 2012. "Complex Gene Regulatory Networks—from Structure to Biological Observables: Cell Fate Determination." In *Computational Complexity*, edited by R. A. Meyers, 527–60. New York: Springer.
Ideker, T., T. Galitski, and L. Hood. 2001. "A New Approach to Decoding Life: Systems Biology." *Annual Review of Genomics and Human Genetics* 2:343–72.
Kelley, R., and T. Ideker. 2005. "Systematic Interpretation of Genetic Interactions Using Protein Networks." *Nature Biotechnology* 23:561–66.
Kitano, H. 2002. "Systems Biology: A Brief Overview." *Science* 295:1662–64.
Konopka, R. J., and S. Benzer. 1971. "Clock Mutants of *Drosophila melanogaster*." *Proceedings of the National Academy of Sciences* 89:2112–16.
Krakauer, D. C., J. P. Collins, D. Erwin, J. C. Flack, W. Fontana, M. D. Laubichler, S. J. Prohaska, et al. 2011. "The Challenges and Scope of Theoretical Biology." *Journal of Theoretical Biology* 276:269–76.
Leloup, J. C., and A. Goldbeter. 2008. "Modeling the Circadian Clock: From Molecular Mechanism to Physiological Disorders." *BioEssays* 30:590–600.
Levy, A., and W. Bechtel. 2013. "Abstraction and the Organization of Mechanisms." *Philosophy of Science* 80:241–61.
Machamer, P., L. Darden, and C. F. Craver. 2000. "Thinking about Mechanisms." *Philosophy of Science* 67:1–25.
Mitra, K., A. R. Carvunis, S. K. Ramesh, and T. Ideker. 2013. "Integrative Approaches for Finding Modular Structure in Biological Networks." *Nature Reviews Genetics* 14:719–32.
Palade, G. E. 1987. "Cell Fractionation." In *The American Association of Anatomists, 1888–1987. Essays on the History of Anatomy in America and a Report on the Membership—Past and Present*, edited by J. E. Pauly. Baltimore: Williams and Wilkins.

Prokop, A., and B. Csukás. 2013. *Systems Biology: Integrative Biology and Simulation Tools*. Vol. 1. Dordrecht: Springer.

Ravasz, E., A. L. Somera, D. A. Mongru, Z. N. Oltvai, and A. L. Barabasi. 2002. "Hierarchical Organization of Modularity in Metabolic Networks." *Science* 297:1551–55.

Schwann, T. 1839. *Mikroskopische Untersuchungen über die Uebereinstimmung in der Struktur und dem Wachstum der Thiere und Pflanzen*. Berlin: Sander.

Tversky, B. 2011. "Visualizing Thought." *Topics in Cognitive Science* 3:499–535.

Vitaterna, M. H., D. P. King, A. M. Chang, J. M. Kornhauser, P. L. Lowrey, J. D. McDonald, W. F. Dove, et al. 1994. "Mutagenesis and Mapping of a Mouse Gene, *Clock*, Essential for Circadian Behavior." *Science* 264:719–25.

Yu, Michael K., M. Kramer, J. Dutkowski, R. Srivas, K. Licon, Jason F. Kreisberg, Cherie T. Ng, et al. 2016. "Translation of Genotype to Phenotype by a Hierarchy of Cell Subsystems." *Cell Systems* 2:77–88.

Zhang, E. E., A. C. Liu, T. Hirota, L. J. Miraglia, G. Welch, P. Y. Pongsawakul, X. Liu, et al. 2009. "A Genome-Wide RNAi Screen for Modifiers of the Circadian Clock in Human Cells." *Cell* 139:199–210.

CHAPTER 14

UPDATING COWDRY'S THEORIES

THE ROLE OF MODELS IN CONTEMPORARY EXPERIMENTAL AND COMPUTATIONAL CELL BIOLOGY

Fridolin Gross

What is the relationship between theory and experiment in cell biology? In physics, for instance, the division of labor seems clear-cut: theoreticians propose theories that are subsequently put to test by experimentalists. It is not obvious, by contrast, where to look for theory in contemporary cell biology, given that experimental cell biologists often seem to be getting along well without referring to the work of theoretical biologists and without speaking of theories. This was evidently different in Cowdry's times: In *General Cytology* (1924) one counts 248 instances of the word *theory*. Compare this to the 2008 edition of Alberts's *Molecular Biology of the Cell*, in which *theory* is used only 24 times, even though the volume has more than twice as many pages (Alberts et al. 2008).

In *General Cytology* we find the cell theory, the reticular theory of protoplasm, the chromosome theory of heredity, theories of irritability, theories of permeability, theories of stimulation, theories of fertilization, and more. These theories are presumably not understood in the logical empiricist's sense of a set of sentences that can be deduced from specific axioms. Taking a closer look at these examples, one gets the impression that theories were understood mostly as speculative hypotheses to explain and unify a collection of known facts and phenomena about the cell. The theories in *General Cytology* were typically grounded in basic chemical or physical principles and often concerned levels of scale that were difficult to investigate with the experimental methods available at the time. This explains why various competing theories associated with a particular phenomenon are often presented, and in fact large passages in *General Cytology* are concerned with the discussion and evaluation of alternative theories. Even though some of the contributors are very cautious about theoretical speculations, many think of them as necessary for scientific progress. For example, Jacobs writes in his chapter on the permeability of the cell,

> With so many of the facts regarding the penetration of the cell by diffusing substances still in uncertainty, the time is not yet ripe for

attempting a comprehensive theoretical explanation of the process itself. Nevertheless, hypotheses are so necessarily and so inextricably connected with the acquisition of new facts that a review such as the present one would not be complete without some mention of several of the chief theories of cell permeability which have been suggested in the past, together with a brief criticism of some of the deficiencies of each. (Jacobs 1924, 149)

The abundance of theories can thus be explained by the fact that, as Garland Allen points out in chapter 8 of this volume, one motivation behind Cowdry's *General Cytology* was to concentrate on those areas of cell biology that were less understood and to which future research should be directed.

But where are the theories in contemporary cell biology? One possible answer is that molecular techniques have enabled biologists to directly observe and intervene on the relevant causal factors involved in producing cellular phenomena. Accordingly, there would be no room for theory, at least in the sense of speculative theory, in current cell biology. But clearly cell biologists are still engaged in hypothetical reasoning, and a more plausible interpretation is that *models* have taken over the epistemic role of theories. In fact, while the term *model* can be found not more than 11 times in Cowdry's book, it appears 348 times in Alberts's contemporary equivalent. This suggests that, instead of reaching for general and unifying theories of cellular behavior, cell biologists are now more concerned with specific explanatory models of cellular mechanisms at the molecular level. However, one conceptual difficulty immediately arises: the term *model* is used with different meaning by theoretical scientists who investigate biological processes with mathematical and computational tools and by biologists performing experiments. In the former case a model usually involves a set of equations describing a cellular process, while in the latter case it refers to the sketch of a hypothesized mechanism—for example in the form of a "cartoon" diagram. What then is the epistemic relationship between the models of the theoretical biologist and the models of the experimental biologist?

Evelyn Fox Keller (2002) has argued that the specificity of a disciplinary culture significantly shapes its ideas of knowledge, theory, and successful explanation. On the basis of her collection of historical case studies in the context of developmental biology, she concludes that theoretical and experimental approaches have had fundamentally different conceptions of what it means to "understand" a biological process. While it is possible that development represents a special case, Keller's analysis suggests that a similar gap exists more generally between theoretical and experimental approaches in biology. Therefore, experimental and computational cell biology may represent different epistemological cultures that are concerned with objects and phenomena of the same domain but pursue different epistemological goals with different means.

A glance at the history of the theoretical modeling of cellular processes seems to confirm this picture. Theoreticians initially were not necessarily drawn to problems that were considered relevant by other biologists; rather, they were attracted by "interesting" phenomena involving, for instance, oscillatory or chaotic behavior. The discovery of the Belousov-Zhabotinsky reaction in 1958, a chemical reaction that gives rise to oscillations and waves, inspired many theoreticians in the following decades to study phenomena of oscillation and pattern formation in living systems as well (Goldbeter 1997). The observation of ordered patterns emerging from the nonlinear kinetic properties of a chemical reaction provided an entry point for theoreticians into the realm of cell biology, but at the same time it suggested that these theoretical studies remained confined to the explanation of "emergent" or otherwise puzzling phenomena.

More recently, many biologists have envisioned a closer interaction of computational and experimental approaches under the label of *systems biology*, and the ideal of this interaction is often described as an iterative loop in which models are built on experimental results and then in turn used to propose new experiments (e.g., Kitano 2002). However, this iterative approach is rarely specified in detail. In what follows I look more closely at the relationship between computational and experimental approaches in cell biology. I argue that one gets a better idea of the interaction of theory and experiment in cell biology by considering the different kinds of models as complementary tools for the discovery of mechanisms. Models, whether computational or not, are used for hypothetical reasoning and thus play a role similar to the theories in Cowdry's *General Cytology*.

The chapter is structured as follows. In the second section, I propose a general way of thinking about the differences between computational and experimental approaches in cell biology. For this purpose I analyze scientific discovery in terms of heuristic strategies. These are strategies that make the search for mechanisms tractable by introducing certain assumptions about the organization and complexity of the system under study. In the third section, I describe a set of assumptions that are characteristic of research in experimental cell biology in the second half of the twentieth century. These assumptions can be derived from the way in which experimentalists represent their mechanistic models, which allow them to do without formal quantitative methods. In the fourth section, I show how computational modeling manages to overcome some of the limitations of this qualitative and informal approach. However, modelers must introduce heuristic assumptions of their own in order to generate tractable and well-constrained research problems. I discuss case studies from research about cell cycle regulation to illustrate and substantiate my claims.

Heuristic Strategies in Cell Biology

At the root of many misconceptions regarding the status of computational approaches in cell biology lies the idea that these approaches should provide the theory that is allegedly missing in experimentalists' work. But contemporary experimental cell biology clearly also has theoretical elements of its own. Starting from the way in which *theory* is understood in *General Cytology*, I focus on aspects of hypothetical reasoning in the discovery of biological mechanisms. As mentioned before, experimental cell biologists seem to prefer the term *model* to *theory*, yet unless they are referring to model organisms or other material models, they usually have something similar in mind: an abstract representation of a hypothetical biological process or mechanism whose accuracy can subsequently be put to test in experiments. (For a more general discussion of the different ways in which biologists use models, see Laubichler and Müller 2007).

The language of some philosophers of biology leads to the impression that hypothetical reasoning plays only a minor role in experimental biology. By avoiding the term *model* in the context of mechanisms altogether and instead speaking of mechanism "sketches" that are to be completed, or "schemas" that are abstracted from finished accounts, Machamer, Darden, and Craver (2000) seem to imply that an accurate representation of a mechanism emerges straightforwardly in the process of collecting experimental data and filling in black boxes. However, this clearly downplays the role of hypothetical reasoning and of the assumptions and preconceptions that enter into the construction of candidate mechanistic models. I suggest that by taking experimental cell biologists' use of *model* in the sense of a heuristic tool for the discovery of mechanisms seriously, one gets a clearer picture of the theoretical elements that are involved in their research activities and of how these activities relate to the strategies of computational biologists. To substantiate this claim, I argue that research in experimental cell biology involves many aspects that are typically considered hallmarks of theoretical approaches. Just like the computational models of theoreticians, the models of experimentalists are based on strategies of simplification, abstraction, and idealization. Interestingly, as will become clear in the next section, it is precisely those elements that justify the absence of formal analytical tools in the construction and description of their models. Their main difference from the computational biologists is thus that they work with *informal* models.

Following Wimsatt (2007), I adopt the concept of heuristics to analyze scientific strategies of approaching complex problems. I want to make plausible that experimentalists in cell biology are, explicitly or implicitly, making use of various heuristic strategies in order to make their scientific tasks manageable. In general, heuristics are rules of thumb that facilitate the discovery process by restricting or directing the search through the problem space. They rely on certain background assumptions about the complexity and organization of the system under study and thereby achieve their aim of reducing the complexity of the research problem.

One common strategy in biology in general is the heuristic of decomposition and localization (Bechtel and Richardson [1993] 2010; see also Bechtel's chapter 13 in this volume). Faced with the task of understanding the complex behavior of a system, scientists often advance by considering the behavior as produced by simpler subactivities that can be assigned to structural components of the system. In this way they arrive at a successful mechanistic explanation of the phenomenon, provided that the assumption of structural and functional modularity can be justified. Bechtel and Richardson ([1993] 2010) show in detail how this strategy was applied by biochemists in the first decades of the twentieth century to figure out the mechanism of fermentation. Subactivities were conceived as intermediate steps in the conversion of alcohol to sugar and localized by identifying them with basic chemical reactions that could be studied in isolation.

A characteristic feature of heuristics is that they are not error-free strategies, but may fail if some of the underlying assumptions are not justified. Decomposition and localization, for instance, may turn out to be misleading if the organization of a system is too complex and integrated. The heuristic of decomposition and localization is not the only strategy that is used in cell biology, and in order to analyze the differences between experimental and computational approaches, more specific heuristics have to be discussed. As I argue in the third section, these strategies have enabled experimentalists to explain a wide range of biological phenomena without having to resort to quantitative or formal methods. However, as increasing amounts of detailed and system-wide data about the molecular features underlying cellular processes become available, many of the assumptions underlying these strategies are called into question in certain contexts. In the fourth section, I argue that computational approaches introduce alternative heuristics with the potential to overcome some of the limitations of the cognitive strategies of the classical experimental approach in cell biology. Computational modeling is thus not primarily used to understand or explain complex behavior that appears puzzling, but it allows researchers to draw on more efficient tools for the discovery of mechanisms. What makes computational models useful is their ability, facilitated by styles of formal reasoning and representation, to impose additional constraints that hypothetical accounts of a mechanism must fulfill. Importantly, they can contribute to the discovery process, even if the final account of a mechanism that is produced with their help is later described in a qualitative and informal way.

Strategies of Discovery in Experimental Cell Biology

In this section I analyze the heuristic strategies that have been and are applied in experimental cell biology. This set of heuristics is to some extent representative of experimental science in general, but it also includes some that are characteristic of research in cell biology. In particular, we must acknowledge that not only the experimental techniques but also the cognitive strategies of discovering cellular mechanisms in the second half of the twentieth century (and beyond) have been significantly shaped by the advent of molecular biology, which is reflected in the importance that is assigned to genes and a general "informational" vision of biological processes (see also Reynolds, chapter 3 in this volume). I suggest that these strategies are often implicit in and can be derived from the ways in which experimentalists represent their ideas about cellular mechanisms. As mentioned before, one fundamental premise is that experimental biologists, like computational biologists, work with *models*, and an important part of my argument is to spell out the differences between the informal models used by experimentalists and the formal models that are used in computational approaches.

The models put forward by experimental cell biologists are sometimes presented verbally, but most commonly they are depicted in visual form as interaction diagrams.[1] Either way, they appear to be products of *informal reasoning*, which means that they are not based on a well-identified set of premises or assumptions and inferential rules. To put it differently, without further specification, it is not obvious how to translate such a description

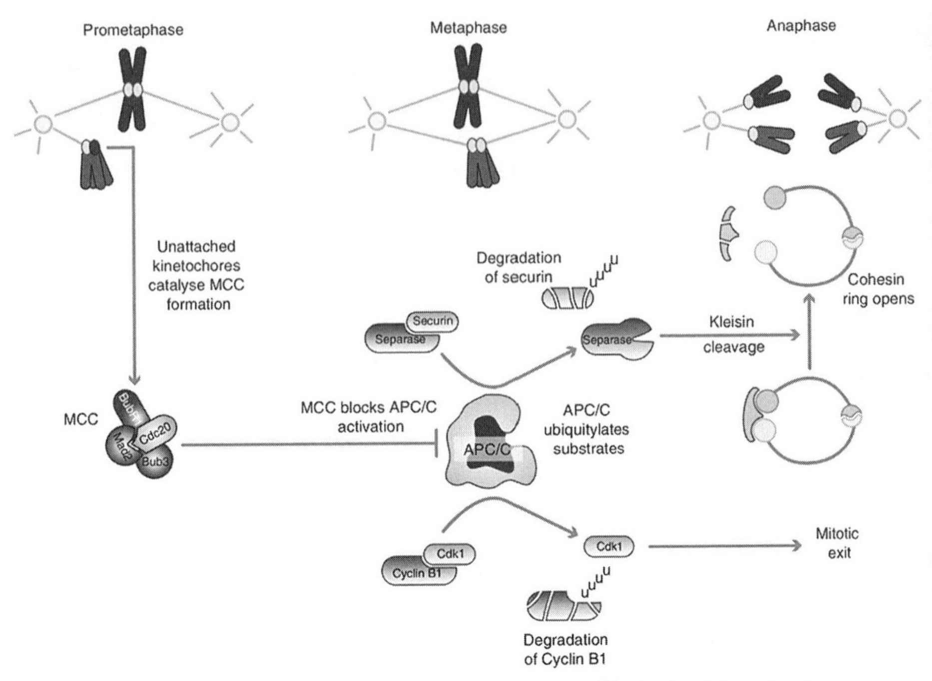

Figure 14.1. Cartoon representation of the spindle assembly checkpoint mechanism. From Lara-Gonzales, Westhorpe, and Taylor 2012. Reprinted by kind permission of Elsevier Science.

into a rigorous logical argument (Evans and Thompson 2004). Experimental cell biologists obviously do not believe that biological processes literally proceed as depicted in their cartoon representations, but the cartoons are considered sufficient to convey the relevant information that is needed to understand the workings of a mechanism. So even though scientists are aware of the simplifying nature of their cartoons and do not take them as accurate portrayals of biological processes, the particular features of those representations are still informative about the kinds of assumptions underlying their strategies of discovery.

Figure 14.1 shows a typical example of a cartoon model from contemporary experimental cell biology. It is a proposed explanation of the spindle assembly checkpoint mechanism, a process that ensures the reliable segregation of the genetic material during cell division. This mechanism works by arresting the cell cycle machinery until all chromosomes are correctly captured by the mitotic spindle. Chromosomes that are not yet attached are able to catalyze the formation of an inhibitory molecular complex (MCC) that in turn blocks the activity of another complex (APC/C) that would otherwise promote progression toward cell division.

There are several aspects of cell biologists' heuristic strategies that become immediately obvious from the way in which the model is represented. To begin with, it illustrates one fundamental assumption underlying most of discovery in cell biology: phenomena of interest are not expected to be produced by the cell *as a whole*. Instead, cell biologists usually look for a spatially confined or more manageable subsystem that underlies the phenomenon of interest. They assume, in other words, that the "locus of control" (Bechtel and Richardson [1993] 2010) of the mechanism can be assigned to a specific structure or set of structural components within the cell.

In the present case the observed phenomenon of cell cycle arrest has been localized within the nucleus and traced back to a handful of molecular complexes. Furthermore, each of these components is assigned a specific causal role that contributes to the phenomenon at the level of the whole mechanism, thus illustrating the heuristic of decomposition and localization. This strategy is well-suited for the discovery of mechanisms with a relatively small number of relevant components that work in relative autonomy from the rest of the cell. In this case it is possible to describe the mechanism without taking into account the complexity of the systemic context. It is represented as receiving an *input* from the cellular environment and as generating an *output* that can in turn serve as an input for another part of the system.

Another feature of the model descriptions of experimental cell biologists is that, while relatively rich in detail, they usually represent mechanisms as simple in terms of organization. Mechanisms are usually depicted as sequentially organized, that is, as stepwise processes. In the example shown in figure 14.1, individual steps are highlighted by arrows representing activation or inhibition: the unattached kinetochores on the chromosomes catalyze the formation of MCC; MCC in turn inhibits the activity of APC/C; when this inhibition is released, APC/C degrades securin and Cyclin B1, which leads to separation of chromosomes and mitotic exit.

Thus, the main explanatory task for the experimental biologist is not to uncover the particular organization of a mechanism, but to figure out the individual steps of the process. The underlying assumption is that important parts of cellular biology can be understood as processes of information transfer: sequential organization seems natural if the core function of a mechanism is framed as the transmission of a signal. The complexity of the research task is reduced if one thinks of a process as a linear chain

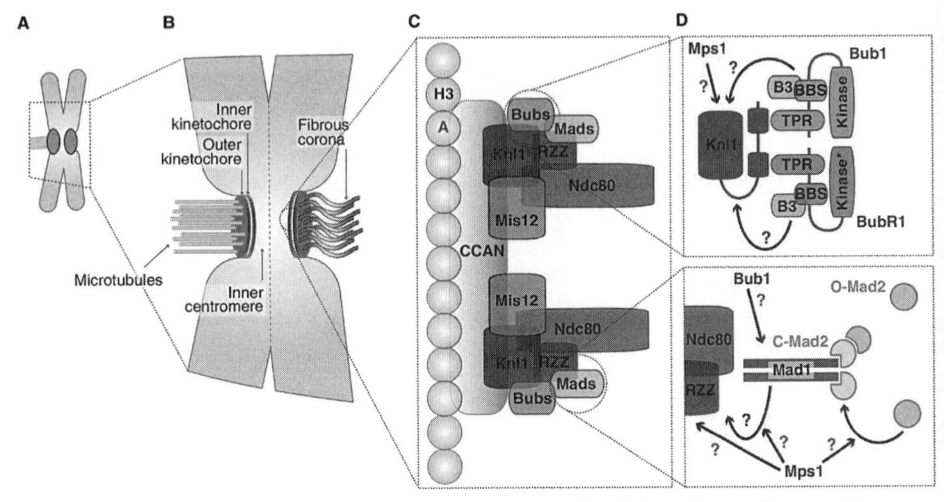

Figure 14.2. The amount of detail that has been accumulated about the spindle assembly checkpoint components is shown with increasing zoom levels. The two panels on the right (*D*) represent hypothetical models for the recruitment of checkpoint proteins. Question marks indicate interactions that have not yet been established. From Lara-Gonzales, Westhorpe, and Taylor 2012. Reprinted by kind permission of Elsevier Science.

of events, because one can zoom in once more and focus on the individual links of the chain.[2] Once the essential steps of a mechanism are figured out, progress is expected in terms of ever more detailed structural investigations, as figure 14.2 illustrates for the case of the spindle assembly checkpoint. The explanation of suboperations requires biologists to go to lower scales and, in particular, to investigate the inner structure of macromolecular complexes. This usually entails a division of labor: different research groups investigate different links of the chain.

The assumption that mechanistic explanations can be given in terms of small subsystems with relatively simple organization can justify an experimental approach that is productive without techniques of quantitative and formal reasoning. Two additional, and more specific, heuristic strategies support the idea that qualitative descriptions suffice for the purpose of explaining cellular phenomena. Both of them rely, even though in different ways, on the idea of molecular mechanisms as processes of information transfer.

The first strategy is based on a conceptual detachment of the organization of a mechanism from certain aspects of the underlying biochemical processes. Experimental cell biologists want to explain how one step in the informational chain leads to the next, but the kinetic features of the biochemical reactions mediating these steps are not considered relevant for the explanation of a mechanism. While biochemistry plays an important role in describing the structure of macromolecular complexes and individual reactions, these investigations are used exclusively to fill in the black boxes corresponding to the single steps in the signaling cascade. The cascade itself is represented in purely qualitative terms and does not rely on any detailed kinetic information about the occurring biochemical reactions.

It seems that with the advent of molecular biology, the complex metabolic reaction schemes that biochemists had studied in the early twentieth century, such as the system responsible for fermentation mentioned earlier, did not seem to provide the right exemplars to illuminate the information-transmitting mechanisms of molecular biology. Instead, the role of biochemistry was largely reduced to the study of the specific reactions occurring in individual steps within such processes. It had no bearing on the general route of the signal and its significance for the rest of the system. Jacques Monod captured this independence of the informational pathways from the chemical nature of the underlying signals with his concept of *gratuité* (gratuity): "Physiologically useful or 'rational,' this relation is chemically arbitrary—'gratuitous,' one might say" (Monod 1971, 77). Monod's view relies on the assumption that evolutionary processes, even though using chemical "bricks," have the freedom to "engineer" physiological systems in a largely unconstrained way: "The very gratuitousness of these systems, giving molecular evolution a practically limitless field for exploration and experiment, enabled it to elaborate the huge network of cybernetic interconnections which makes each organism an autonomous functional unit, whose performances appear to transcend the laws of chemistry if not to ignore them altogether" (Monod 1971, 78). The assumption of gratuity is thus what actually enables biologists to heuristically investigate the individual links in a sequence independently from one another: there is no dependency of the single steps in the process on the overall organization of the system.

The second strategy of the qualitative approach of experimental cell biology is based on the assumption that processes that actually involve ensembles of hundreds or thousands of molecules can be represented as interactions of individual representatives. In other words, "population effects" are largely disregarded in the cell biologists' informal models. As exemplified by the cartoon shown in figure 14.1, biologists describe processes in terms of what happens to individual molecules, even though they are aware that the actual causally efficient factors are typically large collections of identical or similar molecules. This habit relies on the tacit assumption that there is a simple relationship between the activity of the individual molecule and the activity of the population. For instance, if a molecule of type *A* inhibits the activity of another molecule of type *B* by binding to it, the expectation is that the activity of a population of *A*s inhibits the activity of a population of *B*s. However, the effect of one population on another cannot in general simply be equated with the effect of an individual member, as shown by examples of elementary population dynamics, such as the Lotka-Volterra model (Volterra 1926). This model describes the dynamics of two interacting ecological species, one a predator and one its prey. At the level of individual members, predation implies one organism eliminating another, but the interactions at the population level can be far more complex. The prey population is not simply eliminated but depleted at a certain rate, depending on the size of the predator population. Moreover, the model can exhibit complex behavior, such as oscillations, that can only be explained when quantitatively describing the process at the population level.

In general, quantitative aspects, concentrations, kinetic parameters, and so forth are of crucial importance in many applications of biochemistry as well. Take as an example the well-known Michaelis-Menten model of enzyme kinetics. It describes the process in which an enzyme converts a substrate by forming an intermediate complex. Even though a qualitative account of how one single molecule of substrate binds to one molecule of enzyme, and how the former is subsequently converted, may partly illuminate the process, it completely neglects the kinetic aspects of the reaction at the population level. In order to explain, for example, how the presence of the substrate affects the amount of product, one has to apply subtle mathematical methods, and in order to make predictions, one needs precise quantitative measurements of the required kinetic parameters (Gunawardena 2012).

The explanatory schemes of experimental cell biology, by contrast, typically do without any quantitative features. It is assumed, for example, that to understand the relevant aspects of the spindle assembly checkpoint, one does not have to know how many molecules are turned over or what their initial concentrations are, and that one can explain the process by restricting the description to the level of individual molecules. A fundamental assumption underlying the mechanistic schemes of molecular biology, therefore, is that *the individual molecule is sufficient to represent the population*.

My analysis shows that experimentalists' mechanistic models are much closer to computational models than is usually assumed. They are obviously abstractions, since they omit a lot of molecular detail. But at the same time, as we have seen, they can be understood as idealizations in the sense of *distorted* representations of reality. The terminology of mechanism *sketches* or *schemata* (Machamer, Darden, and Craver 2000) highlights only the aspect of abstraction or incompleteness but not the many ways in which reasoning with accounts of mechanisms is based on idealizing assumptions. The difference between computational and experimental cell biologists is not that one works with models and the other performs experiments. The difference is rather that one works with *formal* models, while the other works with *informal* models.

In summary, cell biologists work with representations of molecular processes that are tractable and powerful as heuristic tools for discovery. The assumptions that make this approach so productive go beyond the framework of decomposition and localization and involve specific heuristic strategies that justify a qualitative and informal approach to cell biology. The next section shows different ways in which computational models can contribute to the discovery of mechanisms in contexts where such an approach reaches its limits.

Strategies of Discovery in Computational Cell Biology

Computational biology is not one homogeneous endeavor; rather, it is a large collection of different approaches that have their historical roots in various traditions of theoretical biology or other theoretical fields studying complex systems. What this section shows is that one of the main roles of mathematical models in computational cell biology is to facilitate the discovery of mechanisms. In spite of increasing amounts of molecular data, most areas in cell biology still lack knowledge about the underlying causal structures. Computational models can be used as heuristic tools to restrict the set of candidate mechanisms (i.e., experimentalists' models) that are proposed for the explanation of a particular phenomenon, as I explain in this section. The following quotation, which is taken from an article about the modeling of complex signaling networks, emphasizes this point:

> We believe that modeling these important biological systems cannot wait until all the rates are reliably measured, or even until all the various players and interactions are discovered. Indeed, the most important role of modeling is to identify missing pieces of the puzzle. It is as useful to falsify models—identifying which features of the observed behavior cannot be explained by the experimentalists' current interaction network—as it is to successfully reproduce known results. (Brown et al. 2004, 185)

In particular, computational biology can contribute to discovery in cell biology by overcoming some of the limitations and biases of the approach of experimental cell biology described in the previous section. In what follows, I give examples of ways in which modeling can relax some of the assumptions of the classical approach but at the same time introduce additional constraints. These constraints reduce the size of the problem space and thereby simplify the task of identifying the causal structure underlying the phenomenon of interest. It is important to emphasize that there are many different types of computational modeling, and that no single type allows modelers to overcome *all* limitations. On the contrary, as we will see, each approach has to make additional simplifying assumptions in order to arrive at well-constrained modeling problems. Modelers have to find the right grain of resolution in order to make a meaningful contribution. So, effectively, some of the more specific heuristic strategies of the experimental approach are replaced with alternative heuristics.

I start by presenting a quantitative perspective on the spindle assembly checkpoint mechanism that was discussed in the previous section. A recent review article explains the particular interest in the spindle assembly checkpoint as a target of computational modeling:

> The high fidelity and robustness of this process have made it a subject of intense study in both the experimental and computational realms. A significant number of checkpoint proteins have been identified, but how they orchestrate the communication between local spindle attachment and global cytoplasmic signalling to delay segregation is not yet understood. Here, we propose a systems view of the spindle assembly checkpoint to focus attention on the key regulators of the dynamics of this pathway. These regulators in turn have been the subject of detailed cellular measurements and computational modelling to connect molecular function to the dynamics of spindle assembly checkpoint signalling. (Ciliberto and Shah 2009, 2162)

Thus, in spite of the amount of accumulated molecular detail, the authors think that the mechanism is not yet sufficiently understood. Unlike most experimental biologists, however, they do not see the main problem as missing molecular data, but as a missing link between "molecular function" and "the dynamics of spindle assembly checkpoint signalling." They go on to clarify what motivates the role of modeling in this context:

> Given its role, it is not surprising, but yet striking, that the spindle assembly checkpoint can delay anaphase in response to a single uncaptured chromosome, exhibiting excellent sensitivity. Once this last chromosome attaches, the spindle assembly checkpoint disengages and rapidly promotes anaphase onset. High fidelity and speed are usually competing design constraints in manmade machines, and as such the underlying logic and quantitative mechanisms of the spindle assembly checkpoint are of interest to life scientists and physical scientists alike. (Ciliberto and Shah 2009, 1262)

Therefore, the checkpoint mechanism is interesting for quantitative modeling because it represents a solution to a "design problem" that would provide a challenge for human engineers. On the one hand, it has to work reliably, because the fidelity of chromosome segregation is of crucial importance for the cell (it must be extremely sensitive to the signal produced by a single unattached kinetochore). On the other hand, the inhibition must be released very quickly, because it has been observed that anaphase onset occurs in a matter of minutes after the last chromosome attaches (e.g., Rieder et al., 1995; Howell et al. 2000). The existence of competing constraints and the possible ways of how the biological system may solve this design problem may be seen as clues to the underlying mechanism. But taking these constraints into account requires a quantitative and dynamic perspective on the system. In order to understand whether a proposed mechanism can produce reliable inhibition, even when the signal emanates from only one chromosome, one has to consider both the rate of the putative reaction that produces the inhibitory signal and the diffusion rate of the signal through the cytosol. Similarly, to understand whether the checkpoint can be relieved fast enough, one has to take into account the rate of disassembly of the inhibitory complex as well as the time it takes for the APC/C to carry out its activating function. These specific quantitative features are not taken to be relevant in the experimentalists' account discussed in the previous section.

Ciliberto and Shah use the analogy of a washbasin to illustrate the role of

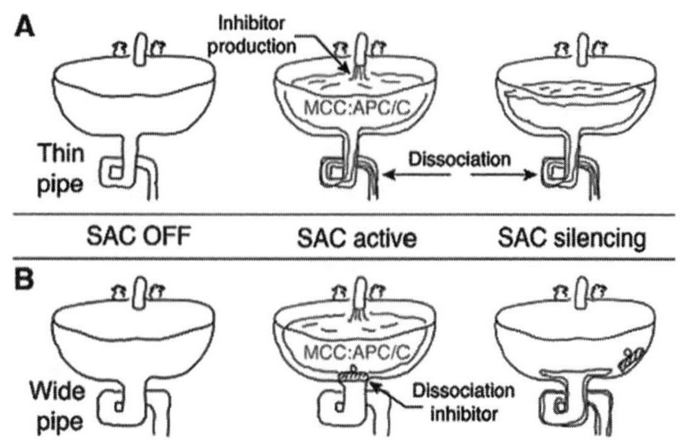

Figure 14.3. Two scenarios for the dynamic regulation of the spindle assembly checkpoint. For explanation, see text. From Ciliberto and Shah 2009. Reprinted by kind permission of John Wiley and Sons.

quantitative reasoning (fig. 14.3). The production of the inhibitor is represented by a faucet filling up the sink, while its dissociation corresponds to the outflow through the drain pipe. In the scenario represented in figure 14.3*A*, the inhibitor is constantly flowing out—that is, dissociated. This dissociation must be slow enough to allow for reliable inhibition while the checkpoint is active. In other words, a thin pipe is needed to guarantee that the outflow does not exceed the inflow. As a result, it takes a long time to drain the sink: the silencing of the checkpoint is slow. Figure 14.3*B* proposes a possible solution to the design problem. This time, the dissociation rate is high, corresponding to a wide pipe, but the checkpoint involves an additional "dissociation inhibitor" that plugs the pipe. As soon as the last kinetochore attaches to the spindle, the faucet is closed and the plug is removed. In this way, the silencing of the checkpoint can be fast. Thus, with a slightly different checkpoint schema, both constraints can be fulfilled.

The analogy shows the possible contribution of a dynamic perspective to the mechanism. At the same time, it illustrates how some of the experimentalists' heuristic assumptions are dropped. The activities of the mechanism are not represented as actions of individual molecules, but in terms of changing quantities, which correspond to the concentrations, or copy numbers, of different molecular species. It also becomes clear that a dynamic vision must pay more attention to the organization of the mechanism: different steps in the process cannot be modeled independently since their dynamic features depend on each other. This kind of interdependence can be taken into account by reintroducing the kinetic aspects of biochemistry. While figure 14.3*A* illustrates the dependency of the processes of inhibition and release, the slightly more complicated model in *B* illustrates how this dependency can be relieved by introducing an additional component into the schema. In order to deal with the dynamic nature of the mechanism, Ciliberto and Shah propose an approach that interprets its basic activities in terms of signaling *modules*:

> These activities, inhibition on the one hand and release of that inhibition on the other, must support the widespread observation of a single unattached kinetochore delaying the onset of anaphase. Moreover, the coupling of these activities and their relative dominance must be controlled entirely through kinetochore attachment to permit the rapid transition to anaphase on kinetochore attachment. Each of these activities: inhibitor generation, release from inhibition, and kinetochore attachment are themselves complex signalling pathways involving a myriad of molecular components. A systems view of spindle assembly checkpoint signalling focuses our attention onto the communication between signalling modules that are likely to govern the quantitative dynamics of this pathway. (Ciliberto and Shah 2009, 2163)

As we have seen in the previous section, the conceptualization of a mechanism in terms of functional modules is implicit also in the traditional approach of cell biology. There I discussed how experimentalists conceptually decompose a mechanism into separate activities, which allows a reduction of epistemic complexity, since each step in the process can be addressed independently. This strategy requires, however, that the interaction between the modules be straightforward. When investigating each activity as an independent step of a sequential process, one can ignore the ways in which the properties of different modules might dynamically depend on each other. The idea of the quantitative approach is to focus instead on the communication between the modules. Complexity is reduced in this approach as well, but this time by *black-boxing molecular detail within each module*. As the authors of the review explain, they "modularize the complexity of the components into the key communicating elements" (Ciliberto and Shah 2009, 2162).

The motivation for using a coarse-grained perspective in terms of modules is not necessarily based on the belief that these modules represent the "real" components of the mechanism, or that the project of figuring out the details of the underlying molecular structure is misguided. Mainly, the modular strategy serves to make the task well-constrained as a modeling problem and is therefore heuristic in nature. Even though the research on the spindle assembly checkpoint "has amassed a substantial amount of quantitative data" (Ciliberto and Shah 2009, 2166), this does not automatically enable scientists to build useful quantitative models at the molecular level. The reason for this is connected to what systems biologist Jeremy Gunawardena calls "the parameter problem" (Gunawardena 2010). The essence of this problem is captured by the famous expression attributed to John von Neumann: "With four parameters I can fit an elephant, and with five I can make him wiggle his trunk" (quoted in Dyson 2004, 297). In an ideal world every parameter of a model would be determined by independent and accurate measurement, but in biological practice most properties of interest cannot be directly measured. Virtually every quantitative model in biology involves a number of unknown parameters and unwarranted idealizations. If a model has many free parameters, the fact that the model accounts for the data might largely be due to mathematical reasons, and not to the model's correspondence to the target system.

There are two strategies to cope with the parameter problem, corresponding to what Gunawardena calls "thin" and "thick" models (Gunawardena 2010, 26). Thin models include only what are assumed to be the essential causal features of the system. These models are typically tested against a small set of mostly qualitative observations and generic physical constraints. Thick modeling, by contrast, is acceptable when enough empirical data are available. In this case one tries to bring the assumptions of the model as close to reality as possible by explicitly including all known components and processes. However, in spite of the available information, biologists must accept a large number of unknown parameters in thick models. This is because it is rarely possible to determine the parameters occurring in the model, such as rates of synthesis or degradation of proteins, directly. Instead, unknown parameters must be inferred indirectly by optimizing the fit of the model simulation to the data. Often scientists use only part of the available data for parameter estimation, and afterward try to reproduce or predict other data. Deviations between predicted and observed behavior can then be exploited to modify the structure of the model and learn about the underlying mechanism.

An example of thin modeling is provided by the study of Doncic, Ben-Jacob, and Barkai (2005). It consists of a comparison of three different models of the spindle assembly checkpoint mechanism in budding yeast. These models are evaluated with respect to reliability of inhibition and time of release, the two properties that were identified above as competing design constraints. Eventually, the authors find that only one of the three proposed models passes the test of properly fulfilling the requirements. All three model variants are loosely based on molecular knowledge, but there is no strict identification of model components with specific proteins, and the main focus is on the role of physical constraints. The cell nucleus is modeled as a sphere with one single kinetochore located in the center as a subsphere with significantly smaller radius. The molecular processes are characterized by a set of reaction-diffusion equations that describe both spatial and temporal changes of the molecular concentrations as well as the chemical interactions. Geometrical scale, reaction rates, and diffusion constants are chosen in agreement with known general properties of cellular systems. As the authors explain, "We did not simulate the full complexity of the network underlying the checkpoint but, rather, compared classes of mechanisms. Each class may be realized by a range of molecular machineries, but its essence can be summarized by a simple model, composed of just a few components" (Doncic, Ben-Jacob, and Barkai 2005, 6336). By abstracting from underlying molecular detail, they are able to cover a large set of possible causal structures with a small number of models. Each model is simple enough that fairly general claims may be derived about its behavior under varying parameter values.

Importantly, the goal of this work is not to *explain* the behavior of the spindle assembly checkpoint mechanism, even though, as a by-product, it might contribute to elucidating how certain causal structures bring about certain behaviors. Instead, the authors' main interest is to compare different models with respect to their ability to fulfill the design constraints. Even though one of the proposed models meets these constraints, it is not proclaimed as the actual mechanism of the checkpoint. The more important result is negative: Certain types of causal structures are *not* able to account for the observed behavior, which raises the bar for the evaluation of proposed molecular mechanisms.

The authors cannot guarantee that their selection of models exhausts all possible checkpoint mechanisms, and whether the actual checkpoint mechanism is in line with the successful candidate can be established only on the basis of additional molecular knowledge. However, the authors can motivate the exclusion of some proposed mechanisms by taking into account constraints that do not appear in the mechanistic models of experimental biologists. First, they quantify observed behavior: it is not enough that the checkpoint is released after attachment, but it must be released *within a certain time*. Information about upper limits of the rates of chemical reactions are important to evaluate whether a proposed mechanism can fulfill this time constraint. Similarly, it is not sufficient to show that one type of molecule is able to inhibit another, but inhibition must be strong enough in terms of the fraction of inhibited molecules. Inhibition has different meanings depending on whether one talks about populations or about individual molecules. The interaction between an inhibiting and an inhibited species is a chemical reaction that produces a dynamical equilibrium in which there always remain a number of uninhibited molecules. The activity of a single molecule is inhibited if it is bound to its inhibitor, whereas the activity of the population is inhibited if the number of uninhibited molecules is below a certain threshold. The chemical perspective, therefore, implies reasoning in terms of *populations* of molecules. A further consequence of quantitatively accounting for the chemical reactions is that the strength of inhibition is connected to the timing for the release from inhibition. This is exactly what was illustrated in Ciliberto and Shah's washbasin model (fig. 14.3). Aside from this, the models of Doncic and colleagues take into account spatial properties of the system: inhibition must be strong everywhere in the nucleus, and not only near the kinetochore. Limits of possible diffusion rates of proteins, therefore, set further important constraints on the possible signaling mechanism.

Note, however, that the strategy of thin modeling involves important trade-offs. It seems that in order to serve as powerful heuristic tools, the proposed models must be of rather low complexity. In the work of Doncic and his colleagues, we can find many steps of idealization. First of all, they lump a whole network of interactions into a minimal number of effective reactions. This is not an assumption they want to test, but it is a requirement of their strategy. Moreover, they make simplifying assumptions, such as the idealized spherical geometry of the system or the conservation of the numbers of all interacting particles. Especially this latter assumption is problematic, since it has been shown that some of the components of the checkpoint mechanism are actively degraded during and after the mitotic arrest. Obviously, their models can accurately represent only mechanisms that approximately fulfill these underlying assumptions. To the extent that the assumptions are unrealistic, the overall strategy cannot amount to a strict criterion to exclude candidate mechanisms.

A case of thick modeling can be found in the detailed analysis of cell cycle regulation presented in Chen et al. (2004). Based on the wiring diagram

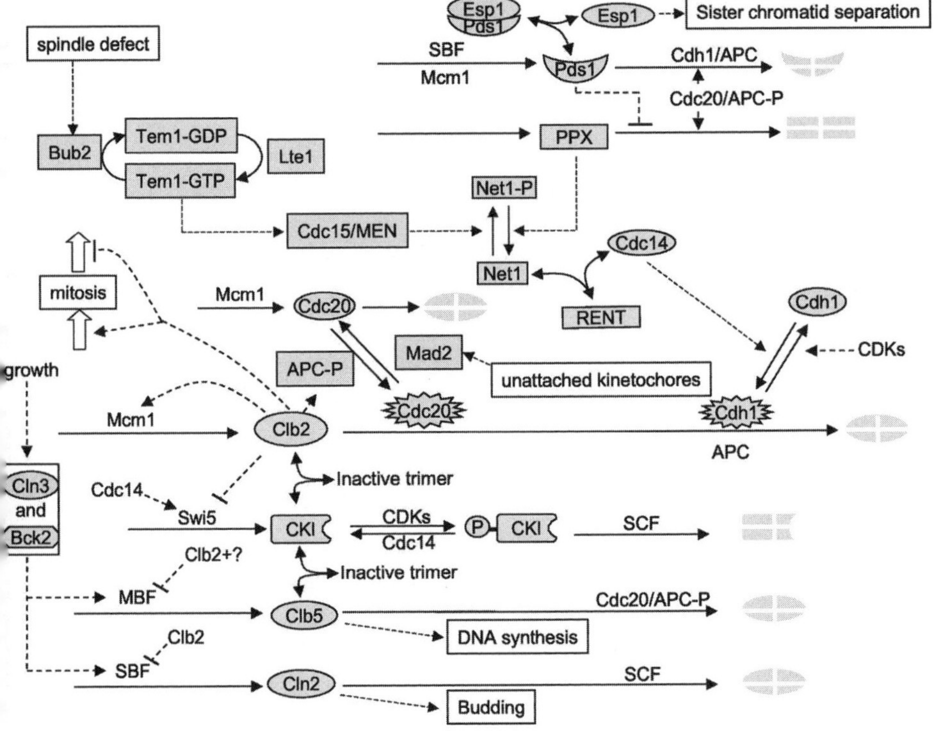

Figure 14.4. Wiring diagram of the detailed model of cell cycle regulation. From Chen et al. 2004. Reprinted by kind permission of the American Society for Cell Biology.

shown in figure 14.4, this model is claimed to present a "realistic" picture of the cell cycle machinery in budding yeast. The model consists of fifteen ordinary differential equations and twelve algebraic equations that together depend on more than one hundred free parameters. The equations summarize experimental results collected from various publications, but they also incorporate specific hypotheses that are based on a quantitative and dynamic perspective of cell cycle regulation: The model is designed to conform to the idea of two stable steady-states generated by the antagonism of two groups of factors: Clb-kinases and G1-stabilizers. During a normal cycle, the cell periodically traverses these two states, driven by growth and division. Checkpoints, such as the spindle assembly checkpoint, act as signals that delay these transitions by stabilizing the different stages of the cell cycle.

After the model was set up, the equations were solved numerically and compared to experimental observations. According to the authors, the model manages to accurately describe division and growth in wild-type yeast cells. Moreover, it reportedly agrees with observations for 120 of 131 tested mutant strains. The large set of empirical information that is used to test the model appears to justify the approach of building a large and detailed model of cell cycle regulation.

Note, however, that, in spite of the level of incorporated detail, thick models typically also rely on many simplifying assumptions. As Chen et al. admit, "There is no unique correspondence between a wiring diagram and a set of mathematical equations" (Chen et al. 2004, 3845). Thus even though many of the equations are directly based on fundamental principles of biochemical kinetics, certain pragmatic choices and assumptions are required to make the modeling problem well-constrained and tractable. For example, proteins with redundant functions, such as the cyclins Cln1 and Cln2, are represented by only one variable. Moreover, the use of ordinary differential equations implies that noise effects due to fluctuations in the copy numbers of molecules and their spatial localization are neglected. Yet, unlike the informal way in which a diagram like figure 14.4 is usually interpreted, setting up a quantitative model forces researchers to make all of their assumptions explicit.

Most of the model's parameters could not directly be quantified by experimental measurements. Values for these parameters were not determined by a systematic search; they were "selected by a painstaking process of trial-and-error to provide a suitable fit to the full data set" and "represent a compromise of many, often competing, observations" (Chen et al. 2004, 3850). Some of the rate constants could be estimated from measured protein concentrations in cell culture, but in general modeling relied on so-called auxiliary variables to connect the temporal evolution of concentrations to measurable quantitative data (e.g., the timing of bud emergence, the onset of DNA synthesis, or cell separation).

Given that most of the data used to evaluate the model already entered into its construction, it is difficult to maintain that the model is confirmed or validated in a strong sense by empirical evidence. Instead, the authors argue that its main role consists in checking the "sufficiency and consistency of the mechanism" (Chen et al. 2004, 3859). The model clearly retains a hypothetical character and is used as a heuristic reasoning tool that enables biologists to uncover gaps in their intuitive way of understanding the mechanism of cell cycle regulation. Note, however, that modeling is required not because the mechanism is too complex to be understood intuitively at all, but because often our intuitive reasoning cannot be trusted. In fact, the original caption of figure 14.4 in Chen et al.'s article contains a detailed verbal and qualitative description of the processes displayed in the figure, showing that it is entirely possible to get a qualitative understanding of the mechanism by reasoning along the arrows of the diagram. However, due to the interconnectedness of the system, such a description is not reliable. In the author's own words, our intuition has to be "disciplined by precise numerical simulation" (Chen et al. 2004, 3851). Whenever deviations between model predictions and empirical observations are discovered, one must either check the model's assumptions or question the completeness and adequacy of the hypothesized mechanism.

Just as in the example of thin modeling, an important contribution of the model is to raise the standard that a candidate explanation must meet. The model is not a finished product used for explanatory purposes but a tool that contributes to the search for the actual mechanism. This picture is confirmed by the authors in another article that presents an earlier version of the model:

> One can learn as much from the failures of the model as from its successes. Where there are inconsistencies between the model and experiment, we are prompted, first of all, to look for a better parameter set. If that fails, we consider slight changes in the mechanism, which might bring the model in accord with observations. If that fails, and if the experimental community is convinced that the observations are reliable and significant, then we have identified an area that deserves closer scrutiny to resolve the discrepancies. If the mechanism proves insufficient, that does not invalidate our approach. Mathematical modeling, as a tool, is no more "falsifiable" than gel electrophoresis. The tool tells us what a mechanism can and cannot explain. When the model fails, the fault lies with the mechanism, not the tool. (Chen et al. 2000, 385)

Conclusion

In this chapter I have discussed the relationship between experimental and computational approaches in contemporary cell biology. Computational biologists and experimentalists often share the same epistemic goals and use models to represent their hypotheses about cellular mechanisms. However, the models are different in kind and indicate different underlying heuristics. The main difference from computational approaches is that experimentalists usually work with *informal* models.

The assumptions underlying the heuristic strategies of experimentalists justify an approach that can do without sophisticated quantitative techniques and restricts itself to informal and qualitative reasoning. This approach has been successful, but it reaches its limits whenever these assumptions are not warranted.

Computational biologists, by contrast, use formal and quantitative models to overcome some of the limitations of the experimentalists' approach. The examples presented in the previous section exemplify that computational modeling is often used as a strategy for discovery. In both cases the starting point is a candidate mechanism that is consistent with established molecular knowledge and represents an intuitive explanation of observed behavior. Computational modeling is used to decide between possible causal structures and to reveal gaps in our current understanding. Computational biologists thus directly build on the findings of cell biologists, and they are interested in the solution of the same epistemic puzzle, but they propose a different strategy for solving it. Describing the phenomenon and the hypothesized causal structure quantitatively allows them to detect discrepancies between proposed mechanisms and experimental observations. Moreover, the introduction of physical and biochemical constraints can lead to the exclusion of mechanistic models, even if they are considered plausible candidates by traditional molecular biologists.

More generally, I have tried to present models as tools of hypothetical reasoning in contemporary cell biology. Experimentalists formulate their hypotheses about specific mechanisms in terms of qualitative and informal models. Computational biologists use the tools of formal and quantitative modeling to facilitate the process of testing and revising these models. The models in cell biology can therefore be thought of as the updated equivalents of the speculative theories in Cowdry's *General Cytology*. The prevalent use of the term *model* suggests that the current focus of cell biologists is less on general and unifying accounts of cellular behavior and more on specific mechanisms at the molecular level. While there are also contemporary approaches that aim at more general or unifying accounts by identifying "organizational principles" or "design principles" found in many different systems (see Bechtel, chapter 13 in this volume), my aim in this chapter was to highlight the productive interaction of experimental and computational approaches in the discovery of mechanisms.

Notes

1 Here I have in mind the typical "cartoon" diagrams used by experimentalists, and not, for instance, the network diagrams that are based on graph-theoretical analysis and can be seen as products of formal reasoning (see Bechtel, chapter 13 in this volume).

2 *Linear* in this context means that there are no branchings or loops within the chain. This kind of linearity can be applied both to spatial and to temporal chains. In static diagrams temporally linear processes are often represented as spatially linear chains.

References

Alberts, Bruce, Alexander Johnson, Julian Lewis, Martin Raff, Keith Roberts, and Peter Walter. 2008. *Molecular Biology of the Cell.* 5th ed. New York: Garland Science.

Bechtel, William, and R. C. Richardson. (1993) 2010. *Discovering Complexity: Decomposition and Localization as Strategies in Scientific Research.* Cambridge, MA: MIT Press. Originally published in 1993. Princeton, NJ: Princeton University Press.

Brown, K. S., C. C. Hill, G. A. Calero, C. R. Myers, K. H. Lee, J. P. Sethna, and R. A. Cerione. 2004. "The Statistical Mechanics of Complex Signaling Networks: Nerve Growth Factor Signaling." *Physical Biology* 1:184–95.

Chen, K. C., L. Calzone, A. Csikasz-Nagy, F. R. Cross, B. Novak, and J. J. Tyson. 2004. "Integrative Analysis of Cell Cycle Control in Budding Yeast." *Molecular Biology of the Cell* 15: 3841–62.

———, A. Csikasz-Nagy, B. Gyorffy, J. Val, B. Novak, and J. J. Tyson. 2000. "Kinetic Analysis of a Molecular Model of the Budding Yeast Cell Cycle." *Molecular Biology of the Cell* 11:369–91.

Ciliberto, Andrea, and Jagesh V. Shah. 2009. "A Quantitative Systems View of the Spindle Assembly Checkpoint." *EMBO Journal* 28:2162–73.

Cowdry, Edmund V., ed. 1924. *General Cytology: A Textbook of Cellular Structure and Function for Students of Biology and Medicine.* Chicago: University of Chicago Press.

Doncic, Andreas, Eshel Ben-Jacob, and Naama Barkai. 2005. "Evaluating Putative Mechanisms of the Mitotic Spindle Checkpoint." *Proceedings of the National Academy of Sciences* 102: 6332–37.

Dyson, Freeman. 2004. "A Meeting with Enrico Fermi: How One Intuitive Physicist Rescued a Team from Fruitless Research." *Nature* 427:297.

Evans, Jonathan St. B. T., and Valerie A. Thompson. 2004. "Informal Reasoning: Theory and Method." *Canadian Journal of Experimental Psychology* 58(2): 69–74.

Goldbeter, Albert. 1997. *Biochemical Oscillations and Cellular Rhythms: The Molecular Bases of Periodic and Chaotic Behaviour.* Cambridge: Cambridge University Press.

Gunawardena, Jeremy. 2010. "Models in Systems Biology: The Parameter Problem and the Meanings of Robustness." In *Elements of Computational Systems Biology*, edited by Huma M. Lodhi and Stephen H. Muggleton, 21–47. Hoboken, NJ: John Wiley & Sons.

———. 2012. "Some Lessons about Models from Michaelis and Menten." *Molecular Biology of the Cell* 23: 517–19.

Howell, B. J., D. B. Hoffman, G. Fang, A. W. Murray, and E. D. Salmon. 2000. "Visualization of Mad2 dynamics at Kinetochores, along Spindle Fibers, and at Spindle Poles in Living Cells." *Journal of Cell Biology* 150: 1233–50.

Jacobs, Merkel H. 1924. "Permeability of the Cell to Diffusing Substances." In Cowdry 1924, 97–164.

Keller, Evelyn F. 2002. *Making Sense of Life: Explaining Biological Development with Models Metaphors and Machines*. Cambridge, MA: Harvard University Press.
Kitano, Hiroaki. 2002. "Systems Biology: A Brief Overview." *Science* 295:1662–64.
Lara-Gonzalez, Pablo, Frederick G. Westhorpe, and Stephen S. Taylor. 2012. "The Spindle Assembly Checkpoint." *Current Biology* 22: R966–R980.
Laubichler, Manfred D., and Gerd B. Müller, eds. 2007. *Modeling Biology: Structures, Behaviors, Evolution*. Cambridge, MA: MIT Press.
Machamer, Peter, Lindley Darden, and Carl F. Craver. 2000. "Thinking about Mechanisms." *Philosophy of Science* 67:1–25.
Monod, Jacques. 1971. *Chance and Necessity: An Essay on the Natural Philosophy of Modern Biology*. New York: Alfred A. Knopf.
Rieder, Conly L., Richard W. Cole, Alexey Khodjakov, and Greenfield Sluder. 1995. "The Checkpoint Delaying Anaphase in Response to Chromosome Monoorientation Is Mediated by an Inhibitory Signal Produced by Unattached Kinetochores." *Journal of Cell Biology* 130: 941–48.
Volterra, Vito. 1926. "Fluctuations in the Abundance of a Species Considered Mathematically." *Nature* 118: 558–60.
Wimsatt, William C. 2007. *Re-Engineering Philosophy for Limited Beings: Piecewise Approximations to Reality*. Cambridge, MA: Harvard University Press.

ACKNOWLEDGMENTS

Visions of Cell Biology is the outcome of two workshops called "Updating Cowdry at the MBL" held at the Marine Biological Laboratory (MBL) in Woods Hole. The first workshop was supported by funds from the University of Chicago to promote collaborative research at the MBL. The second was supported by a very timely grant from the Edwin S. Webster Foundation, facilitated by Ms. Suzanne H. Sears, for which we are most grateful. Related research has been supported by several grants from the National Science Foundation for the MBL History Project, with considerable financial resources also provided by Arizona State University.

Our thanks go out to all biologists, historians, philosophers, and students who participated in the workshops. In particular, Bill Wimsatt, Everett Mendelsohn, and Christie Henry from the University of Chicago Press provided considerable advice and encouragement along the way. We are also grateful to Ruth Crawford at the University of Chicago and Barbara Burbank at the MBL for administrative support, to Michelle Sullivan at Arizona State for editorial help, and to the entire MBL team for helping make the workshops run flawlessly.

Former directors and presidents Gary Borisy and Huntington Willard have strongly encouraged exploring history in connection with current scientific research. This volume is the first in a series of such explorations at the intersections of science, history, and philosophy.

CONTRIBUTORS

Garland E. Allen, PhD
Professor of Biology, Emeritus
Washington University in St. Louis
Campus Box 1137
One Brookings Drive
St. Louis, MO 63130
gallen@wustl.edu
314-935-6808

William Bechtel, PhD
Distinguished Professor of Philosophy
Department of Philosophy and Center for Circadian Biology
University of California, San Diego
9500 Gilman Drive
La Jolla, CA 92093-0119
bechtel@ucsd.edu
858-822-4461

Fridolin Gross, PhD
University of Kassel
Nora-Platiel-Straße 1
34127 Kassel
Germany
fridolin.gross@uni-kassel.de
+49 561 804 3614

Lijing Jiang, PhD
Haas Postdoctoral Fellow
Beckman Center for the History of Chemistry
Chemical Heritage Foundation
315 Chestnut Street
Philadelphia, PA 19106
LJiang@chemheritage.org
215-629-5189

Manfred D. Laubichler, PhD
President's Professor, Arizona State University
Professor, Santa Fe Institute
School of Life Sciences
Arizona State University
PO Box 4501
Tempe, AZ 85287-4501
manfred.laubichler@asu.edu
480-965-5481 / 480-965-6214

Daniel Liu, PhD
Andrew W. Mellon Post-Doctoral Fellow in the Biohumanities
Illinois Program for Research in the Humanities
University of Illinois at Urbana-Champaign
Levis Faculty Center, Suite 400
919 West Illinois Street
Urbana, IL 61801
liud@illinois.edu
714-723-1377

Kate MacCord, PhD
McDonnell Fellow
Marine Biological Laboratory
7 MBL Street, Woods Hole, MA 02543
kmaccord@asu.edu
508-289-7513

Jane Maienschein, PhD
University Professor, Regents' Professor, and President's Professor Fellow, Marine Biological Laboratory
School of Life Sciences
Arizona State University
427 East Tyler Mall
Tempe AZ 85287-4501
maienschein@asu.edu
480-965-6105

Karl S. Matlin, PhD
Professor of Surgery
Member, Committee on Conceptual and Historical Studies of Science
The University of Chicago
5841 S. Maryland Ave., MC 5032
Chicago, IL 60637-1470
kmatlin@uchicago.edu
773-834-2242

Rudolf Oldenbourg, PhD
Senior Scientist
Marine Biological Laboratory
7 MBL Street
Woods Hole, MA 02543
rudolfo@mbl.edu
508-289-7426

Andrew Reynolds, PhD
Professor of Philosophy
Cape Breton University
1250 Grand Lake Road
Sydney, Nova Scotia
Canada B1P 6L2
andrew_reynolds@cbu.ca
902-563-1301

Jan Sapp, PhD
Professor of Biology and History
York University
Biology Department
Faculty of Science
4700 Keele St.
Toronto M3J 1P3
jsapp@yorku.ca
416-736-2100

Jutta Schickore, PhD
Professor
Department of History and Philosophy of Science and Medicine
Indiana University
Ballantine Hall 647
Bloomington, IN 47405
jschicko@indiana.edu
812-855-9728

Beatrice Steinert, BA
Brown University
Box G
Providence, RI 02902
beatrice_steinert@brown.edu
(347) 604-0222

William C. Summers, MD, PhD
Professor
Yale University
450 Saint Ronan St.
New Haven, CT 06511
william.summers@yale.edu
203-887-9224

INDEX

Page numbers in italics refer to figures.